Molecular Theory of Water and Aqueous Solutions

Part II: The Role of Water in Protein Folding, Self-Assembly and Molecular Recognition

Molecular Theory of Water and Aqueous Solutions

Part II: The Role of Water in Protein Folding, Self-Assembly and Molecular Recognition

Arieh Ben-Naim

The Hebrew University of Jerusalem, Israel

World Scientific

NEW JERSEY · LONDON · SINGAPORE · BEIJING · SHANGHAI · HONG KONG · TAIPEI · CHENNAI

Published by

World Scientific Publishing Co. Pte. Ltd.

5 Toh Tuck Link, Singapore 596224

USA office: 27 Warren Street, Suite 401-402, Hackensack, NJ 07601

UK office: 57 Shelton Street, Covent Garden, London WC2H 9HE

Library of Congress Cataloging-in-Publication Data
Ben-Naim, Arieh, 1934–
    Molecular theory of water and aqueous solutions / Arieh Ben-Naim.
     v. cm.
    Includes bibliographical references and index.
    Contents: pt. 2. The role of water in protein folding, self-assembly and molecular recognition
    ISBN-13: 978-981-4350-53-2 (pt. 2 : hardcover : alk. paper)
    ISBN-10: 981-4350-53-2 (pt. 2 : hardcover : alk. paper)
    ISBN-13: 978-981-4350-54-9 (pt. 2 : pbk : alk. paper)
    ISBN-10: 981-4350-54-0 (pt. 2 : pbk : alk. paper)
    1. Solution (Chemistry) 2. Water. 3. Molecular theory. I. Title.
    QD541.B459 2009
    546'.22--dc22

                                                  2009013992

British Library Cataloguing-in-Publication Data
A catalogue record for this book is available from the British Library.

Typeset by Stallion Press
Email: enquiries@stallionpress.com

Printed in Singapore.

This book is dedicated to
Walter Kauzmann

Painting by Alexander Vaisman
(Based on pictures provided by Peter Kauzmann.)

In favor of the hydrophobic paradigm:

"Hydrogen bonds, taken by themselves, give a marginal stability to ordered structures... The hydrophobic bond is probably one of the more important factors involved in stabilizing the folded configuration in many native proteins." Kauzmann (1959)

"Let us examine the basis of hydrophobic interactions, which are a major driving force in the folding of macromolecules." Stryer (1988)

"This preference of non-polar atoms for non-aqueous environments has come to be known as the hydrophobic interaction. It is a major factor in the stabilities of proteins, nucleic acids and membranes, and it has some unusual characteristics." Creighton (1993)

"Hydrophobic interaction has a profound effect on the structure of proteins. When the polypeptide chain of a protein folds... the non-polar amino acids... are in the interior... while the polar amino acid residues are on the exterior." Chang (2005)

"The importance of the hydrophobic effect was for a long time underestimated. Charged interactions and hydrogen bonds are not strong intra molecular forces because water molecules compete significantly with these effects." Whitford (2005)

"An important non-covalent force that causes a polypeptide to fold into its native conformation are the hydrophobic interaction forces... Hydrogen bonds... although they contribute to the thermodynamic stability of protein's conformation, their formation may not be a major driving force for folding." Delvin (2006)

"The hydrophobic effect, which causes non-polar substances to minimize their contacts with water, is the major determinant of native protein structure... hydrogen bonds, which are central features of protein's structures, make only minor contributions to protein stability." Voet *et al.* (2008)

"One important aspect of protein folding is the hydrophobic effect... The hydrophobic effect stabilizes protein structures primarily due to a favorable entropy contribution." Liljas *et al.* (2009)

My doubts about the hydrophobic paradigm:

"...in spite of my researches in this field over almost ten years, I cannot confirm that there is at present either theoretical or experimental evidence that unequivocally demonstrates the relative importance of the hydrophobic interactions over other types of interaction in aqueous solutions." Ben-Naim (1980)

And my conviction about the hydrophilic paradigm:

"This finding leads us to conclude that intramolecular hydrophilic interactions are quite strong, highly dependent on orientation, and very specific to the properties of the solvent. For these reasons, we believe that these interactions are probably more important than hydrophobic interactions, especially in highly specific processes such as protein folding and molecular recognition." Ben-Naim (1989)

List of Abbreviations

BB: Backbone
BE: Binding energy
CN: Coordination number
EEC: Entropy-Enthalpy Compensation
HS: Hard sphere
HB: Hydrogen bond
HϕI: Hydrophilic
HϕO: Hydrophobic
i.g.: Ideal gas
KB: Kirkwood-Buff
LJ: Lennard Jones
MM: Mixture Model
1-D: One-dimensional
2-D: Two-dimensional
3-D: Three-dimensional
PCP: Pseudo Chemical potential
PDB: Protein Data Bank
PFP: Preferential folding pathways
PMF: Potential of mean force
RDF: Radial distribution function
SMI: Shannon's measure of information
SPT: Scaled particle theory

Preface

Aqueous solutions of macro-biomolecules such as proteins and nucleic acids are extremely complicated systems. They are difficult to study experimentally, very demanding computationally and almost impossible to deal with using purely theoretical methods. Yet, because of their crucial importance for understanding biological processes, we must pursue the search for answers even though we can find only partial answers, and even then, to only a fraction of the problems.

In a recent editorial of *Science* (2005), the editors listed 125 unsolved "big questions" in science.

Three of those questions which are relevant to this book are:

I. What is the structure of water?
II. Can we predict how proteins will fold?
III. How do proteins find their partners?

None of these questions are well defined. The first depends on how one defines the "structure of water." In my opinion, the most important aspect of liquid water is not its structure, however one chooses to define it, but the understanding of the outstanding properties of water. At present, these are well understood and I do not believe they should be considered as "unknowns of science." This question, as well as its answer, is discussed in Part I of this series.

Questions II and III are discussed in Chapters 3 and 4, respectively, of this book. I believe that a major obstacle that has hindered the quest for answers to these questions was the adherence to the $H\phi O$ paradigm. As I shall discuss throughout the book, switching to the $H\phi I$ paradigm brings us as close as one can hope to answering both of these questions.

The main purpose of this book is to convey a single message: The dogma that the *hydrophobic effect* is the dominant driving force in processes such as protein folding and protein–protein association has now totally collapsed.

For me personally, the hydrophobic ($H\phi O$) dogma lost much of its power during the 1980s. Unfortunately, even today, many books and review articles still propagate the $H\phi O$ dogma. Here is a typical quotation from a recent book by Whitford (2005):

> "The importance of the hydrophobic effect was for a long time underestimated. Charged interactions and hydrogen bonds are not strong *intra* molecular forces because water molecules compete significantly with these effects."

My view is diametrically opposite. I shall deal with the different parts of this quotation throughout the book. Here, I will only make the statement that the $H\phi O$ effect was, for a long time — almost 50 years — overestimated, not underestimated.

The quotation from Whitford should not be understood as criticism of the book. On the contrary, I found that the book presents an excellent introduction to protein structure and function. It is beautifully illustrated, rich in scope and clear in style. I brought up this quotation only as an example of a very widespread view expressed in most of the biochemical literature. These views have been lingering for over 20 years after the evidence favoring the $H\phi O$ effect was shown to be faulty and all the available evidence is now overwhelmingly in favor of hydrophilic ($H\phi I$) effects (see also Sec. 3.5).

I should add here that I am referring to the relative importance of $H\phi O$ versus $H\phi I$ effects in processes such as protein folding and protein–protein association. There is no doubt that $H\phi O$ effects are dominant in the formation of large aggregates of $H\phi O$ groups or molecules, such as micelles and membranes.[1]

The general methodology advocated in this book is based on the idea that a grand problem can be broken into smaller, more manageable problems, and that the so-called "driving-forces" for a biochemical process may be split into a *direct* and an *indirect* part. The indirect part is further broken down into smaller ingredients.

In this book, we focus mainly on the indirect part of the "driving forces," or the solvent-induced contribution to the Gibbs energy change for a biochemical process involving macromolecules. Of course, one must eventually take into account both the direct, and the indirect parts of the driving force. However, I believe that the study of the solvent-induced contribution to the driving force is the more intriguing part, and the more difficult one to understand. The belief that the solvent has a major role in determining both the strength of the driving force and the specificity of the end product has long been recognized. However, in my opinion the role played by the so-called hydrophobic effects has been overemphasized — sometimes even over-overemphasized — in processes such as protein folding, protein association, etc. Instead, I believe that presently, the overwhelming evidence is in favor of various hydrophilic effects. In a way, we are witnessing here a closing of a circle that started with the recognition of the importance of hydrogen bonds (HBs) in biochemical processes. Then came the $H\phi O$ effects, which dominated the field for a half of a century. And finally, hydrogen bonds have made a comeback, not only as direct HB, but as a rich repertoire of solvent-induced effects involving *HBs* between

---

[1] For an excellent review of this topic, see Tanford (1973) and Reynolds (2001).

solvent molecules and hydrophilic groups that are exposed to the solvent.

The ideas expressed in this book are highly personal and in fact against the mainstream views featuring in the literature. These views are based on evidence which I will discuss throughout the book. The evidence, both experimental and theoretical, shows that the various $H\phi I$ effects are much stronger, and the repertoire of $H\phi I$ effects much richer, than the corresponding $H\phi O$ effects.

Presently, it is still impossible to say which factor is the most important — $H\phi O$, $H\phi I$ or perhaps other factors that have not even been identified. However, it is clear to me that in processes such as protein folding, protein–protein association and molecular recognition, the strength of the evidence in favor of $H\phi I$ effects is more compelling vis-à-vis the evidence in favor of $H\phi O$ effects.

This book is organized into four chapters. In Chapter 1, we discuss the solvation of macromolecules and the method for dissecting the solvent Gibbs energy of a protein into small contributions, each of which may be studied using either experimental or theoretical methods.

Chapter 2 is concerned with the solubility of globular proteins. It is argued that the main reason for the high solubility of certain proteins is the occurrence of water bridges connecting $H\phi I$ groups on the surface of the protein.

In Chapter 3, we discuss the role of solvent-induced contributions to the stability of the native form of the protein, as well as their effect on the selection of preferential folding pathways. We conjecture that $H\phi I$ effects are probably the main source of the *"driving force"* for the folding process, and that *$H\phi I$ forces* are likely to determine the preferential pathways for the folding process.

In Chapter 4, we turn to discuss a few association processes and show how various $H\phi I$ effects contribute to both the stability and the specificity of the binding process. We also show how $H\phi I$ effects can radically change the mechanism of molecular recognition.

In most parts of the book, I have tried to present the argument in, as far as possible, a qualitative descriptive style. The supporting mathematical arguments are relegated to the Appendices.

There, I have included a discussion of a few terms and concepts that appear quite frequently in the biochemical literature. Terms like "the thermodynamic hypothesis," "reversibility" and "cooperativity" feature in many discussions of protein folding. I find that these concepts are sometimes misapplied, and are sometimes irrelevant to the process of protein folding. Another topic discussed in the Appendices is the so-called hydrogen bond inventory argument. To me it is both puzzling and disappointing to see that the HB inventory argument is still invoked in spite of the fact that it was proved to be fundamentally wrong almost 20 years ago.

In fact, much of the material in this book was published almost 20 years ago. Unfortunately, it was largely ignored. I shall not dwell on the reasons for that in this book. I strongly encourage the readers to communicate to me their views. I will publish all of them along with my own views in a forthcoming book.

The book is addressed primarily to biochemists, biophysicists and molecular biologists. I also hope that it will be of interest to anyone interested in the role of water in biological system.

Contents

Acknowledgements

I am dedicating this book to the memory of Walter Kauzmann, who for many years was a source of inspiration and encouragement for me. I believe that the main impact of Kauzmann's work was to produce a strong motivation and impetus to study the properties of water and aqueous solutions, as well as the role of water in biochemical processes.

For over 20 years, I was fascinated by Kauzmann's idea of using the hydrophobic effect in the process of protein folding. In one of the Gordon conferences on water way back in the 1970s, I discussed with Kauzmann the term he used, "hydrophobic bond." I suggested replacing this term with the more specific terms "hydrophobic solvation" and "hydrophobic interaction," a suggestion he gladly and enthusiastically accepted.

Throughout the years, Kauzmann's support and appreciation for my work were evident in his personal letters to me, his reviews of my articles, as well as his comments on my book *Hydrophobic Interaction*, published in 1980, and his more recent article (Henn and Kauzmann, 2003). His positive views about my work have instilled pride in me and given me a great deal of encouragement.

Over the past 20 years, I became convinced that the Kauzmann model for the role of the hydrophobic effect was not adequate in representing the contribution of hydrophobic groups in the processes of protein folding. Furthermore, I found that various hydrophilic effects might be far more important to biochemical processes such as protein folding, protein–protein association and molecular recognition. The results of these findings are summarized in this book. Notwithstanding these findings, I still maintain a deep and unwavering appreciation for Kauzmann the man, and Kauzmann the scientist.

I am not aware about his views or opinions on my recent works on hydrophilic effects, but I would like to believe that although the results of these works go against the grain of the very ideas that he had espoused, he would have appreciated or even admired the new views as presented in this book. My belief is supported by what Arthur Henn, Kauzmann's last graduate student, wrote[2]:

"Another aspect of Walter Kauzmann that impressed people was his open-mindedness and willingness to listen to… other points of view, even if they might be in opposition to his own."

Although it might sound paradoxical, I feel deeply that my forays into the field of *hydrophilic effects* can be attributed to his encouragement of my earlier work in the field of *hydrophobic effects*. For this reason, I feel that I will forever owe him a debt of gratitude.

Throughout the years, I have learned and benefited from discussions with many friends and colleagues. In particular, I would like to mention Harry Saroff, who introduced me to the problems of protein folding and molecular recognition; to Bob Jernigan, whom I collaborated with during the early stages of the emergence of $H\phi I$ effects; George Rose, whom I've met more recently, and who advocates a backbone-based approach to protein folding, an approach that has some overlaps with my views regarding the relative importance of $H\phi I$ effects.

---

[2]Henn (2003).

I would also like to thank Robert Baldwin, Ralph Dougherty, David Eisenberg, Paul King, Robert Mazo, Mihaly Mezei, Jorge Numata and George Rose for reading parts of, or the entire manuscript, and who offered valuable comments.

As always, I am indebted to my dear wife for her gracious help in typing and stylizing the prose of this book.

I am indebted to Alex Vaisman and Jimmy Low for their gracious help in drawing the cover design of the book. I am also grateful to the Kauzmann's family for providing me with pictures of Walter Kauzmann which served as the basis on which Alex Vaisman drew Kauzmann's portrait which appears in the dedication page.

CHAPTER 1

Introduction, the Main Problem and the Main Tools

This book as a whole reflects the story of my conversion from being an enthusiast, as well as a believer and admirer of hydrophobic (HϕO) effects, into being a firm believer in the relative importance of hydrophilic (HϕI) effects. This chapter presents the "evidence" that led to my conversion. The succeeding chapters present the applications of the various HϕI effects to biochemical process.

My first encounter with the term HϕO interaction (or the HϕO bond) was in the early 1970s, when I was asked to write a chapter for a book edited by R. A. Horne and to include a review about what was then known on HϕO interactions. At that time, HϕO interactions were not well defined and certainly not well understood. Nevertheless, I was fascinated by the idea that water, a liquid possessing so many unusual properties, could cause such strong attraction between two simple non-polar molecules, which in the absence of water interact very weakly.

In fact, HϕO interactions were not considered by Kauzmann. What Kauzmann referred to as the hydrophobic bond was actually the hydrophobic solvation of non-polar solutes. It was only much later[1] that I suggested that he make a clear-cut distinction between HϕO solvation and HϕO interactions. The latter were defined in terms of molecular distribution functions,

---

[1] Ben-Naim (1971, 1974).

1

or equivalently in terms of Gibbs energy changes for the process of bringing two (or more) HϕO solutes to some close separation.

I dedicated almost 20 years to studying various aspects of HϕO solvation and HϕO interactions. The results of these studies were summarized in a book titled Hydrophobic Interactions, which was published in 1980.

In the preface of this book I wrote:

> *...in spite of my researches in this field over almost ten years, I cannot confirm that there is at present either theoretical or experimental evidence that unequivocally demonstrates the relative importance of the hydrophobic interactions over other types of interaction in aqueous solutions.*

In this quoted paragraph, I did not specify the processes in which the HϕO interactions were involved, but it was clear from the context of the preface as a whole that I was referring specifically to the processes of protein folding and protein–protein association. After all, Kauzmann himself did not "discover" the HϕO effect in 1959. They were "hovering in the air" much earlier.[2] However, it was Kauzmann who introduced the idea that the HϕO effect, in the sense of HϕO solvation, could contribute significantly to the stability of the native structure of proteins.

As I have noted in Part I,[3] Kauzmann's idea came at a time when direct hydrogen bonds (HBs) were deemed to be contributing "marginally" to the stability of the proteins. The reason given by Kauzmann and others was simple and convincing. Two HϕI groups, such as the carbonyl and the amine groups of the protein backbone, can form strong HBs. However, when these bonds are formed in aqueous media, there is some kind

---

[2]Tanford and Reynolds (2001).
[3]Ben-Naim (2009), Chapter 4. This book will be referred to as Part I.

of compensation between the total HB energy gained and the total HB energy lost. This compensation led to the so-called HB inventory argument,[4] which may be traced back to the studies of Schellman on the association of urea in aqueous solutions.[5]

During the 1960s and the 1970s, most biochemists were convinced that hydrogen bonding in aqueous media could not contribute significantly to the stability of proteins. On the other hand, the HϕO bond argument as brought forward by Kauzmann was so convincing that most people believed that the HϕO effect was the dominant force in protein folding.[6]

It was at this stage that I entered the scene, accepting both the HB inventory argument and Kauzmann's HϕO bond argument. I accepted these views, but I always had doubts about their validity, as I have expressed in the quotation from my book.

My views changed dramatically in the late 1980s, when I undertook the task of examining the question of the solvent-induced effect on protein folding. The result of this examination was two-fold. First, I found that the HB inventory argument was fundamentally flawed.[7] It followed, therefore, that hydrogen bonding, even in aqueous media, might be more important than previously thought. Second, I found that Kauzmann's model — transferring a HϕO molecule from water into an organic liquid — was not appropriate for estimating the contribution of the loss of solvation of a HϕO group attached to a protein.[8] First, because there is a big difference between the solvation Gibbs energy of a non-polar molecule and the *conditional solvation Gibbs energy of a non-polar group hung on the backbone of*

---

[4]Fersht (1984, 1985), Fersht *et al.* (1985).

[5]Schellman (1955).

[6]Here, I use the term "dominant force" in a qualitative sense. I shall further discuss this term in Sec. 1.2.

[7]Ben-Naim (1991).

[8]Doubts regarding the adequacy of the Kauzmann model were also expressed by Klapper (1973).

the protein. Second, the search for the most appropriate organic liquid to mimic the interior of the protein is futile. The Gibbs energy of solvation of the non-polar group in an organic liquid is simply not needed for estimating the contribution of the non-polar group to the "driving force" of protein folding.

Thus, my doubts regarding the relative importance of HϕO effects turned into a conviction that various HϕI effects, one of which is hydrogen bonding in aqueous media, are probably the more important effects in processes such as protein folding and protein–protein association.

These views were published in a series of papers and summarized in a book published in 1992.[9] The diffusion of these views into the biochemical literature was slow and only partial. Today, it seems to me that many biochemists have regained the view that hydrogen bonding in aqueous media is important. One result of this change in view is the additional weight given to the role of the backbone of the protein rather than to the side chains. However, the intramolecular hydrogen bonding in proteins is only one of a large repertoire of different HϕI effects which might be decisive in stabilizing the structure of proteins.

In this introductory chapter, we shall start by defining the thermodynamic driving force and the solvent-induced part of the driving force for a general process in aqueous solutions. In Sec. 1.2, we shall discuss the relationship between the force and the thermodynamic "driving force." This is necessary to remove much of the confusion that exists in the literature about these terms.

In Sec. 1.3, we introduce the definition of the solvation process and the corresponding thermodynamic quantities associated with this process. The definition applies to any molecule

---

[9]Ben-Naim (1992).

solvated in any liquid. We shall briefly discuss other measures of solvation thermodynamic quantities and their relation to the thermodynamics of the solvation process. In Sec. 1.4, we shall introduce the concept of conditional solvation and the corresponding thermodynamic quantities. The latter quantities appear whenever we dissect the solvation quantities of a large biomolecule into small, manageable quantities. The definition of the thermodynamic "driving force" and the corresponding contribution of the solvent to the "driving force" are also discussed in Sec. 1.4.

In Secs. 1.5 and 1.6, we briefly describe the process of dissecting the solvation Gibbs energy of a globular protein. In Sec. 1.7 we present a few numerical values of the various ingredients of the solvation Gibbs energy of a protein.

The reader is urged to read Sec. 4.12 of Part I, where a historical perspective of the shift in the paradigm from hydrophobic to hydrophilic effects is presented.

1.1. The General Problem

Let CR denote any chemical reaction occurring under well-defined conditions: Fixed temperature T, pressure P and given concentrations of all the reactants and products. We shall usually assume that all the solutes involved in the chemical reaction are very dilute in the solvent so that they form a dilute ideal solution. For such a reaction, we denote by $\Delta G^{l}(CR)$ the Gibbs free energy of a unit reaction[10] in the liquid phase l. Similarly, for the same reaction, but in the absence of the solvent, we denote by $\Delta G^{g}(CR)$ the corresponding Gibbs energy change in an ideal gas phase.

---

[10]Per molecule or per mole, depending on whether we are discussing the reaction in molecular terms, or in macroscopic measurable quantities.

We define the *solvent-induced* contribution to the Gibbs energy change of the reaction CR by the difference

$$\delta G(CR) = \Delta G^l(CR) - \Delta G^g(CR). \qquad (1.1.1)$$

The quantity δG will be our main concern in this book. Roughly speaking, $\Delta G^g(CR)$ is determined by the properties of the molecules involved in the reaction CR (these are referred to as solutes). On the other hand, $\Delta G^l(CR)$ depends on both the properties of the molecules involved in the reaction and the interactions of these solute molecules with the solvent molecules. By taking the difference in (1.1.1), we "extract" from $\Delta G^l(CR)$ that part that depends only on the properties of the solute molecules, and obtain the *solvent-induced* part. The last statement needs some clarification. When the reaction is carried out in an ideal gas phase, *only* the properties of the solute molecules involved in the reaction contribute to $\Delta G^g(CR)$ (given the thermodynamic parameters T, P and all the concentrations). Adding the solvent introduces two major effects: First, the interactions between the solute molecules involved in the reactions and the solvent molecules. Second, the solvent molecules might affect the *internal* properties of the solute molecules.[11]

In most cases, we assume that the second effect is small and therefore negligible. This is a good approximation in many cases. If this approximation is justified, then the quantity δG defined in (1.1.1) depends only on the interactions between the solutes involved in the chemical reaction and the solvent molecules. In this case, it is meaningful to claim that we have "extracted" the part that depends only on the properties of the solutes involved in the reaction.

---

[11] We shall always assume that the translation and the rotation of the entire molecule are not affected by the solute–solvent interactions.

There is another quantity that is also "extracted" in the formation of the difference (1.1.1). This is the *translational* Gibbs energy of all the molecules involved in the reaction. This "extraction" allows us to replace the chemical reaction CR with a new reaction involving the same solute molecules but devoid of *translational* degrees of freedom. With such molecules, it is easier to draw cyclic processes and relate the ΔG of the chemical reaction to the solvation Gibbs energies.

The following example clarifies the simplification achieved by this "extraction." As we shall see, this simplification is rendered possible by introducing the *pseudo-chemical potential*, with which we shall define any *standard* changes in the Gibbs energy of solvation or of a chemical reaction. It also avoids the great confusion that exists in the literature regarding the choice of the standard states.[12]

Consider the following reaction:

$$CR: A + B \rightarrow AB. \tag{1.1.2}$$

We can write the Gibbs energy change for a unit reaction in the liquid phase as

$$\Delta G^l(A + B \rightarrow AB) = \mu_{AB}^l - \mu_A^l - \mu_B^l. \tag{1.1.3}$$

Carrying out the same reaction in an ideal gas phase, the corresponding Gibbs energy change is

$$\Delta G^g(A + B \rightarrow AB) = \mu_{AB}^g - \mu_A^g - \mu_B^g. \tag{1.1.4}$$

We now write the chemical potential of each species as[13]

$$\mu_i^l = \mu_i^{*l} + k_B T \ln \rho_i \Lambda_i^3, \tag{1.1.5}$$

where $\rho_i = N_i/V$ is the *number* density, k_B the Boltzmann constant, Λ_i^3 the de Broglie wavelength, or the momentum partition

---

[12] Ben-Naim (2006) and Appendix B.
[13] See Sec. 1.3 and Appendices A and B.

function, and μ_i^{*l} is referred to as the *pseudo*-chemical potential of the species i.

We shall always assume that the system can be dealt with by classical statistical mechanics, and that the momentum partition function of each molecule is independent of its surroundings.

We now *define* the *solvent-induced* part of the Gibbs energy change for the reaction (1.1.2) as in (1.1.1), i.e.

$$\delta G(A + B \to AB) = \Delta G^l(A + B \to AB) - \Delta G^g(A + B \to AB)$$
$$= (\mu_{AB}^{*l} - \mu_{AB}^{*g}) - (\mu_A^{*l} - \mu_A^{*g}) - (\mu_B^{*l} - \mu_B^{*g})$$
$$= \Delta G_{AB}^* - \Delta G_A^* - \Delta G_B^*. \tag{1.1.6}$$

Thus, we see that the solvent-induced contribution to the Gibbs energy change, or to the "driving force" of the reaction, is determined by the *solvation Gibbs energies* of all the species involved in the reaction.[14] This statement has a far more general validity. It applies to any process carried out in any solvent. Therefore, a large part of this book is devoted to the study of the *solvation* thermodynamics of biomolecules.

The relationship between δG and the solvation quantities can also be deduced from a graphical representation of the process. Figure 1.1 shows the connection between ΔG of the reaction and the corresponding solvation Gibbs energies for the particular reaction (1.1.2). Similar graphical depictions will appear in the succeeding chapters.

It should be noted that in the cyclic process depicted in Fig. 1.1, all the species A, B and AB are devoid of translational degrees of freedom. We have also assumed that all these solutes form dilute ideal solutions; therefore, the pseudo-chemical potentials of these solutes do not depend on the concentrations of the solutes A, B and AB. Thus, in order to

---

[14]See Sec. 1.3.

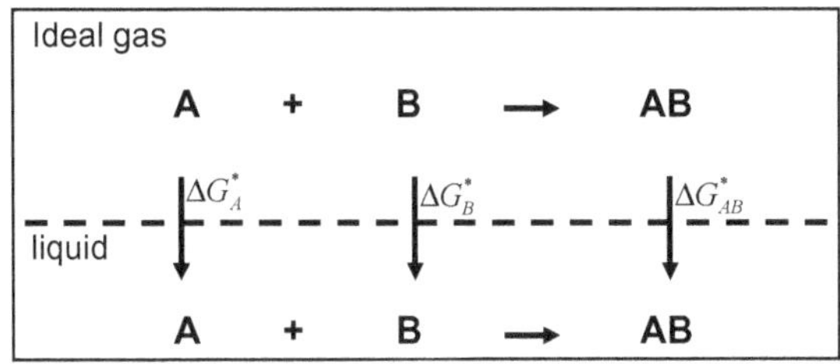

Fig. 1.1　A cyclic process for the reaction (1.1.2). Note here that molecules *A*, *B*, and *AB* are devoid of translational degrees of freedom.

characterize the solvent-induced part of the Gibbs energy of the reaction, all we need to know is the solvation Gibbs energies of the solutes involved. The literature is replete with cyclic processes of the type shown in Fig. 1.1, where the "standard states" of the solutes are not specified, hence rendering the solvation process unspecified. Therefore, the meaning of the corresponding Gibbs energy of "solvation" also becomes unspecified.

1.2. Forces and Driving Forces

In this section, we use a very simple process to define and to distinguish between the concept of *force* and the concept of *driving force* on the one hand, and the concepts of *direct* and *indirect* interactions on the other. It is absolutely necessary for the reader to understand these concepts as applied in a simple process before encountering the same concepts in the more general processes such as protein folding and protein–protein association. The biochemical literature is riddled with confusing, incorrect and misleading statements involving these concepts.

We start with a simple example. Consider the *interaction energy* between two simple spherical atoms, say two atoms of argon. Figure 1.2b shows the general form of the *pair potential* $U(R)$. The pair potential at distance R is simply defined as the

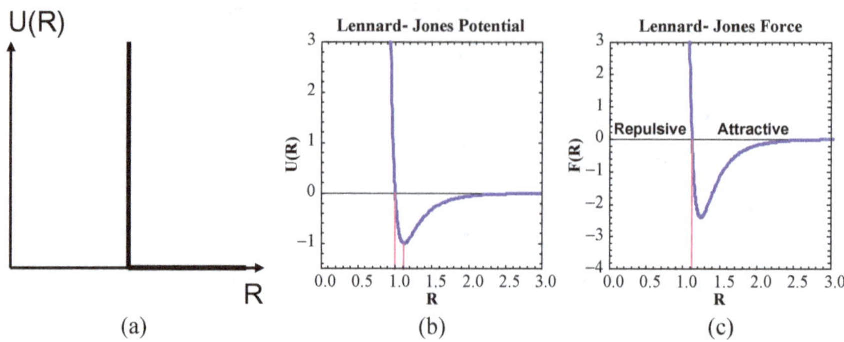

Fig. 1.2   (a) Hard-sphere pair potential, (b) Lennard-Jones pair potential, (c) Lennard-Jones force. (These plots are for illustration only. *R* and *U* are in arbitrary units.)

energy change for the process of bringing two particles from initially at rest and at infinite separation, to the final distance R.

$$U(R) = E(R) - E(\infty). \tag{1.2.1}$$

The interaction energy at $R = \infty$ is, by definition, chosen as zero.

This is, of course, a virtual process. We cannot bring the two motionless particles from infinite separation to some final distance R. However, the function $U(R)$ can either be calculated theoretically, or can be inferred from some experimental data,[15] e.g. from the second virial coefficient, and its temperature dependence.

There are many ways of expressing this function analytically — the most popular one is the Lennard-Jones (LJ) function

$$U(R) = 4\varepsilon \left[\left(\frac{\sigma}{R}\right)^{12} - \left(\frac{\sigma}{R}\right)^{6} \right]. \tag{1.2.2}$$

---

[15]See Hirschfelder *et al.* (1954).

The parameters σ and ε roughly measure the *size* of the particles and the *strength* of the interaction between the two particles.

It is easy to see that σ corresponds to the point where the function $U(R)$ crosses the R-axis, i.e. $U(R = \sigma) = 0$, and $-\varepsilon$ is the depth of the minimum of the function $U(R)$. Thus, the condition for a minimum of $U(R)$ is

$$\frac{dU(R)}{dR} = 4\varepsilon \left[\frac{6\sigma^6}{R^7} - \frac{12\sigma^{12}}{R^{13}} \right] = 0. \qquad (1.2.3)$$

We get

$$R_{\min} = \sigma \sqrt[6]{2}, \qquad (1.2.4)$$

$$U(R_{\min}) = -\varepsilon. \qquad (1.2.5)$$

The interpretation of σ as the diameter of the particles follows from the fact that the function $U(R)$ rises very sharply for distances R smaller than σ. This means that a huge amount of energy must be invested to push the two particles to distances smaller than σ. This conforms to our notion of the size of a *hard* spherical particle, which may be described by a *hard* pair potential depicted in Fig. 1.2a.

Note also that the minimum of the function $U(R)$ is at $R_{\min} = \sigma \sqrt[6]{2}$, slightly to the right of $R = \sigma$.

The *force* acting between the two particles is

$$F(R) = -\frac{dU(R)}{dR}. \qquad (1.2.6)$$

In the region $R_{\min} \leq R \leq \infty$, the slope is positive and the force is negative (Fig. 1.2c). (Note that at the minimum of $U(R)$ at $R_{\min}$, the force is zero.) In this region the particles *attract* each other. In the region $R \leq R_{\min}$ the particles repel each other. Normally, one ignores the difference between σ and $\sigma \sqrt[6]{2}$ and simply refers to $R < \sigma$ and $R \geq \sigma$ as the repulsive and the attractive regions, respectively. For two hard spheres the potential is *infinitely repulsive* at $R < \sigma$ and zero attractive at $R > \sigma$.

Another way of interpreting the slope of the function $U(R)$ is in terms of the direction of motion. Suppose that we start with the two particles at fixed positions and distance R. An attractive force means that if we release the two particles, they will move towards each other, i.e. R will decrease (see green arrow, Fig. 1.3). A repulsive force means that the two particles will move away from each other (red arrow in Fig. 1.3). In this example the *sign* of the force determines the direction of motion. As we shall soon see, this is not exactly correct when we discuss the *average force* between two particles in a solvent.

Suppose now we have n simple spherical particles all at rest at a configuration $\mathbf{R}_1, \ldots, \mathbf{R}_n$. The total potential energy of interaction among all these particles is defined as the energy

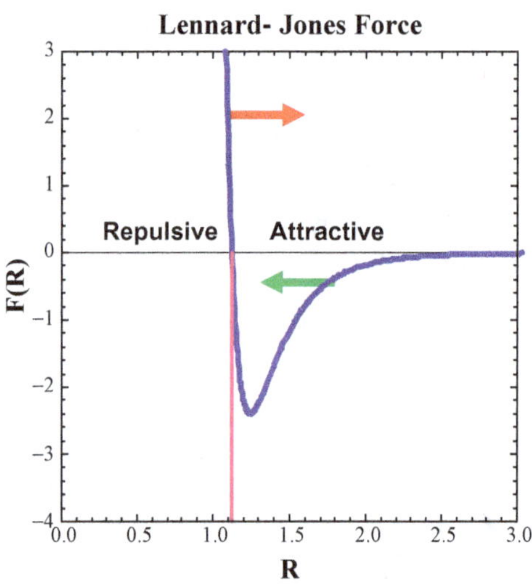

Fig. 1.3   The direction of motion resulting from the direction of the force: attractive (green arrow) and repulsive (red arrow). (This plot is for illustration only. R and $F(R)$ are in arbitrary units.)

difference for the process of bringing all the particles from infinite separation from each other, and at rest, to the final configuration $\mathbf{R}_1, \ldots, \mathbf{R}_n$.

$$U(\mathbf{R}_1, \ldots, \mathbf{R}_n) = E(\mathbf{R}_1, \ldots, \mathbf{R}_n) - E(\text{all } R_{ij} = \infty). \qquad (1.2.7)$$

The force acting on particle i at $\mathbf{R}_i$ is given by

$$\mathbf{F}_i = -\nabla_i U(\mathbf{R}_1, \ldots, \mathbf{R}_n). \qquad (1.2.8)$$

Where the operator ∇_i, the gradient, is defined by

$$\nabla_i = \left(\frac{\partial}{\partial x_i}, \frac{\partial}{\partial y_i}, \frac{\partial}{\partial z_i} \right),$$

i.e. the vector $\nabla_i U$ has the three components:

$$\nabla_i U = \left(\frac{\partial U}{\partial x_i}, \frac{\partial U}{\partial y_i}, \frac{\partial U}{\partial z_i} \right). \qquad (1.2.9)$$

In most cases, we assume that the total potential energy is pairwise additive — this means that $U(\mathbf{R}_1, \ldots, \mathbf{R}_n)$ may be written as a sum of pair interactions

$$U(\mathbf{R}_1, \ldots, \mathbf{R}_n) = \sum_{i<j} U(\mathbf{R}_i, \mathbf{R}_j), \qquad (1.2.10)$$

where the sum on the right-hand side of (1.2.10) is over all the different pairs of particles.

The pairwise-additivity assumption is, of course, an approximation. What it means is that the total work of bringing the n particles from infinite separation to the final configuration is equal to the energy involved in the process of bringing $n(n-1)/2$ pairs of particles, say i and j, from infinity to the final configuration $\mathbf{R}_i, \mathbf{R}_j$. For example, for three particles (Fig. 1.4), we have

$$U(\mathbf{R}_1, \mathbf{R}_2, \mathbf{R}_3) = U(\mathbf{R}_1, \mathbf{R}_2) + U(\mathbf{R}_1, \mathbf{R}_3) + U(\mathbf{R}_2, \mathbf{R}_3). \qquad (1.2.11)$$

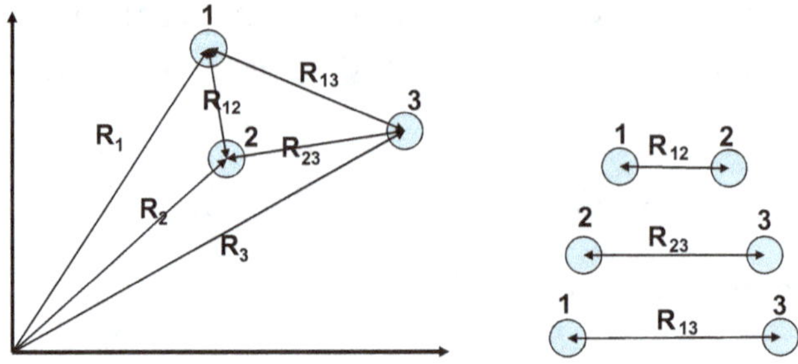

Fig. 1.4 The pairwise additive assumption. The work required to bring the three particles to $\mathbf{R}_1, \mathbf{R}_2, \mathbf{R}_3$ is equal to the sum of the works required to bring the three pairs to the distances R_{ij}.

Thus, for $n = 3$, we have $n(n-1)/2 = 3$ pairs of interacting particles. The assumption (1.1.10) is exact for hard sphere particles. It is a good approximation for simple particles, say argon or methane. It is not a good approximation for molecules like water. Nevertheless, this assumption is made in most of the discussions of force between water molecules.[16]

Thus, for a system of n particles at a specific configuration $\mathbf{R}_1, \ldots, \mathbf{R}_n$, the force acting on particle i is given by

$$\mathbf{F}_i = -\nabla_i U(\mathbf{R}_1, \ldots, \mathbf{R}_n) = -\sum_{j \neq i} \nabla_i U(\mathbf{R}_j, \mathbf{R}_i). \qquad (1.2.12)$$

Note that $\mathbf{F}_i$ is a vector. Each particle $j (j \neq i)$ exerts a force on particle i in different directions. The force $\mathbf{F}_i$ is the vector sum of all the forces exerted by all the particles.

Next, consider the situation depicted in Fig. 1.5. We have two particles at position $\mathbf{R}_1$ and $\mathbf{R}_2$, and a third particle at position $\mathbf{R}_3$. The force on particle 1 is simply

$$\mathbf{F}_1 = -\nabla_1 U(\mathbf{R}_1, \mathbf{R}_2) - \nabla_1 U(\mathbf{R}_1, \mathbf{R}_3). \qquad (1.2.13)$$

---

[16]See Ben-Naim (2009).

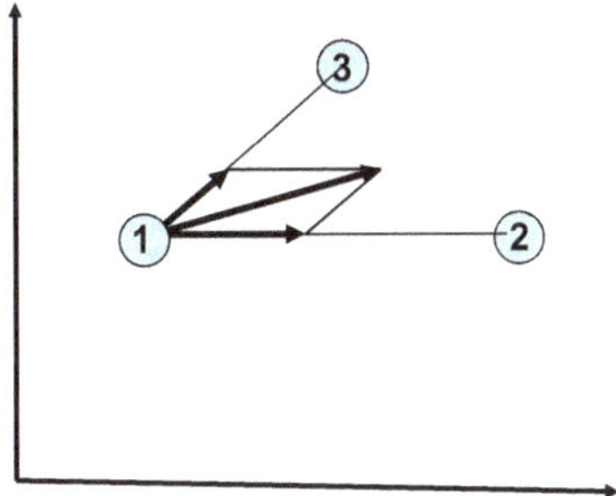

Fig. 1.5   Three particles at $\mathbf{R}_1, \mathbf{R}_2$, and $\mathbf{R}_3$. The force on particle 1 is the resultant of the forces exerted by particles 2 and 3.

Note that both terms on the right-hand side of (1.2.13) are *direct* forces. The first term is the force exerted by 2 on 1; the second term is the force exerted by 3 on 1.

Now suppose that particle 3 is moving around, and we are interested only in the *average* force exerted by this particle, while particles 1 and 2 are at fixed positions $\mathbf{R}_1$ and $\mathbf{R}_2$, respectively. In this case, the force on 1 is different from (1.2.13).

$$\mathbf{F}_1 = -\nabla_1 U(\mathbf{R}_1, \mathbf{R}_2) - \langle \nabla_1 U(\mathbf{R}_1, \mathbf{R}_3) \rangle$$
$$= \mathbf{F}_1^{\mathrm{D}} + \mathbf{F}_1^{\mathrm{SI}}. \tag{1.2.14}$$

We shall refer to the two terms on the right-hand side of (1.2.14), as the direct (D) force exerted by 2 on 1, and an indirect, or "solvent-induced" (SI) force. The reason for these terms will become clear soon. For the moment, the first term on the right-hand side of (1.2.14) is the *direct* force exerted by 2 on 1, and the second is the *average* force exerted by particle 3 (the "solvent") on particle 1, the average being over all positions of particle 3. Of course there are different ways of defining the average force in (1.2.14). It depends on the probability distribution of the *locations* of particle 3. The one particular average that will interest us is[17]

$$\mathbf{F}_1^{SI} = -\int \nabla_1 U(\mathbf{R}_1, \mathbf{R}_3) \Pr(\mathbf{R}_3/\mathbf{R}_1, \mathbf{R}_2) d\mathbf{R}_3$$

$$= -\frac{1}{V} \int \nabla_1 U(\mathbf{R}_1, \mathbf{R}_3) \exp[-\beta U(\mathbf{R}_1, \mathbf{R}_3) - \beta U(\mathbf{R}_2, \mathbf{R}_3)] d\mathbf{R}_3,$$

$$(1.2.15)$$

where $\beta = (k_B T)^{-1}$, with k_B the Boltzmann constant and T the absolute temperature. The average in (1.2.15) is taken with the *conditional* probability of finding particle 3 at the element of volume $d\mathbf{R}_3 = dx_3 dy_3 dx_3$, given that particles 1 and 2 are at positions $\mathbf{R}_1$ and $\mathbf{R}_2$[18]

$$P(\mathbf{R}_3/\mathbf{R}_1, \mathbf{R}_2) d\mathbf{R}_3 = \frac{1}{V} \exp\left[-\beta U(\mathbf{R}_1, \mathbf{R}_3) - \beta U(\mathbf{R}_2, \mathbf{R}_3)\right] d\mathbf{R}_3.$$

$$(1.2.16)$$

The integration in (1.2.15) extends over all possible locations of particle 3.

We may refer to the quantity $-\nabla_1(\mathbf{R}_1, \mathbf{R}_3)$ as the *instantaneous* force exerted by particle 3 on 1. When particle 3 is moving around, we get an average force $\mathbf{F}_1^{SI}$, or a "solvent-induced" force. Here, the solvent consists of one particle only. The total force on 1 is the vector sum of the two forces.

Note that the *average* force is always along the line connecting the centers of particles 1 and 2. Remember that the two particles 1 and 2 are simple spherical particles. Therefore, the conditional density $\rho(\mathbf{R}_3/\mathbf{R}_1, \mathbf{R}_2)$ has axial symmetry about the line connecting 1 and 2. This means that for each force $-\nabla_1(\mathbf{R}_1, \mathbf{R}_3)$ with components (f_x, f_y, f_z), there is another force with components $(-f_x, -f_y, -f_z)$, and these two forces have the same probability. Therefore, the components of all the forces

---

[17]For derivation, see Appendix C and Ben-Naim (2006).
[18]Note here that $d\mathbf{R}_3$ is an element of volume, not a vector.

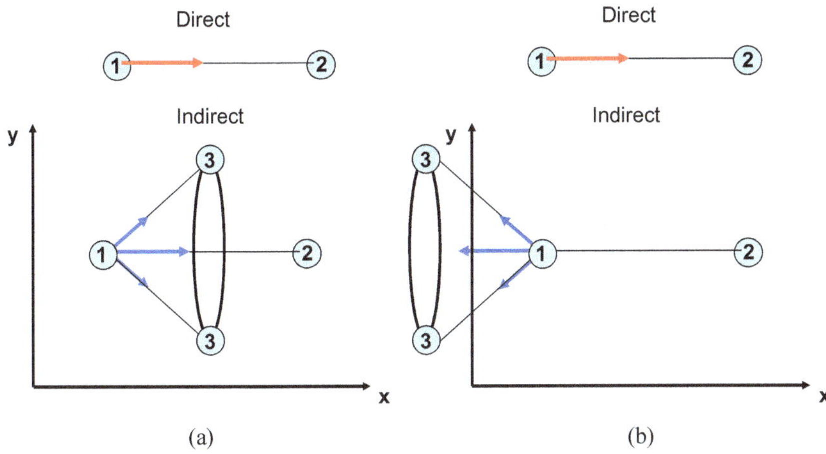

Fig. 1.6   The *direct* (red) and the *indirect* (blue) force exerted on particle 1. Note that the indirect force is also along the x-axis. The direct and indirect forces are (a) in the same direction, (b) in opposite directions.

in the y and z directions cancel out. In the particular example drawn in Fig. 1.6a, the average force is in the same direction as the direct force $-\nabla_1(\mathbf{R}_1, \mathbf{R}_2)$. However, in general, the average force is not necessarily in the same direction as the direct force (see Fig. 1.6b). For more details, see Appendix C.

We can now generalize Eq. (1.2.15) to the case when the two particles 1 and 2 are immersed in a solvent. For simplicity, suppose that the solvent is pure water and the configuration of each water molecule is described by six coordinates, three locations and three angles. We denote by $\mathbf{X}_W$ the vector $(x_W, y_W, z_W, \theta_W, \phi_W, \psi_W)$ and rewrite the force on 1, given the locations of 1 and 2 at $\mathbf{R}_1$ and $\mathbf{R}_2$, respectively as[19]

$$\mathbf{F}_1 = \mathbf{F}_1^D + \mathbf{F}_1^{SI}$$

$$= -\nabla_1 U(\mathbf{R}_1, \mathbf{R}_2) - \int \nabla_1 U(\mathbf{R}_1, \mathbf{X}_W)\rho(\mathbf{X}_W/\mathbf{R}_1, \mathbf{R}_2)d\mathbf{X}_W.$$

$$(1.2.17)$$

---

[19]For derivation, see Appendix C and Ben-Naim (2006).

Here, the first term on the right-hand side of (1.2.17) is as before, the *direct* force exerted by particle 2 on 1, as if the two particles were in a vacuum. The second, the indirect part of the force, is due to the presence of the solvent. Instead of only one particle playing the role of solvent in (1.2.15), we have many water molecules wandering around particles 1 and 2. The term $-\nabla_1 U(\mathbf{R}_1, \mathbf{X}_W)$ is again the *instantaneous* force exerted by a water molecule at $\mathbf{X}_W$ (Fig. 1.7). This force may depend on both the location and the orientation of the water molecule. The term $\rho(\mathbf{X}_W / \mathbf{R}_1, \mathbf{R}_2)$ is the *conditional* density of water molecules at configuration $\mathbf{X}_W$, given two *simple* solute particles at $\mathbf{R}_1$ and $\mathbf{R}_2$. The integral on the right-hand side of (1.2.17) is a conditional average force. It is a conditional average because the term $\rho(\mathbf{X}_W / \mathbf{R}_1, \mathbf{R}_2) d\mathbf{X}_W$ is also the *conditional*

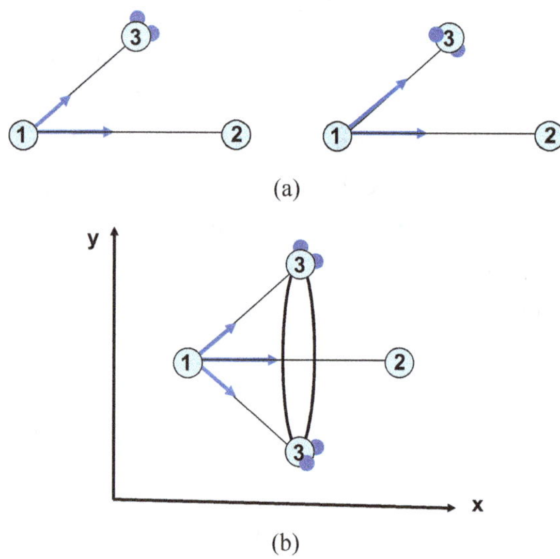

Fig. 1.7 (a) The instantaneous force exerted on 1 by the water molecule at $\mathbf{X}_W$ depends on both the location and the orientation of the water molecule. However, the average force (b) exerted on 1 is still along the line connecting the centers of 1 and 2 (i.e. the x-axis).

probability of finding any water molecules in a small "volume" $d\mathbf{X}_W$ at $\mathbf{X}_W$ given the two particles at $\mathbf{R}_1$ and $\mathbf{R}_2$.

Note again that since the two particles 1 and 2 are presumed to be simple spherical particles, the conditional density $\rho(\mathbf{X}_W/\mathbf{R}_1, \mathbf{R}_2)$ has axial symmetry about the line connecting the centers of 1 and 2, and therefore the average force must also be along the line connecting 1 and 2. The direction of the average force can either be the same or the opposite of the direction of the direct force. See below for an example.

We now introduce the *potential* of *mean force* (PMF), denoted by W. Here, we define the PMF in terms of the Gibbs energy of a system consisting of a solvent, say water, and two solute particles at fixed positions $\mathbf{R}_1$ and $\mathbf{R}_2$.[20]

$$W(\mathbf{R}_1, \mathbf{R}_2) = G(\mathbf{R}_1, \mathbf{R}_2) - G(R_{12=\infty}). \qquad (1.2.18)$$

If the *direct* solute–solute interaction $U(R)$ is independent of the presence of the solvent, then the change in the Gibbs energy for the process of bringing the two solutes from infinite separation to the final configuration $(\mathbf{R}_1, \mathbf{R}_2)$ can be split into two parts

$$\Delta G = G(\mathbf{R}_1, \mathbf{R}_2) - G(\infty) = U(\mathbf{R}_1, \mathbf{R}_2) + \delta G(\mathbf{R}_1, \mathbf{R}_2).$$

$$(1.2.19)$$

Here, ΔG is split into two terms: The direct interaction and the indirect, or *solvent-induced* part of the work associated with the process as depicted in Fig. 1.8.

The quantity δG is actually defined in (1.2.19). The PMF is the potential from which the mean force is obtained. The two components of the force in (1.2.17) correspond to the two terms

---

[20]Traditionally, the PMF is *defined* in terms of molecular distribution functions. Then one can show that the PMF is equal to the Gibbs (or Helmholtz) energy change of the process as described in (1.2.18). For details, see Ben-Naim (1992, 2006).

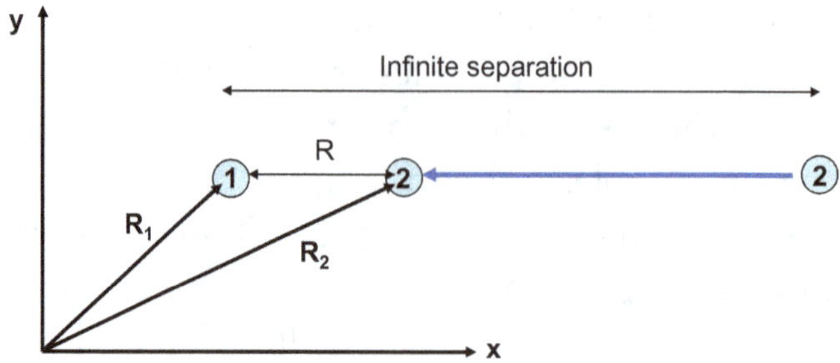

Fig. 1.8 The process of bringing two simple particles from infinite separation to the final configuration $\mathbf{R}_1, \mathbf{R}_2$ and distance $R = |\mathbf{R}_2 - \mathbf{R}_1|$.

in (1.2.19), i.e.

$$F_1^D = -\nabla_1 U(\mathbf{R}_1, \mathbf{R}_2), \qquad (1.2.20)$$

$$F_1^{SI} = -\nabla_1 \delta G(\mathbf{R}_1, \mathbf{R}_2). \qquad (1.2.21)$$

The quantity δG is the central quantity in the study of hydrophobic and hydrophilic interactions. δG may be related to the *solvation* Gibbs energies of the solute molecules. We shall define the solvation Gibbs energy of a solute in the next section. Here, we can deduce the relationship between δG and the solvation quantities from the cyclic process depicted in Fig. 1.9 (but for two solutes s)

$$\delta G(\mathbf{R}_1, \mathbf{R}_2) = \Delta G^*(\mathbf{R}_1, \mathbf{R}_2) - 2\Delta G_s^*, \qquad (1.2.22)$$

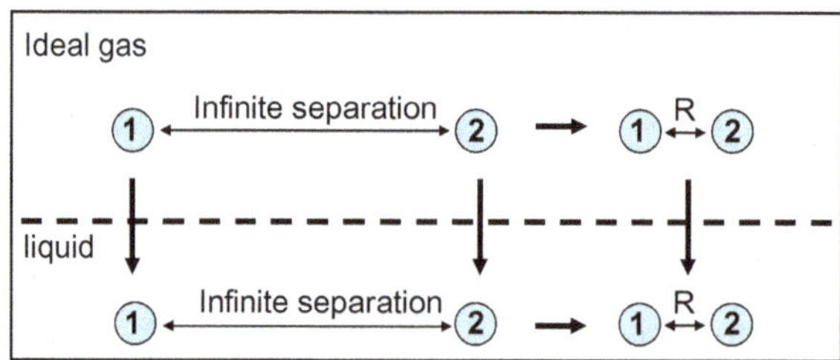

Fig. 1.9 The process of bringing two simple solutes from infinite separation $(R = \infty)$ to some final distance R. The vertical arrows indicate solvation from an ideal gas to the liquid.

where ΔG_s^* is the solvation Gibbs energy of the solute s, and $\Delta G^*(\mathbf{R}_1, \mathbf{R}_2)$ is the solvation Gibbs energy of the pair of solutes at a configuration $\mathbf{R}_1, \mathbf{R}_2$.

In thermodynamics, it is customary to refer to the change in Gibbs energy as a thermodynamic *driving force*. This is not strictly a force, yet the concept of *driving force* is useful. When the ΔG of a process is large and negative in some direction, we can expect that there will be a strong *tendency* for the process to occur in that direction. However, the value of ΔG is not to be confused with the *force*, neither should it be confused with the *speed* at which this process will occur.

We shall now discuss a few important differences between a *driving force* and a *force* on one hand, and the driving force for macroscopic processes and for molecular processes on the other hand.

Figure 1.10 shows two functions $U(R)$ and $G(R)$ (or $W(R)$) for Lennard-Jones particles. Note first that the notation $G(R)$

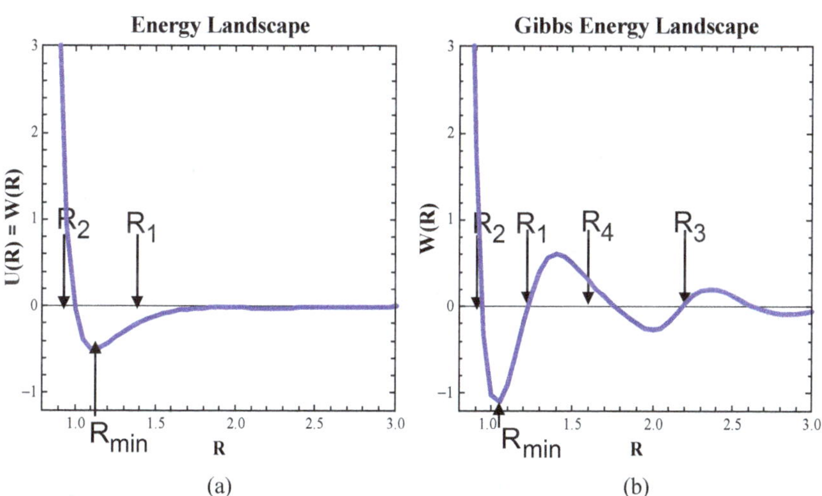

Fig. 1.10   (a) The energy landscape and (b) the Gibbs energy landscape for two simple particles, as a function of the distance R.

is a shorthand notation for $G(T, P, \mathbf{N}; R)$. Thus, whereas $U(R)$ is the energy as a function of R, and is independent of the temperature T, the pressure P, or the composition of the "solvent" $\mathbf{N}$, the function $G(R)$ *depends* on $T, P, \mathbf{N}$. A second important difference between the two functions is revealed when we discuss systems with many particles, e.g. a protein, the conformation of which is described by $\mathbf{R}_1, \ldots, \mathbf{R}_n$. In such cases the *potential* function may be assumed to be approximately *pairwise additive*. However, the corresponding Gibbs energy function is not pairwise additive, even when the energy is exactly pairwise additive.[21]

Turning to Fig. 1.10, we can say that the function $U(R)$ is the *energy landscape* for the system of two simple particles, and $G(R)$ is the free energy, or the *Gibbs energy landscape* for the same system. Note that we are here discussing simple spherical systems, such that both $U(R)$ and $G(R)$ depend on the scalar distance $R = |\mathbf{R}_2 - \mathbf{R}_1|$. Note also that in this particular system, the energy landscape has only one minimum at $R_{\min}$, whereas the Gibbs energy landscape has several minima and maxima (Fig. 1.10b). In Chapter 3, we shall see that having multiple minima and maxima is a feature of both landscapes of a protein.

A more important difference between the energy landscape and the Gibbs energy landscape is the relationship between the direction of the *force* and the direction of the instantaneous motion. On the energy landscape, the force at each point determines the direction of motion at that point. For instance, if we start at the point R_1 in Fig. 1.10a, the force is attractive. This means that if we release the two particles at R_1, they will move *towards* each other. This is simply the result of Newton's law of motion. If we start at the point R_2 in Fig. 1.10a, there is a repulsive force, and the two released particles will move *away*

---

[21]See also Appendix E.

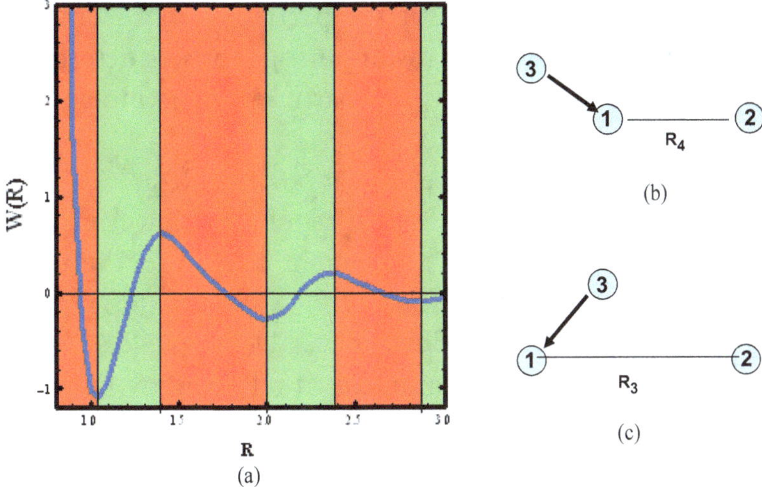

Fig. 1.11 (a) Regions of attraction (green) and repulsion (red) in the Gibbs energy landscape. (b) Instantaneous attraction and (c) instantaneous repulsion between 1 and 2.

from each other. (Note that this is true provided there are no constraints on the motion of the particle.)

This relationship between the direction of the *force* and the direction of *motion* does not necessarily hold in the Gibbs energy landscape. As can be seen in Fig. 1.11, there are several regions of attractive and repulsive force, but an attractive force does not imply instantaneous motion of the particles *towards* each other, and repulsive force does not imply instantaneous motion of the particles *away* from each other. This is a fundamental difference between a direct force derived from the energy landscape and an average force derived from the Gibbs energy landscape. For example, if we start at the point R_3 in Fig. 1.10b, we see that there is an attractive force. However, when the two particles are released at the point R_3, the instantaneous motion can be in *any* direction. It can be towards each other, away from each other, or in any other direction not necessarily along the line connecting the two particles. Similarly, at point R_2 or R_4, there is an average repulsive force. However, this force does not imply instantaneous motion of the particles away from each other. The reason is that the instantaneous force is determined by the collisions of the "solvent" particles at the moment we release the two particles 1 and 2.

As an example, suppose we start at position R_4 in Fig. 1.10b. Here, on average, the force will be repulsive. However, when we release the two particles, there might be a solvent molecule that collides with particle 1 in such a way that the instantaneous motion of 1 will be *towards* particle 2, as illustrated in Fig. 1.11b. Similarly, at the moment of releasing two particles at position R_3, a solvent molecule might hit particle 1 in such a way to cause an instantaneous motion away from particle 2 (Fig. 1.11c).

The unpredictability of the instantaneous direction of motion from the direction of the *average* force is a feature of any Gibbs energy landscape, including the case of a protein molecule in a solvent, which we shall discuss in Chapter 3.

Next, we discuss the difference between the *force* and the *thermodynamic driving force*. Again, care must be exercised when inferring the direction of *force* from the direction of the *driving force*, or vice-versa. On the energy landscape (Fig. 1.10a) the force is related to the slope at each point. The thermodynamic driving force is related to the *difference in the energy* $U(R_{min}) - U(R_i)$ for any R_i. This means that releasing the two particles at *any* point R_i, there will be an overall tendency to move *towards* R_{min}, independently of whether the force is attractive or repulsive. We call this a thermodynamic driving force towards the minimum energy at R_{min}. We can also say that R_{min} is a kind of "attractor" for any initial point on the energy landscape.

On the other hand, on the Gibbs energy landscape, one can say that there exists a thermodynamic driving force to move from any point R_i to R_{min}, but there is no guarantee that the system will actually move towards $R_{\min}$. Furthermore, there is no simple relationship between the direction of the force on the Gibbs energy landscape and the direction of the driving force. For instance, at R_4, the *force* is repulsive, i.e. directing the particles to move away from each other, whereas the *driving force* is towards $R_{\min}$, i.e. directing the particles to move towards each other.

Thus, when we release the two particles at point R_4 in Fig. 1.10b, we cannot tell anything about the instantaneous direction of the motion of the two particles. What is actually happening is that the two particles are bombarded incessantly by the "solvent" particles. At each instance, there is a small movement of the particles. What we can observe (for instance in Brownian motion) is the net movement after a period of time t. The relationship between the instantaneous motion and the resultant motion that we observed is shown schematically in Fig. 1.12.

What we have described is the resultant motion taken after some measurable period of time t. In writing the average force, as in Eq. (1.2.17), we have actually written an ensemble average. Therefore, all one can say is that there is a large probability for

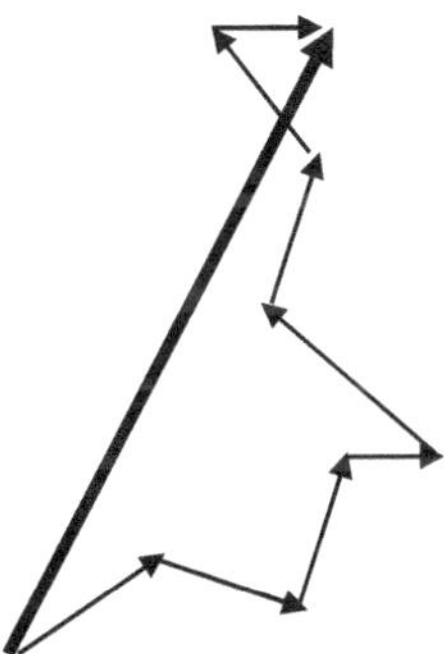

Fig. 1.12 Instantaneous motion of one particle as a result of an instantaneous force of the "solvent" particles. The resultant "average" motion during a measurable time is indicated by the heavy arrow.

the particle to move in the direction of the force, but there is no guarantee that the move will always be in the direction of the force. Certainly, there is no guarantee that the system will be moving according to the dictates of the *driving force.*

Thus, one should be very careful in inferring the direction of motion of the system from the thermodynamic driving force. There is also an essential difference between the thermodynamic driving force when applied to *macroscopic* processes, and when applied to microscopic processes.

Figure 1.13 shows a thermodynamic isolated system. There are two compartments separated by a movable partition. If the pressures in the two compartments are unequal, say $P_1 > P_2$, then the partition will move from left to right. In this particular example, there is a net *force* acting on the partition, which causes it to move to the right. The corresponding *thermodynamic force* is ΔS, which is positive for this process. Note that for macroscopic systems, the entropy function $S(x)$, where x is the location of the partition, has a *single* maximum at x_{eq}, i.e. the value of x at which the pressures in the two compartments are equal, i.e. $P = P_1 = P_2$.

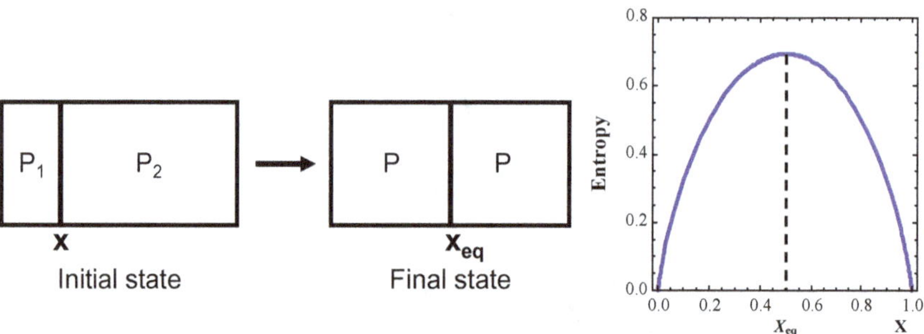

Fig. 1.13   An isolated system with a movable partition and the corresponding entropy as a function of the location of the partition x. Initially, the pressures in the two compartments are different, $P_1 > P_2$. After equilibrium is reached, $P = P_1 = P_2$.

What we have described above is essentially the second law of thermodynamics. When we release a constraint on the system, here the location of the partition x, the system will move with *high probability* towards the equilibrium value x_{eq}, for which the entropy is maximum. Once we reach the equilibrium state, the system will stay there forever. We shall not notice any measurable deviations from the equilibrium state.

In thermodynamics, high probability means *certainty*, or probability one. This is far from being the case when we apply the concept of a driving force to microscopic systems (the two particles at distance R, or a protein in some conformation $\mathbf{R}_1, \ldots, \mathbf{R}_n$). Here, there is usually more than one maximum, and there are finite probabilities for finding the system in any state (here a specific distance R). There are states with a higher probability, but these are far from having probability one.

We have presented above the example of an isolated system to show the difference between a driving force applied to a macroscopic process, and when applied to a microscopic process. The same is true for a system at constant temperature and pressure.

In a macroscopic system characterized by the variables $T, P, \mathbf{N}$ (where $\mathbf{N}$ is the composition vector $(N_1, \ldots, N_c)$), the Gibbs energy will reach a *minimum* with respect to any internal parameter, say ξ, the extent of the reaction, or the mole fraction of one component in an equilibrium between two isomers of the same compound. In such a case, there is again a *single minimum* of the function $G(P, T, \mathbf{N}; \xi)$ with respect to ξ, provided that we keep $P, T, \mathbf{N}$ constant.[22] If we start with any initial state and release the constraint on ξ, the system proceeds to the state for

---

[22]Note that $\mathbf{N}$ constant refers to the number of *independent* components, i.e. we refer to a closed system. When a chemical reaction occurs in a closed system, the number of independent components is unchanged.

which the Gibbs energy is minimum, i.e.

$$\left(\frac{\partial G}{\partial \xi}\right)_{P,T,\mathbf{N}} = 0. \tag{1.2.23}$$

This is equivalent to the second law, but stated for a closed system at constant P and T.

As an example, consider a system at constant P, T and with total number of molecules N. Each molecule can be in one of two states, say A and B. Suppose we prepare the system with an arbitrary composition x_A (and $x_B = 1 - x_A$) and add a catalyst to initiate the conversion between A and B. If we keep T, P, N constant, the system will move towards the state of minimum G having the equilibrium composition $x_{A,\text{eq}}$, determined by the condition

$$\mu_A(P, T, x_{A,\text{eq}}) = \mu_B(P, T, x_{A,\text{eq}}). \tag{1.2.24}$$

Figure 1.14a shows the initial and the final states of the system and the Gibbs energy as a function of x_A (with P, T, N held constant). A similar example is shown in Fig. 1.14b. Here in the initial system, all the particles are in the left compartment, which

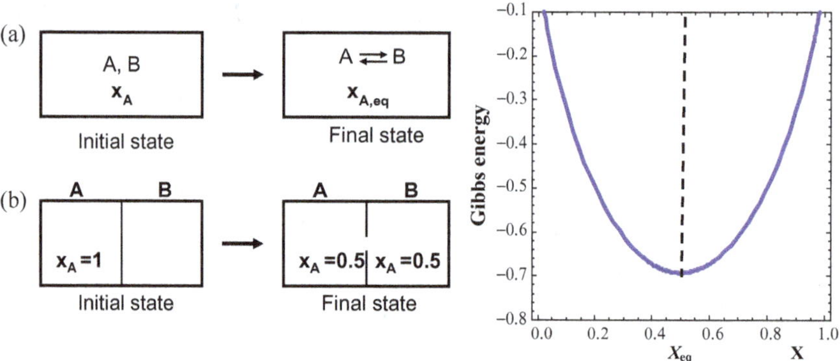

Fig. 1.14   (a) The system at the initial composition x_A moves to the final composition $x_{A,\text{eq}}$, at which the Gibbs energy has a minimum. (b) Process of expansion. Initially, all the particles are in compartment A. After the expansion, the particles are spread uniformly in A and B.

we denote by the A-compartment. The initial mole fraction of particles in the A-compartment is $x_A = 1$. When we open a small window between the two compartments, the Gibbs energy of the system will change in that direction for which the *driving force* is negative, i.e. $\Delta G < 0$. When the system reaches the equilibrium value of x_{eq} (here $x_A = 1/2$), the system will stay there forever. The reader is urged to examine carefully the two examples shown in Figs. 1.14a and 1.14b, and perhaps invent other examples to clarify the role of the driving force in a thermodynamic process and compare that to its role in a microscopic process.

Note carefully that in this case the system is macroscopic — there is only *one minimum* to the function $G(x_A)$, and that from any initial point x_A, the system will move monotonically towards the equilibrium state at $x_{A,\text{eq}}$. Strictly speaking, the motion to the equilibrium state is not perfectly monotonic. However, for macroscopic systems we shall not be able to observe any deviations from such monotonic downward motion towards the minimum of G.

The situation is quite different when we consider the function $W(R)$, or $G(R)$ where R is the distance between the two particles. Figure 1.10b shows that the function $G(R)$ has more than one minimum. Starting from any point R, the system has a thermodynamic driving force to move towards the location R_{min}, i.e. $\Delta G = G(R_{\text{min}}) - G(R) < 0$. However, there is no guarantee that if we release the system at a distance R, it will move monotonically towards R_{min}. Figure 1.10 shows two possible initial states, R_3 and R_4. In both, we have $G(R_{\text{min}}) - G(R_3) < 0$ and $G(R_{\text{min}}) - G(R_4) < 0$. However, at R_3 the average *force* is in the same direction towards the minimum at R_{min}, but at R_4 the average force is in the opposite direction. In both cases, the initial movement of the system is not necessarily along the average force, nor towards the minimum of G at R_{min}.

The distinction between *force* and *thermodynamic driving force* on one hand, and between direct and indirect "forces" on the other, will be important in the study of protein folding in Chapter 3. Unfortunately, the biochemical literature is quite confusing on these concepts. For instance, in a review article titled "Dominant Forces in Protein Folding,"[23] there is nothing about actual *forces* acting on the protein, or any discussion of the different meanings of *forces* and *thermodynamic forces*. We shall also see in Chapter 3 that what the author refers to as "dominant forces" are far from being dominant!

In a more recent review by Levy and Onuchic,[24] one finds a schematic curve of $W(r)$, referred to as "the potential energy function," similar to Fig. 1.15. The reader should be careful to note that what is referred to as the "potential energy function" by Levy and Onuchic, as well as many others, is referred to in this book as the PMF or the Gibbs energy function. Furthermore, the configurations A and B in Fig. 1.15 are referred to as "direct interaction" and "solvent-mediated interaction,"

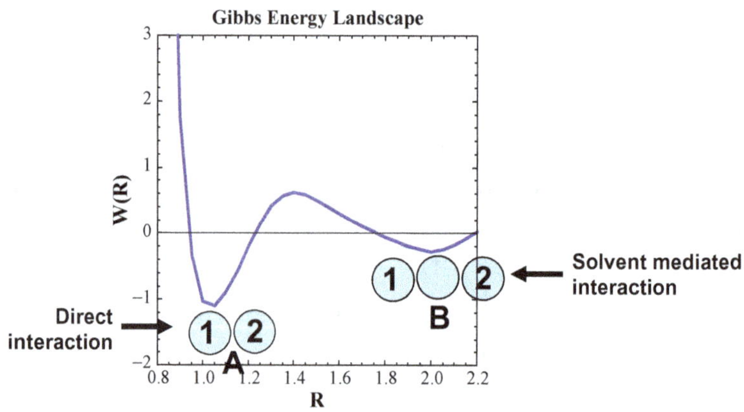

Fig. 1.15   A schematic form of the PMF sometimes referred to as the "potential energy function." The two configurations of particles 1 and 2, A and B are referred to by Levy and Onuchic (2006), as "direct" and "solvent-mediated" interactions.

---

[23] Dill (1990).
[24] Levy and Onuchic (2006).

respectively. In this book, when we talk about the energy landscape, $U(R)$ in Fig. 1.10a, *at any point* on this landscape, the interaction is *direct*. On the other hand, when we talk about the Gibbs energy landscape (Figs. 1.10b or 1.15) again *at any point* (including A and B of Fig. 1.15), we have both direct and *indirect* "interactions," i.e. at each distance R, the PMF has two components, the direct and the indirect parts.

1.3. Definition of the Solvation Process and the Corresponding Thermodynamic Quantities

In this section we turn to another notoriously confusing concept that is essential to understanding the role of the solvent in any biochemical process. This is the concept of solvation and the associated thermodynamic quantities of solvation.

Traditionally, the solvation of solutes was studied within the context of thermodynamics. In this context, it could be studied only at the limit of very dilute solutions, i.e. in the concentration range where Henry's law is obeyed.

In 1978, a new process of solvation was introduced along with the corresponding thermodynamic quantities.[25] With this new concept, the study of solvation became a powerful tool for probing the extent of the interaction between the solute and the solvent and the effect of the solute on various molecular distribution functions in the solvent.

We *define* the process of solvation using the following simple thought experiment. We transfer a molecule α from a *fixed* position in an ideal gas phase into a *fixed* position in the liquid phase (l) (Fig. 1.16).[26]

---

[25] Ben-Naim (1978), and for more details, Ben-Naim (2006).

[26] Note that this definition of the solvation process applies to any *molecule α* being solvated by *any* liquid l.

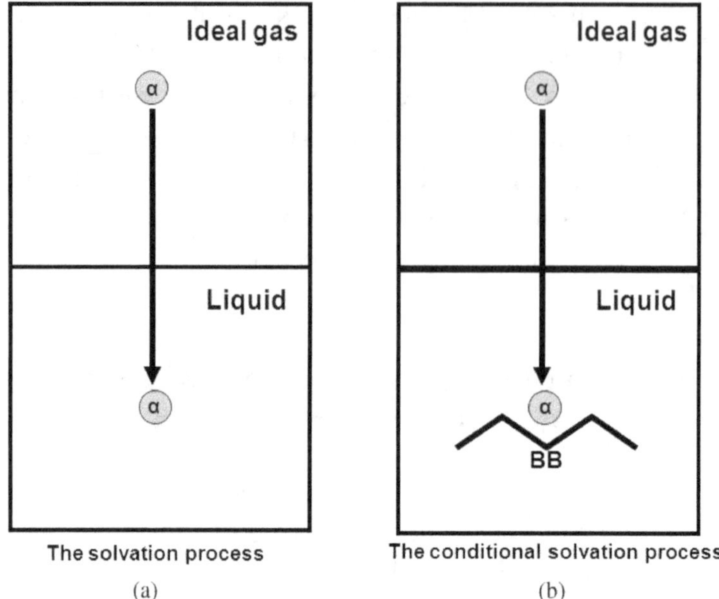

Fig. 1.16 The process of (a) solvation and (b) the conditional solvation of a molecule α.

If α is a spherical particle such as argon, then we place the center of α at a fixed position. When α is not spherical, then we may choose the center of mass of the molecule, or any other convenient point in the molecule, to be placed at a fixed position. The important thing is not where we choose the "center" of the molecule, but the lack of translational degrees of freedom of the molecule as a whole.

Clearly, the process defined above cannot be carried out experimentally, but since we consider a classical system, it is meaningful to think of a particle lacking any translational degrees of freedom, i.e. at *fixed* position and *zero* velocity. Fortunately, all the thermodynamic quantities associated with this *solvation process* may be obtained from experimentally *measurable* quantities.

The general expression for the chemical potential of α in any liquid system can be written as[27]

$$\mu_\alpha^l = \mu_\alpha(T, P, \mathbf{N}) = \mu_\alpha^{*l}(T, P, \mathbf{N}) + k_B T \ln \rho_\alpha \Lambda_\alpha^3. \qquad (1.3.1)$$

We use here the independent variables $T, P, \mathbf{N}$.[28] These are the most common variables that are controlled in solution chemistry. In writing (1.3.1), we have already committed ourselves to systems where the translational degrees of freedom are classical (Λ_α^3 arises from the integration over the classical Maxwell–Boltzmann distribution of momenta). Any other internal degrees of freedom that α might have can be treated either classically or quantum mechanically. The corresponding *internal* partition function of α denoted q_α is absorbed by μ_α^{*l}.

It is convenient to interpret the two terms in (1.3.1) as follows (Fig. 1.17). The process of adding one α to the system is performed in two steps. First, we place α at a fixed position in the system. The corresponding change in Gibbs energy is μ_α^{*l}, referred to as the pseudo-chemical potential of α. Second, we release the particle from the constraint of being at a fixed

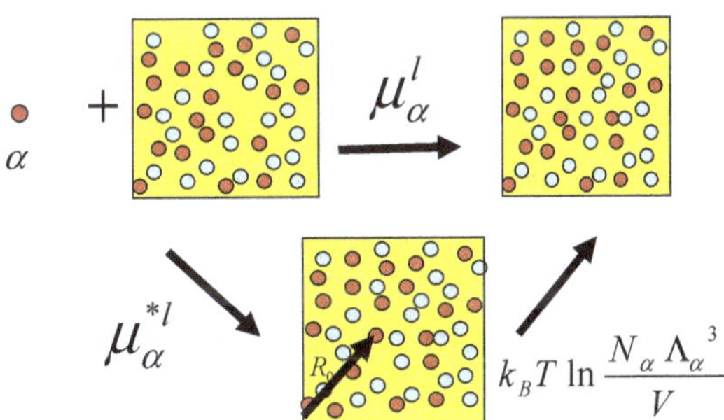

Fig. 1.17 The process of introducing molecule α into the liquid l is split into two steps: First, the molecule is placed at a fixed position at $\mathbf{R}_0$, then it is released.

---

[27] For more details, see Appendix A and Ben-Naim (2006).
[28] $\mathbf{N}$ is understood to contain the entire composition of the liquid phase l.

position. We call this the *liberation* process. The corresponding change in Gibbs energy is $k_B T \ln \rho_\alpha \Lambda_\alpha^3$. For classical particles, this term is always negative.[29]

If α is in an ideal gas phase, then the chemical potential of α has the form

$$\mu_\alpha^{ig} = \mu_\alpha^{*ig} + k_B T \ln \rho_\alpha \Lambda_\alpha^3$$
$$= k_B T \ln q_\alpha^{-1} + k_B T \ln \rho_\alpha \Lambda_\alpha^3. \qquad (1.3.2)$$

Again, we may interpret the first term on the right-hand side of (1.3.2) as the change in Gibbs energy for the process of placing α at a fixed position in the ideal gas phase. Clearly, in this process the particle carries with it all its internal degrees of freedom. By definition of the ideal gas system, there are no interactions between α and other particles in the system. We now define the solvation Gibbs energy associated with the solvation process by the difference

$$\Delta\mu_\alpha^* = \Delta G_\alpha^*(ig \to l) = \mu_\alpha^{*l} - \mu_\alpha^{*ig}. \qquad (1.3.3)$$

Note that we have assumed that Λ_α^3 is the same in the two phases. Again, classically speaking, the momentum partition function depends only on the temperature and is not affected by the intermolecular interactions. Any other degrees of freedom might or might not be affected by the interactions. For small rigid molecules, we also assume that q_α is unaffected by the interactions. Hence, in this case, (1.3.3) reduces to[30]

$$\Delta\mu_\alpha^* = \Delta G_\alpha^*(ig \to l) = W(\alpha|l) = -k_B T \ln\langle\exp[-\beta B_\alpha]\rangle_0.$$
$$(1.3.4)$$

---

[29]See Hill (1956) and Ben-Naim (2006).

[30]This expression results directly from the statistical mechanical definition of the chemical potential. See Ben-Naim (2006) and Appendices A and B.

Thus, the solvation Gibbs energy is the *coupling work* of α to the rest of the system. It is the average of the quantity $\exp[-\beta B_\alpha]$, where B_α is the binding energy of α to all the molecules in the system at some fixed configuration.

The binding energy of α to the rest of the system is simply defined by

$$B_\alpha = \sum_{i=1}^{N} U(\mathbf{R}_\alpha, \mathbf{X}_i), \qquad (1.3.5)$$

where $\mathbf{R}_\alpha$ is the locational vector of the "solute" α[31] and $\mathbf{X}_i$ is the configuration (location and orientation) of the ith "solvent"[32] molecule. In writing (1.3.5) we assume that α is a spherical particle and that the solvent molecules are rigid molecules described by the location of the center of mass and three orientational angles. We also assume for simplicity that the total interaction energy in the system is pairwise additive. This assumption is not essential for the definition of the solvation process.[33]

The average in (1.3.4) is taken over all the configurations of all the molecules in the system, in the T, P, N ensemble, using the distribution function of configurations of all the particles in the system *before* the addition of α.[34] This average is indicated by the symbol $\langle \ \ \rangle_0$.

In the more general case, when α is inserted into the system, some internal degrees of freedom might be affected. Hence $\Delta\mu_\alpha^*$ in general includes both the coupling work and the effect of

---

[31] Note that the molecule α is, in general, a solute molecule. However, in this definition, α can be any molecule in the system.

[32] Here, "solvent" includes all the molecules in the system except the specific α-molecule at $\mathbf{R}_\alpha$.

[33] See Ben-Naim (2006, 2009).

[34] Note again that we do not require that α be diluted in the system. By "all the particles" in the *system,* we include any number of α molecules, except the particular molecule placed at $\mathbf{R}_\alpha$.

the solvation process on the internal degrees of freedom. In this section we shall assume that q_α (as well as Λ_α^3) is not affected by the process of solvation.

A detailed comparison between $\Delta\mu_\alpha^*$ and the conventional standard Gibbs energy of solvation has been discussed elsewhere.[35] It has been shown that none of the conventional quantities of solvation is identical to the solvation quantities as defined above. Nevertheless, some confusion still lingers concerning the meaning of the various solvation quantities. We therefore add the following two comments:

First, $\Delta\mu_\alpha^*$ is, in general, *not* the excess chemical potential of α, neither with respect to ideal gas, nor with respect to dilute ideal solutions.[36] Unfortunately, $\Delta\mu_\alpha^*$ is sometimes referred to as the excess chemical potential of the solute.[37] These two excess chemical potentials are defined as (see Eq. (1.3.2))

$$\mu_\alpha^{EX,IG} = \mu_\alpha^l - k_B T \ln \rho_\alpha \Lambda_\alpha^3 q_\alpha^{-1}, \qquad (1.3.6)$$

$$\mu_\alpha^{EX,DI} = \mu_\alpha^l - [\mu_\alpha^{0,\rho} + k_B T \ln \rho_\alpha]. \qquad (1.3.7)$$

where $\mu_\alpha^{0,\rho}$ is the *standard* chemical potential of α defined by

$$\mu_\alpha^{0,\rho} = \lim_{\rho_\alpha \to 0} [\mu_\alpha^l - k_B T \ln \rho_\alpha]. \qquad (1.3.8)$$

If q_α is *unaffected* by the interactions among the molecules in the system, we have

$$\mu_\alpha^l = W(\alpha|l) + k_B T \ln \rho_\alpha \Lambda_\alpha^3 q_\alpha^{-1} = \mu_\alpha^{*l} + k_B T \ln \rho_\alpha \Lambda_\alpha^3. $$
$$(1.3.9)$$

---

[35] Ben-Naim (2006).

[36] See, for example, Arthur and Haymet (1999).

[37] Note also that "excess" chemical potential may also mean an excess with respect to symmetrical ideal solutions [see Ben-Naim (2006)].

Hence, from (1.3.2), (1.3.6), (1.3.7) and (1.3.8), we get for this case

$$\mu_\alpha^{EX,IG} = W(\alpha|l) = \Delta\mu_\alpha^{*l}, \tag{1.3.10}$$

$$\mu_\alpha^{EX,DI} = W(\alpha|l) - \lim_{\rho_\alpha\to 0} W(\alpha|l)$$

$$= \mu_\alpha^{*l} - \lim_{\rho_\alpha\to 0} \mu_\alpha^{*l} = \Delta\mu_\alpha^{*l} - \lim_{\rho_\alpha\to 0}\Delta\mu_\alpha^{*l}. \tag{1.3.11}$$

Thus, only when q_α is unaffected by the solvation process, $\mu_\alpha^{EX,IG}$ becomes identical to the solvation Gibbs energy, whereas $\mu_\alpha^{EX,DI}$ is the difference between the solvation Gibbs energy of α in the actual system, and in the limit of the dilute ideal system with respect to α.

The second comment is more subtle. As we can see from (1.3.10), when q_α is unaffected by the solvation process, $\mu_\alpha^{EX,IG}$ becomes *numerically* identical to $\Delta\mu_\alpha^{*l}$. However, these two quantities refer to two *different* processes. $\Delta\mu_\alpha^{*l}$ is the change in the Gibbs energy of the solvation process as defined above. On the other hand, $\mu_\alpha^{EX,IG}$ may be interpreted as the change in the Gibbs energy for transferring one α molecule from an ideal gas to the liquid at the same *number density*, i.e.

$$\mu_\alpha^{EX,IG} = \mu_\alpha^l - \mu_\alpha^{ig}\Big|_{\rho_\alpha^l=\rho_\alpha^{ig}} = \mu_\alpha^{*l} - \mu_\alpha^{*ig} = \Delta\mu_\alpha^{*l}. \tag{1.3.12}$$

Clearly, the latter process is different from the solvation process, yet the change in the Gibbs energies in the two processes is the same. This has led some to confuse the solvation process with the conventional *standard* free energy of solvation based on the number density.[38] To clarify the issue, let us denote by ρ-process the process of transferring one α molecule from an ideal gas to the liquid at the same number density. Thus, the *numerical values* of the Gibbs energy change for the *solvation process*

---

[38]For more details, see Ben-Naim (2006).

and the ρ-process are the same. However, since the two processes are different, they might have different thermodynamic quantities of solvation, such as entropy, enthalpy, etc.

To see this, consider the entropy change for the two processes. The solvation entropy is simply defined by

$$\Delta S_\alpha^{*l} = -\left(\frac{\partial \Delta \mu_\alpha^{*l}}{\partial T}\right)_P. \tag{1.3.13}$$

On the other hand, the entropy change for the ρ-process is

$$\Delta S(\rho\text{-process}) = -\left(\frac{\partial \Delta \mu_\alpha^{*l}}{\partial T}\right)_P - \frac{\partial}{\partial T}\left(k_B T \ln \frac{\rho_\alpha^l}{\rho_\alpha^{ig}}\right)_P\bigg|_{\rho_\alpha^l = \rho_\alpha^{ig}}$$

$$= \Delta S_\alpha^{*l} + k_B T \alpha_P^l - k_B, \tag{1.3.14}$$

where

$$\alpha_P^l = \frac{1}{V^l}\left(\frac{\partial V^l}{\partial T}\right)_P. \tag{1.3.15}$$

Thus, when studying the entropy of solvation of a solute α, the proper quantity is ΔS_α^{*l}, as defined in (1.3.13). However, since the thermal expansion coefficients at constant pressure for the liquid and the gaseous phases are different, the quantity $\Delta S(\rho\text{-process})$ is, in general, not equal to the entropy change for the solvation process. Similar comments apply to the enthalpy, the volume, etc., of the solvation process. The temperature and the pressure derivatives of the solvation Gibbs energy $\Delta \mu_\alpha^{*l}$ give the correct thermodynamic quantities of solvation. On the other hand, the temperature and pressure derivatives of the Gibbs energy change of the ρ-process (or the excess quantity $\mu_\alpha^{EX,IG}$) do not provide an appropriate measure of the relevant solvation thermodynamic quantities.

Table 1.1 shows some values of the thermodynamic quantities of solvation of inert gases in water. Table 1.2 shows the

Table 1.1.   Values of ΔG_s^*, ΔS_s^*, ΔH_s^*, and $\Delta C_{p,s}^*$ for Neon Argon, Krypton and Xenon in H_2O and D_2O[a]

$t\,(^\circ C)$		ΔG_s^* (kJ mol^{-1})	ΔS_s^* (J/K mol)	ΔH_s^* (kJ/mol)	$\Delta C_{p,s}^*$ (J/K mol)
Neon					
25	(H_2O)	11.19	−45.4	−2.3	
	(D_2O)	10.99	−43.6	−2.0	
50	(H_2O)	12.14	−33.2	1.4	150
	(D_2O)	11.90	−32.2	1.5	140
75	(H_2O)	12.81	−23.1	4.8	130
	(D_2O)	12.56	−22.9	4.6	120
100	(H_2O)	13.26	−14.7	7.8	120
	(D_2O)	13.02	−15.2	7.3	110
Argon					
25	(H_2O)	8.40	−65.1	−11.0	
	(D_2O)	8.15	−68.5	−12.26	
50	(H_2O)	9.76	−47.8	−5.7	210
	(D_2O)	9.60	−50.8	−6.8	220
75	(H_2O)	10.74	−33.6	−0.9	190
	(D_2O)	10.64	−36.0	−1.9	200
100	(H_2O)	11.39	−21.8	3.3	170
	(D_2O)	11.34	−23.5	2.6	180
Krypton					
25	(H_2O)	6.94	−70.5	−14.1	
	(D_2O)	6.99	−68.4	−13.4	
50	(H_2O)	8.40	−51.5	−8.2	230
	(D_2O)	8.43	−50.8	−8.0	220
75	(H_2O)	9.46	−36.8	−3.3	200
	(D_2O)	9.48	−36.7	−3.3	190
100	(H_2O)	10.21	−25.4	0.7	160
	(D_2O)	10.23	−25.4	0.75	160
Xenon					
25	(H_2O)	5.62	−81.9	−18.8	
	(D_2O)	5.64	−70.9	−15.5	
50	(H_2O)	7.28	−56.7	−11.0	310
	(D_2O)	7.18	−55.2	−10.7	200
75	(H_2O)	8.41	−38.6	−5.0	240
	(D_2O)	8.34	−41.0	−5.9	190
100	(H_2O)	9.17	−25.4	−0.3	190
	(D_2O)	9.17	−28.3	−1.4	180

[a]Based on data from Crovetto *et al.* (1982), taken from Ben-Naim (1987).

Table 1.2.   Values of the Solvation Gibbs Energy, Entropy, Enthalpy, and Heat Capacity of Methane in Water and in some Non-aqueous Solvents at Two Temperatures[a]

Solvent	t (°C)	ΔG_s^* (kJ mol^{-1})	ΔS_s^* (J mol^{-1} K^{-1})	ΔH_s^* (kJ mol^{-1})	$\Delta C_{p,s}^*$ (J mol^{-1} K^{-1})
Water	10	7.31	−76.6	−14.3	} 227
	25	8.37	−64.8	−10.9	
Methanol	10	1.43	−10.8	−1.63	{ −88
	25	1.63	−15.4	−2.97	
Ethanol	10	1.38	−13.4	−2.38	{ −22
	25	1.59	−14.6	−2.72	
1-Propanol	10	1.44	−18.0	−3.68	{ 105
	25	1.67	−12.5	−2.10	
1-Butanol	10	1.54	−11.7	−1.76	{ −136
	25	1.80	−3.81	−3.81	
1-Pentanol	10	1.67	−13.8	−2.43	{ −13
	25	1.88	−15.1	−2.63	

[a]Data from Yaacobi and Ben-Naim (1974).

solvation quantities of methane in water and in a few other solvents.

1.4. The Conditional Solvation Process

The solvation process was defined in Sec. 1.3 as the process of transferring one molecule α from a fixed position in an ideal gas phase to a fixed position in the liquid phase. The statistical mechanical expressions for the solvation Gibbs energy of a molecule α is[39]

$$\Delta\mu_\alpha^* = -k_B T \ln\langle\exp[-\beta B_\alpha]\rangle_0, \qquad (1.4.1)$$

---

[39]Recall that this is true for molecules for which the internal partition function is not affected by the solvation process (see Sec. 1.3 and Appendix B).

where the average is over all the configurations of all the molecules in the system in the T, P, N ensemble with the probability distribution $P_0(\mathbf{X}^N)$.[40]

The conditional solvation process is defined similarly to the solvation process, except that the molecule α is placed at a fixed point near a backbone (BB) molecule (Fig. 1.16b). The corresponding expression for the conditional Gibbs energy of solvation is[41]

$$\Delta\mu^*_{\alpha/\mathrm{BB}} = -k_B T \ln\langle\exp[-\beta B_\alpha]\rangle_{\mathrm{BB}}. \qquad (1.4.2)$$

Note that in (1.4.2), B_α is the same as in (1.4.1), i.e. this is the total interaction energy of α with all the other molecules (the "solvent") in the system, but excluding the backbone molecule. The average in (1.4.2) is taken with the conditional probability density $P_0(\mathbf{X}^N/\mathrm{BB})$, which means the probability of finding the configuration $\mathbf{X}^N$ of all the molecules in the system (excluding the specific α molecule and the BB), given that the BB is at some fixed position in the liquid.

As we shall see later in this chapter and in Chapters 3 and 4, it is the *conditional* solvation quantities rather than the solvation quantities that feature in the building up of the solvent-induced part of the driving force in both protein folding and protein–protein association.

It is important to understand the qualitative difference between the solvation process (Fig. 1.16a) and the conditional solvation process (Fig. 1.16b). In both, the factor $\exp[-\beta B_\alpha]$ is the same. However, the distribution function is different. In one,

---

[40]$P_0(\mathbf{X}^N)$ means the distribution of the configurations of all molecules in the system except the specific molecule α placed at the fixed position. The subscript zero is used to indicate that the distribution refers to all molecules in the system before placing one α molecule at a fixed position. Likewise, $P_0(\mathbf{X}^N/\mathrm{BB})$ is the same as $P_0(\mathbf{X}^N)$, except that now the backbone (BB) is present at some fixed position in the system (Fig. 1.16b).

[41]See also Appendix D.

we have the distribution of the configurations of all "solvent" molecules before adding the specific α molecule. In the second, we have a *modified* distribution, i.e. a distribution of "solvent" molecules that is affected by the presence of the backbone. In a sense, the conditional solvation is similar to the solvation process of α in a *different* solvent; a solvent, the properties of which were modified by the presence of the BB. This effect is similar to the effect produced by the addition of a new co-solvent or a new solute into the system.

1.5. Some Numerical Values of Solvation Thermodynamics

In this section, we present some numerical values for the solvation and conditional solvation thermodynamic quantities of a few molecules or groups in water.

1.5.1. *Solvation of hydrophilic molecules or groups*

We start with the solvation of a water molecule in liquid water. Table 1.3 shows the solvation quantities of water in water (both H_2O and D_2O) at different temperatures. The solvation Gibbs energies are calculated from the densities of water in the liquid and in the gaseous phases at equilibrium, and the assumption is made that the vapor phase is an ideal gas.

$$\Delta G_W^* = k_B T \ln(\rho_W^g / \rho_W^l)_{eq} \approx k_B T \ln(P_W / k_B T \rho_W^l)_{eq}. \quad (1.5.1)$$

The reader is urged to compare the values of ΔG_W^* in Table 1.3 with those in Tables 1.1 and 1.2. Note in particular that in the case of solvation of a hydrophobic group (Table 1.1), the solvation Gibbs energy is dominated by the entropy term

Table 1.3. Solvation Thermodynamic Quantities of H_2O and D_2O as a Function of Temperature. The Pressure Corresponds to the Vapor Pressure at Equilibrium with the Liquid at Each Temperature[a]

	Temperature (°C)	ΔG_W^* (kJ mol^{-1})	ΔS_W^* (JK^{-1} mol^{-1})	ΔH_W^* (kJ mol^{-1})	$\Delta C_{p,W}^*$ (JK^{-1} mol^{-1})
H_2O	0	−27.79	−54.56	−42.69	
H_2O	5	−27.52	−53.92	−42.52	35.4
D_2O		−27.94	−57.8	−44.02	43.3
H_2O	10	−27.25	−53.28	−42.34	36.6
D_2O		−27.65	57.01	−43.80	44.7
H_2O	15	−26.99	−52.62	−42.15	37.4
D_2O		−27.37	−56.22	−43.57	45.6
H_2O	20	−26.73	−51.97	−41.96	38.0
D_2O		−27.09	−55.43	−43.34	46.2
H_2O	25	−26.47	−51.33	−41.77	38.4
D_2O		−26.82	−54.64	−43.11	46.4
H_2O	30	−26.21	−50.68	−41.58	38.6
D_2O		−26.54	−53.87	−42.87	46.5
H_2O	35	−25.96	−50.05	−41.38	38.6
D_2O		−26.28	−53.11	−42.64	46.3
H_2O	40	−25.71	−49.43	−41.19	38.5
D_2O		−26.01	−52.37	−42.41	46.0
H_2O	45	−25.47	−48.83	−41.00	38.2
D_2O		−25.75	−51.64	−42.18	45.6
H_2O	50	−25.22	−48.23	−40.81	37.9
D_2O		−25.50	−50.94	−41.96	45.1

[a]Data from Marcus and Ben-Naim (1985).

(i.e. $|T\Delta S_s^*| > |\Delta H_s^*|$). On the other hand, the solvation Gibbs energy of water in water is dominated by the enthalpy term (i.e. $|T\Delta S_s^*| < |\Delta H_s^*|$). It should be noted that these values in themselves will not be used in the discussions of the solvent-induced effects in biochemical processes. What is more important is the *conditional* solvation thermodynamic quantities of hydrophilic *groups* such as amine (NH), carbonyl (C=O) and hydroxyl (O–H). The reason for this will become clear from the analysis made in Sec. 1.8, and from the rest of the book.

Here, we shall "extract" from the solvation Gibbs energies in Table 1.3 a quantity to which we shall refer to as the *conditional solvation Gibbs energy of one arm of a water* molecule. This quantity will be used in the rest of the book to assess the relative importance of hydrophilic vs. hydrophobic effects.

To achieve that, we assume that we can solvate a water molecule in two steps. We assume that the total binding energy of a water molecule may be split into two parts

$$
B_W(\mathbf{X}_0) = \sum_{i=1}^{N} U(\mathbf{X}_0, \mathbf{X}_i)
$$

$$
= \sum_{i=1}^{N} U^{LJ}(\mathbf{X}_0, \mathbf{X}_i) + U^{HB}(\mathbf{X}_0, \mathbf{X}_i)
$$

$$
= B^{LJ} + B^{HB}. \tag{1.5.2}
$$

The idea is that the interaction between a water molecule at $\mathbf{X}_0$ and any "solvent" molecule at $\mathbf{X}_i$ may be split into two contributions, a Lennard-Jones (LJ) part and a hydrogen bond (HB) part.[42] Corresponding to the split of the binding energy in (1.5.2), we can write the solvation Gibbs energy of a water molecule as

$$
\Delta G_W^* = \Delta G_{LJ}^* + \Delta G_{HB/LJ}^* \approx \Delta G_{Ne}^* + \Delta G_{HB/LJ}^*. \tag{1.5.3}
$$

The split of ΔG_W^* into two parts corresponds to the process of solvation of a water molecules in two steps. First, we "solvate" the LJ part, which may be approximated by the solvation

---

[42]We also include in the HB part all interactions due to electric multi-poles. For more details on the HB part of the water–water interaction, see Ben-Naim (2009), Chapter 2.

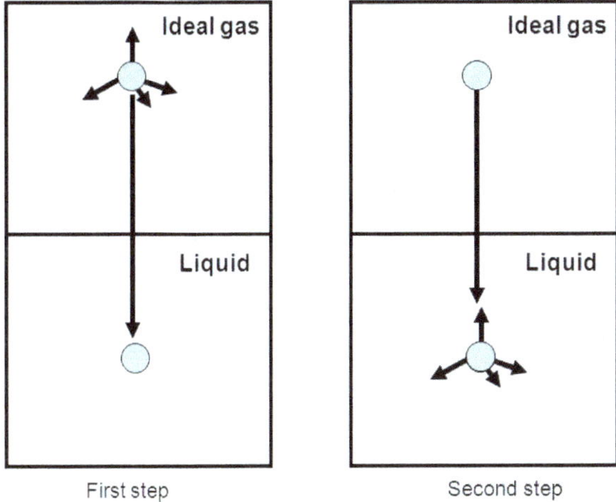

Fig. 1.18   The solvation of a water molecule in two steps. First, we solvate the hard and soft parts of the interactions, then we solvate the four "arms" of the water molecule.

Gibbs energy of a neon atom. Second, we "turn on" the HB part of the interaction, i.e. we "solvate" the four "arms" of the water molecule. This process is shown schematically in Fig. 1.18. Furthermore, we assume that the HB part of the solvation Gibbs energy is the sum of four equal parts, each due to the conditional solvation Gibbs energy of one "arm" of the water molecule. The idea underlying this assumption is that a water molecule can form four HBs with other water molecules along four "arms," as shown in Fig. 1.19[43] and that the four arms are independently solvated.[44] The theoretical background for these assumptions is discussed in Part I, Chapter 2.

---

[43] One can make a distinction between an O–H "arm" and a lone-pair "arm." However, when a HB is formed, the energy of the HB is the same for these two kinds of "arms."

[44] This means that the conditional solvation of one arm is approximately independent of whether other arms are solvated or not. Strictly, one should turn on one arm at a time; the conditional solvation of "arm" k might be different from that of "arm" $k - 1$. There is evidence that these differences are negligible. For more details, see Part I, Chapter 2, and Predota *et al.* (2003).

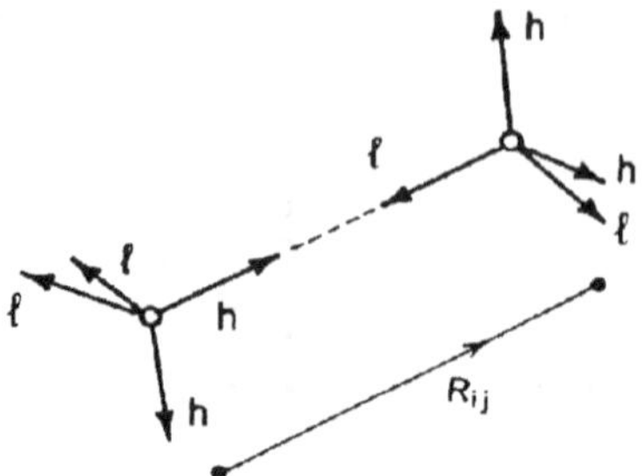

Fig. 1.19   The HB interaction between two water molecules is realized when one "arm" of one water molecule is directed towards an "arm" of the second water molecule. *h* and *l* refer to the hydrogen and the lone pair of electron directions. For the mathematical description of this interaction, see Ben-Naim (2009), Chapter 2.

With these assumptions we can estimate that the conditional solvation Gibbs energy of one "arm" of water at 25°C is about (see Tables 1.1 and 1.3)

$$\Delta G^*_{HB}(\text{one arm of H}_2\text{O/LJ}) = \frac{\Delta G^*_W - \Delta G^*_{Ne}}{4}$$

$$= \frac{26.47 - 11.19}{4} = -9.42 \, \text{kJ/mol}. \tag{1.5.4}$$

For D_2O the corresponding value is slightly larger, i.e.

$$\Delta G^*_{HB}(\text{one arm of D}_2\text{O/LJ})$$

$$= \frac{-26.82 - 10.99}{4} = -9.45 \, \text{kJ/mol}. \tag{1.5.5}$$

We also add here the value of the conditional solvation entropy and enthalpy of one arm of water at 25 °C.

$$\Delta S^*_{HB}(\text{one arm of H}_2\text{O/LJ}) = \frac{\Delta S^*_W - \Delta S^*_{Ne}}{4}$$

$$= \frac{-51.33 + 45.4}{4} = -1.48 \, \text{JK}^{-1}/\text{mol}, \tag{1.5.6}$$

$$\Delta H^*_{HB}(\text{one arm of } H_2O/LJ) = \frac{\Delta H^*_W - \Delta H^*_{Ne}}{4}$$

$$= \frac{-41.77 + 2.3}{4} = -9.87\,\text{kJ/mol}. \qquad (1.5.7)$$

Thus, we see that the value of ΔG^*_{HB} (either per arm or per molecule) is dominated by the *enthalpy* (or the energy) of solvation. The entropy contribution to ΔG^*_{HB} is minor.

It is of interest to note that if we repeat the same type of calculation for the solvation Gibbs energies of H_2S and H_2Se in water we get much smaller values, thus

$$\Delta G^*_{HB}(\text{one arm of } H_2S/LJ)$$

$$= \frac{\Delta G^*_{H_2S} - \Delta G^*_{Ar}}{4} \approx -2.7\,\text{kJ/mol}, \qquad (1.5.8)$$

$$\Delta G^*_{HB}(\text{one arm of } H_2Se/LJ)$$

$$= \frac{\Delta G^*_{H_2Se} - \Delta G^*_{Kr}}{4} \approx -2.17\,\text{kJ/mol}. \qquad (1.5.9)$$

We next turn to some data on the conditional solvation Gibbs energies of hydrophilic groups next to some hydrocarbon molecules.

A hydroxyl group

To estimate the conditional Gibbs energy of a hydroxyl group, we can take one of the two differences:

$$\Delta G^*_{HB}(\text{of } OH/CH_3CH_2)$$

$$= \Delta G^*_{CH_3CH_2OH} - \Delta G^*_{CH_3CH_3}$$

$$= -27.82\,\text{kJ/mol}, \qquad (1.5.10)$$

$$\Delta G_{HB}^*(\text{of OH/LJ of } CH_3CH_2CH_3)$$

$$= \Delta G_{CH_3CH_2OH}^* - \Delta G_{CH_3CH_2CH_3}^*$$

$$= -28.36\,\text{kJ/mol.} \tag{1.5.11}$$

These are two different estimates.[45] In the first, we take an ethane molecule and replace a hydrogen atom with an OH group to obtain ethanol. This is exactly the meaning of a conditional solvation of an OH group next to an ethane molecule. In the second case, we start with a propane molecule. We view the methyl group as an approximate "neon group," then we replace it with a hydroxyl group. Therefore, what we have added is only the conditional solvation of the hydrogen-bonding "arms" of the hydroxyl group. Both of these involve approximations, but the values obtained are very similar. If we take the value of $-9.42\,\text{kJ/mol}$ as the conditional solvation Gibbs energy of one arm of water, see (1.5.4), then the values obtained in (1.5.10) and (1.5.11) are very nearly the same as the value estimated for the conditional solvation of *three* arms, i.e.

$$\Delta G_{HB}^*(\text{three arms of } H_2O/LJ)$$

$$= 3\Delta G_{HB}^*(\text{one arm of } H_2O/LJ) \approx 3(-9.42)$$

$$= -28.26\,\text{kJ/mol.} \tag{1.5.12}$$

This is consistent with what is known about the solvation of an OH group, i.e. that OH has three "arms" along which HBs may be formed (Fig. 1.20).[46]

Tables 1.4 and 1.7 show some more values of the conditional solvation Gibbs energies of a hydroxyl group next to different "backbones."

---

[45] See Appendix D.
[46] See also Thanki *et al.* (1988).

Table 1.4. Conditional Solvation Gibbs Energy of a Hydrophilic Group next to a Saturated Hydrocarbon (in kJ/mol)

$$\Delta G^*_{BB-OH} \;-\; \Delta G^*_{BB-H} = \Delta G^*_{OH/BB}$$

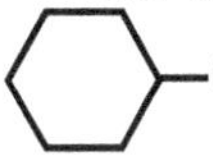 ⬡–OH	– ⬡	=	 -28.05
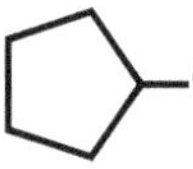 ⬠–OH	– ⬠	=	 -28.01
C–OH	– C	=	-29.75
C–C–OH	– C–C	=	-28.62
C–C–C–OH	– C–C–C	=	-28.37
C–C–C (with OH on middle C)	– C–C–C	=	-28.08
C–C–C–C (with OH on second C)	– C–C–C–C	=	-27.86

*Based on data from Cabani *et al.* (1981).

Table 1.5. Conditional Solvation Gibbs Energy of a Hydrophobic Group next to a Saturated Hydrocarbon (in kJ/mol)

$$\Delta G^*_{BB-CH_3} \;-\; \Delta G^*_{BB-H} = \Delta G^*_{CH_3/BB}$$

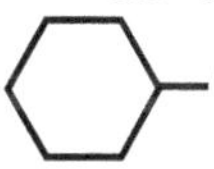

⬡–CH₃	– ⬡	=	1.88
⬠–CH₃	– ⬠	=	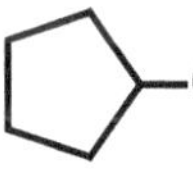  1.63
C–C	– C	=	-0.67
C–C–C	– C–C	=	0.88
C–C–C–C	– C–C–C	=	0.72
C–C–C (with C on middle C)	– C–C–C	=	1.30

*Based on data from Cabani *et al.* (1981).

Table 1.6. Conditional Solvation Gibbs Energy of a Hydrophobic Group next to a Unsaturated Hydrocarbon (in kJ/mol)

$$\Delta G^*_{BB-CH_3} - \Delta G^*_{BB-H} \;=\; \Delta G^*_{CH_3/BB}$$

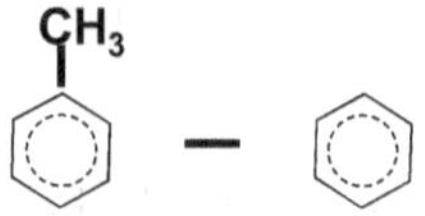

structure difference		value
(CH$_3$–benzene ring − benzene ring)	=	0.05
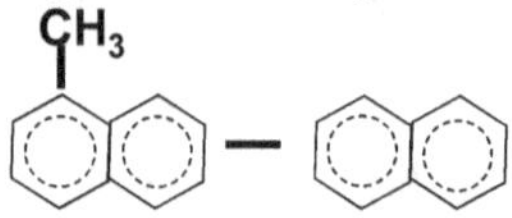 (CH$_3$–naphthalene − naphthalene)	=	0.1
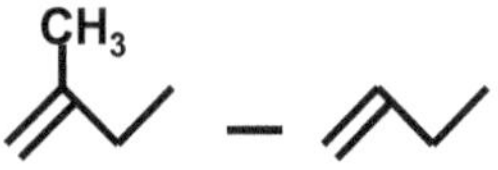	=	0.04
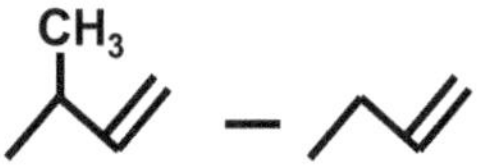	=	1.88
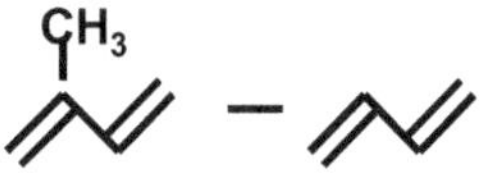	=	0.25

*Based on data from Cabani *et al.* (1981).

A carbonyl group

The conditional Gibbs energy of solvation of a carbonyl group is obtained from the following differences in the solvation Gibbs energies of ketones

$$\Delta G^*_{HB}(\text{of } C{=}O/CH_3CH_2CH_3)$$

$$= \Delta G^*_{CH_3CH_2CH_2CHO} - \Delta G^*_{CH_3CH_2CH_3}$$

$$= -21.55 \,\text{kJ/mol}, \tag{1.5.13}$$

$$\Delta G^*_{HB}(\text{of } C{=}O/CH_3CH_2CH_2CH_3)$$

$$= \Delta G^*_{CH_3CH_2CH_2CHO} - \Delta G^*_{CH_3CH_2CH_2CH_3}$$

$$= -21.97 \,\text{kJ/mol}. \tag{1.5.14}$$

Table 1.7. Conditional Solvation Gibbs Energy of a Hydrophobic and Hydrophilic Groups next to an Unsaturated Hydrocarbon (in kJ/mol)

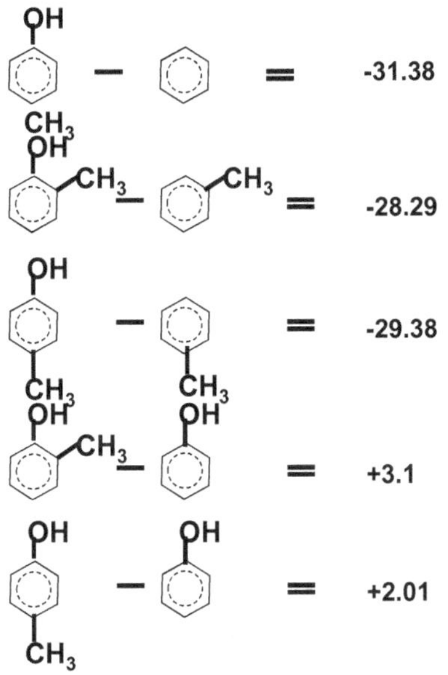

phenol − benzene	= −31.38
cresol − toluene	= −28.29
benzyl alcohol − toluene	= −29.38
cresol − phenol	= +3.1
cresol − phenol	= +2.01

*Based on data from Cabani *et al.* (1981).

Again, we see that the two values are very similar, though the process of adding the carbonyl group is different.[47] Assuming that a carbonyl group can form two HBs[48] through two arms (Fig. 1.20) we could have estimated from (1.5.4)

$$\Delta G^*_{HB}(\text{two arms of } H_2O/LJ)$$

$$= 2\Delta G^*_{HB}(\text{one arm of } H_2O/LJ) \approx 2(-9.42)$$

$$= -18.84 \, \text{kJ/mol}. \tag{1.5.15}$$

---

[47] See Appendix D.
[48] See also Thanki *et al.* (1988).

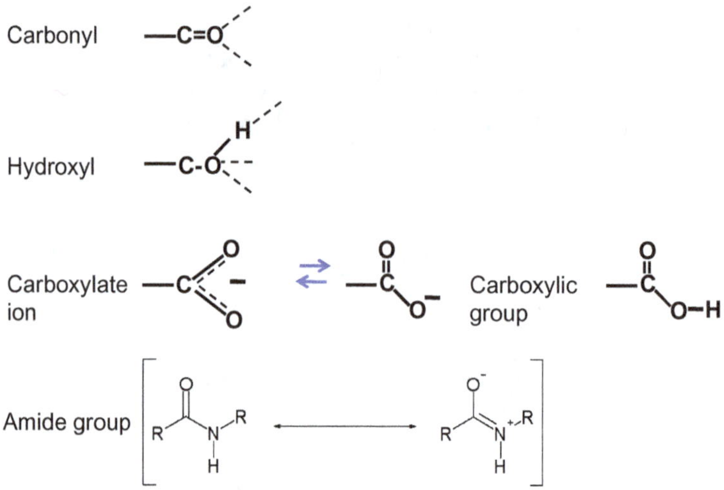

Fig. 1.20   The possible hydrogen-bonding arms of carbonyl, hydroxyl and carboxyl groups.

We see that this value is slightly smaller than the values in (1.5.13) and (1.5.14). Note however, that these values are consistent with the fact that a carbonyl group can form only two HBs. The difference between the values in (1.5.15) and those in (1.5.13) and (1.5.14) is probably due to the fact that the strength of the HB of an arm of water is different from that of an arm of a carbonyl group.

A carboxylic group

The conditional Gibbs energy of solvation of a carboxylic group may be estimated from the following differences in the solvation Gibbs energies

$$\Delta G^*_{HB} \left(\text{of COOH/CH}_3 \right) = \Delta G^*_{CH_3COOH} - \Delta G^*_{CH_4}$$

$$\approx -36.4 \, \text{kJ/mol}, \qquad (1.5.16)$$

$$\Delta G^*_{\text{HB}}(\text{of COOH/CH}_3\text{CH}_3)$$

$$= \Delta G^*_{\text{CH}_3\text{CH}_2\text{COOH}} - \Delta G^*_{\text{CH}_3\text{CH}_3}$$

$$\approx -34.73 \,\text{kJ/mol}, \tag{1.5.17}$$

$$\Delta G^*_{\text{HB}}(\text{of COOH/CH}_3\text{CH}_2\text{CH}_3)$$

$$= \Delta G^*_{\text{CH}_3\text{CH}_2\text{CH}_2\text{COOH}} - \Delta G^*_{\text{CH}_3\text{CH}_2\text{CH}_3}$$

$$\approx -34.73 \,\text{kJ/mol}. \tag{1.5.18}$$

These values correspond to the solvation of about *four* arms, i.e.

$$4\Delta G^*_{\text{HB}} \left(\text{one arm of H}_2\text{O/LJ}\right) \approx -37.7 \,\text{kJ/mol}. \tag{1.5.19}$$

Clearly, if the carbonyl and the hydroxyl groups were independently solvated, we would have expected the solvation of *five* arms. However, because of the proximity of the two groups (Fig. 1.20), we can expect some interference of the solvation of the two groups, thus the conditional solvation Gibbs energy is approximately equal to the solvation of *four* arms.

An amine group

For an amine group, we can estimate the conditional solvation Gibbs energy as

$$\Delta G^*_{\text{HB}}(\text{of NH}_2/\text{CH}_3) = \Delta G^*_{\text{CH}_3\text{NH}_2} - \Delta G^*_{\text{CH}_4} \approx -27.5 \,\text{kJ/mol}, \tag{1.5.20}$$

$$\Delta G^*_{\text{HB}}(\text{of NH}_2/\text{CH}_3\text{CH}_3) = \Delta G^*_{\text{CH}_3\text{CH}_2\text{NH}_2} - \Delta G^*_{\text{CH}_3\text{CH}_3}$$

$$\approx -27.2 \,\text{kJ/mol}. \tag{1.5.21}$$

This corresponds roughly to the solvation of three arms. In the protein backbone, we have the group NH in the amide group (Fig. 1.20). We do not have the relevant data on the solvation Gibbs energies of molecules with or without a NH group. It is reasonable to assume that the NH group can form only one hydrogen bond,[49] and that the conditional solvation Gibbs energy of NH is similar in magnitude to that of one arm of water, i.e. about 9.42 kJ/mol.

A peptide group

Although there exists no direct way of extracting the conditional solvation Gibbs energy of the two FGs for each peptide group, we can safely estimate that the two FGs, the carbonyl and the amine group are solvated through three arms.[50]

1.5.2. *Solvation of hydrophobic molecules or groups*

Tables 1.5 and 1.6 show some values of the conditional solvation Gibbs energies of methane near a saturated and unsaturated hydrocarbon molecule.

First, note the difference between the solvation Gibbs energy and the *conditional* solvation Gibbs energy (Table 1.8). This is an important observation which is very relevant to the role of the hydrophobic solvation in biochemical processes. The reader is referred to Chapter 4 of Part I for details. Briefly, Kauzmann's idea was to use the solvation process of a *HϕO molecule* as a model for transferring a *HϕO group* from being exposed to the solvent, into the interior of the protein. As we shall see in Sec. 1.9 and in Chapter 3, it is not the *solvation* quantities but the *conditional* solvation quantities that contribute to the process of protein folding. As can be seen from Tables 1.5 and 1.8, the conditional solvation Gibbs energies are much smaller than the corresponding solvation Gibbs energies.

---

[49] See Thanki *et al.* (1988).
[50] Gong, Perskie and Rose (preprint 2010).

Table 1.8. Values of the Solvation and the Conditional Solvation Gibbs Energy of a Hydrophobic Group, and the Conditional Solvation Gibbs Energy of Hydrophilic Group

$$\Delta G^*_{CH_3/BB} \approx 1.46\,\text{kJ/mol}$$

$$\Delta G^*_{CH_4} \approx 7.95\,\text{kJ/mol}$$

$$\Delta G^*_{OH/BB} \approx -27.2\,\text{kJ/mol}$$

Second, note the large effect of the backbone (BB) on the conditional Gibbs energy of solvation. The rationale for this large effect is the following:

A *HϕO* molecule placed in water will be surrounded by *pure water*. The distribution of water molecules around the solute will be the same as in pure water.[51] On the other hand, the same solute placed at a point next to a backbone will be solvated by water molecules, the distribution of which has been perturbed by the presence of the BB. As can be seen from Figs. 1.16 and 1.21, the surroundings of the BB are not "pure water," but water that was affected by the presence of the BB. The BB has two effects. First, the geometrical effect; the BB partially "shields" the solute from being solvated from all sides. Second, the polarity effect; the distribution of the water molecules can be affected by the presence of the BB. The latter effect could be quite large if the BB is polar, as is the backbone of a protein (Fig. 1.21).

Therefore, one can say that a *HϕO* group next to a BB is solvated by a solvent which is *different* from pure water. This fact alone casts a shadow of doubt on the adequacy of Kauzmann's

---

[51]This is true for a dilute solution of the *HϕO* molecule. For more details, see Part I and Appendix D.

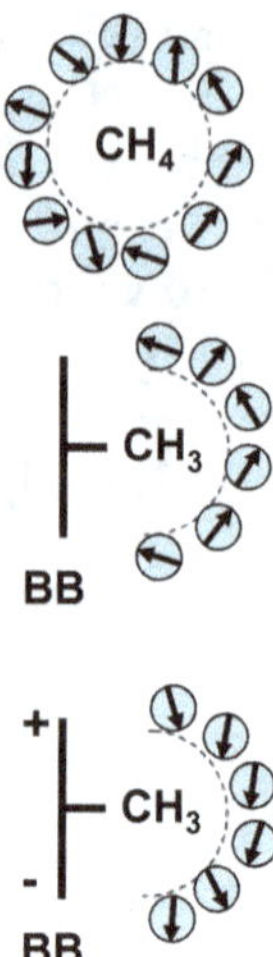

Fig. 1.21   Distribution of solvent molecules around (a) a methane molecule, (b) a hydrocarbon BB and (c) a polar backbone. The solvent molecules are represented as dipoles pointing from the negative to the positive charges.

model for the transferring of $H\phi O$ groups into the interior of the protein.

Table 1.6 shows some data on the conditional solvation Gibbs energy of a methyl group next to an aromatic molecule. As we can see, an aromatic BB reduces the conditional solvation Gibbs energy of the methyl group by more than an order of magnitude compared to the effect of a saturated hydrocarbon.

From Tables 1.5 and 1.6, we can conclude that a backbone has a significant effect on the conditional solvation Gibbs energy. We can expect that the BB of a protein would also have a similar effect. Unfortunately, there exists no relevant data for such a backbone.

1.6. Hydrophobicity Scales

Kauzmann's idea of using the solvation Gibbs energies of nonpolar solutes as a measure of the tendency of different side chains to penetrate into the interior of the protein has spawned a number of "hydrophobicity scales."

The first table of Gibbs energies of transferring amino acids from water to ethanol was published by Nozaki and Tanford.[52] Following Nozaki and Tanford, there have been many attempts to construct "hydrophobicity scales." In 1987, Cornett *et al.* listed at least 38 such scales.[53] As we shall see in Chapter 3, none of these scales reflect the tendencies of the various side chains to enter into the interior of the protein. Wolfenden *et al.*[54] correctly argued that what one needs is the absolute affinity of a compound for the aqueous environment. They clearly recognized the fact that a side chain of an amino acid in a protein is flanked by peptide bonds, and these might (should) affect the surroundings of the side chain.

Let us denote a side chain of an amino acid by R.[55] As we shall see, the quantity that contributes to the total Gibbs energy of protein folding is the conditional solvation Gibbs energy of R, given the protein backbone (BB). This is formally defined by

$$\Delta G^*_{R/BB} = \Delta G^*_{BB-R} - \Delta G^*_{BB-H}, \qquad (1.6.1)$$

where ΔG^*_{BB-R} and ΔG^*_{BB-H} are the solvation Gibbs energies of the whole protein (the backbone BB) with and without the side chain R. Unfortunately, we have no information on these solvation Gibbs energies. Therefore, we use an approximation for the backbone. The justification for such an approximation is discussed in Appendix D.

Instead of the whole protein, we take a segment of the protein, which we shall refer to as the model BB. The model must

---

[52]Nozaki and Tanford (1971).

[53]Cornett *et al.* (1987).

[54]Wolfenden *et al.* (1981).

[55]Note that here R denotes a group of "radical" R, in most of the book, *R* is the distance between two groups or molecules.

be large enough to mimic the effect of the whole protein on the surroundings of the side chain R. It must also be small enough so that its solvation Gibbs energy can be measured.[56] Thus, instead of (1.6.1) we write

$$\Delta G^*_{R/M} = \Delta G^*_{M-R} - \Delta G^*_{M-H},   (1.6.2)$$

where M is the model compound replacing the BB.

If the model compound fulfills the two conditions stated above, then the two quantities on the right-hand side of (1.6.2) can be measured, and $\Delta G^*_{R/M}$ is the required measure of the hydrophobicity of the side-chain R.

To the best of my knowledge, such an "ideal" hydrophobicity scale is not available.[57] The closest to this ideal scale is the one that was suggested by Fauchere and Pliska.[58] These authors suggested using the *distribution* coefficient of N-acetyl-amino-acid-amide (NAAAA) between water and octanol. Table 1.9 shows the hydrophobicity scale suggested by Fauchere and Pliska. Note carefully that the choice of octanol to mimic the environment of the interior of the protein is not essential. The reason, as will become clear in Chapter 3, is that the interaction of the side chain with its neighbors in the interior of the protein is part of the *direct interaction* energy between the various groups of the protein and not part of the solvent-induced contribution to the process of protein folding. Therefore, the "ideal" hydrophobicity scale should be based on the *conditional* solvation Gibbs energy of the side chains, i.e.

$$\Delta G^*_{R/NAAAA} = \Delta G^*_{NAAAA-R} - \Delta G^*_{NAAAA-H},   (1.6.3)$$

where each ΔG^* on the right-hand side of (1.6.3) is a solvation Gibbs energy of a solute in water.

---

[56] More details in Appendix D.
[57] Ben-Naim (1990).
[58] Fauchere and Pliska (1983).

Table 1.9.   Hydrophobicity Scale suggested by
Fauchere and Pliska

Trp	1.35	Thr	0.156
Ile	1.08	His	0.078
Phe	1.07	Gly	0
Leu	1.02	Ser	−0.024
Cys	0.924	Gln	−0.132
Met	0.738	Asn	−0.36
Val	0.732	Glu	0.384
Tyr	0.58	Asp	−0.462
Pro	0.43	Lys	−0.594
Ala	0.186	Arg	−0.606

Relative values of δG^{*}R/N-acetyl-amino-acid-amide in water and in octanol.

The Fauchere–Pliska scale is essentially the difference

$$\Delta G^{*}_{R/NAAAA}(\text{in water}) - \Delta G^{*}_{R/NAAAA} \ (\text{in octanol}). \quad (1.6.4)$$

We add here a qualitative argument as to why one does not need the solvation Gibbs energy of the side chain in octanol, or in any organic liquid.[59]

When a side chain R is transferred from being exposed to water into the interior of the protein, there is a *loss* of the *conditional* solvation Gibbs energy of R and a *gain* of the total interaction energy of R with its surroundings in the interior of the protein. The latter may be approximated by taking the average binding energy of R to the rest of the protein in the folded form. This quantity is, in principle, different from the solvation Gibbs energy of a solute, similar to R, in the interior of the protein which has the form

$$\Delta G^{*}_{R} = -k_{B}T \ln\langle\exp[-\beta B_{R}]\rangle. \quad (1.6.5)$$

---

[59]See also Appendix H.

To see this, suppose that the attractive part of the binding energy is very weak, so that one could approximate ΔG_R^* as[60]

$$\Delta G_R^* \approx \Delta G_{HS}^* + \langle B_R \rangle. \qquad (1.6.6)$$

In (1.6.6), the solvation Gibbs energy is approximated by a cavity work ΔG_{HS}^* and an average binding energy of R.[61] The latter could be similar in magnitude to the interaction *energy* of R with its surroundings in the interior of the protein. However, the cavity work in (1.6.6) is not needed to approximate the interaction of R with its surroundings. As an extreme example, suppose that R itself is a hard sphere (HS), in which case the interaction of R with its surroundings in the interior of the protein is zero. The same is true for $\langle B_{HS} \rangle$. But the solvation Gibbs energy of a HS in an organic liquid is not zero. In other words, if we take the Gibbs energy of transfer of a HS from water into an organic liquid, we actually take the *difference* of the cavity works in the two liquids. On the other hand, when we transfer the HS side chain from water into the interior of the protein, we only lose the cavity in water, but we gain no cavity work in the interior of the protein. The cavity work is done only once, and we do not need to dig a cavity within an already existing cavity. A simple model illustrating this argument is discussed in Appendix H.

1.7. Some Numerical Values of the Pairwise Hydrophobic and Hydrophilic Interactions

First, a comment on nomenclature: We shall use the term interaction energy for the *direct* interaction between two (or more)

---

[60] See Appendix K.
[61] See Appendix K.

particles. This interaction is usually assumed to be independent of temperature and of the surrounding molecules. Thus, $\Delta U(R)$, or simply $U(R)$, is defined as the change in energy for bringing two particles from infinite separation to the final distance R.

When the same process is carried out in a solvent, the associated work is the potential of mean force (PMF), which is also the change in the Gibbs (or Helmholtz) energy for the process of bringing the two particles from infinite separation to distance R within the solvent at a given temperature T, pressure P and composition.[62] Thus

$$W(R) = G(R) - G(\infty) = U(R) + \delta G(R). \qquad (1.7.1)$$

Here, $W(R)$ is split into two terms, the *direct interaction* between the two particles $U(R)$, and the indirect part, or the solvent-induced part of $W(R)$, or of the Gibbs energy change (see also Sec. 1.2 and Appendix C).

One should note that there are several important differences between the *direct* and the *indirect* parts of the PMF.

First, the direct interaction energy $U(R)$ is presumed to be independent of temperature and of pressure. On the other hand, the indirect part has the character of free energy, i.e. it can be dependent on the temperature and pressure.

Second, the indirect part, being a solvent-induced part of $W(R)$, is an average over all the configurations of the solvent molecules. Obviously, $\delta G(R)$ depends on the solvent. In general, $\delta G(R)$ depends on both the solute–solute interactions and the solute–solvent and solvent–solvent interactions. When the solute is very dilute in the solvent, $\delta G(R)$ becomes independent of the solute–solute interaction, which is denoted by $U(R)$ in

---

[62]The exact definition is discussed in Appendix C.

(1.7.1). This important property of $\delta G(R)$ has led to the construction of a simple measure of the $H\phi O$ interaction discussed in great detail in Chapter 4 of Part I.

Finally, it should be noted that when we generalize the relation (1.7.1) for many solute particles, there is one more important difference between the direct interaction among all the solute particles and the indirect part of the corresponding PMF. The former, to a good approximation, may be assumed to be pairwise additive. The latter, however, is not pairwise additive. We discuss this property of δG in Appendix E, and its implication for the process of protein folding in Chapter 3.

Thus, whenever we are interested in the solvent-induced effect, we can focus on $\delta G(R)$ and ignore the direct interaction $U(R)$. We shall refer to δG as the pairwise $H\phi O$ interaction when the two solute particles are hydrophobic. Similarly, when the two solute particles are hydrophilic, we shall refer to δG as the pairwise $H\phi I$ interaction.

Following this definition of (pairwise) $H\phi O$ interaction,[63] much work has been done to study, either theoretically or via simulation, the $H\phi O$ interactions.[64]

However, late in the 1980s, a shift of the focus was suggested, from the study of the $H\phi O$ interaction, to the study of the *conditional* $H\phi O$ interaction (as well as the conditional $H\phi I$ interaction).

1.7.1. *Pairwise hydrophobic (HϕO) interaction*

Most of the information we have on pairwise hydrophobic interactions comes from either simulation of aqueous solutions

---

[63]Ben-Naim (1971).
[64]See Chapter 4 of Part I.

of $H\phi O$ solutes such as methane, or from a measure of the strength of the $H\phi O$ interaction, based on experimental data. These quantities were discussed in Part I, Chapter 4. We note here that although these quantities are of interest in the context of the study of the properties of aqueous solutions, they are not relevant to biochemical processes. In biochemistry, we never encounter the "association" of two methane molecules to form a "dimer." What we do encounter is either the association between macromolecules or "association" between two $H\phi O$ groups which are attached to the BB of the protein in the process of protein folding. In such processes, the relevant quantities are always the *conditional* pairwise $H\phi O$ interactions. These quantities are defined similarly to the PMF. The difference between $H\phi O$ interaction and *conditional* $H\phi O$ interaction is depicted in Fig. 1.22. The *condition* here is the same condition we encountered in the study of conditional solvation, i.e. the presence of the protein BB. Thus, in one case we have two solute molecules approaching each other in a "pure solvent," in the second case, the two groups approaching each other while they are attached to a BB. Clearly, since the BB might affect the solvent properties, the solvent-induced interaction between the two

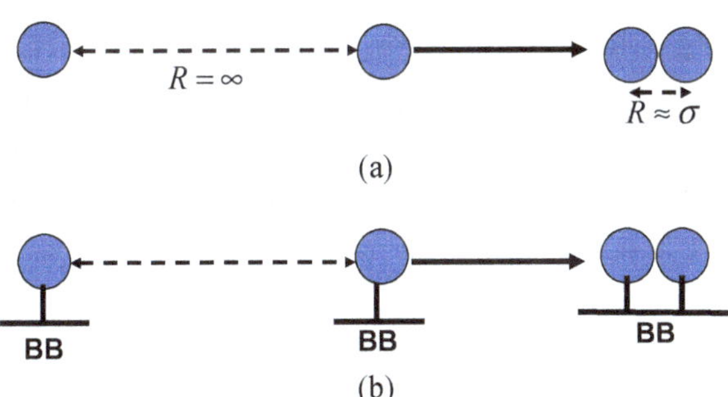

Fig. 1.22   The difference between (a) pairwise $H\phi O$ interaction and (b) conditional pairwise $H\phi O$ interactions.

solutes might be different from the solvent-induced interaction between the two groups.

To the best of my knowledge, there exists no data on the conditional $H\phi O$ interactions derived either from theoretical or from simulation calculations. All we have is some data on the conditional $H\phi O$ interactions between two groups connected to small model compounds.[65] These data were obtained using model BB, similar to the one discussed in Sec. 1.5. The quantity required here is the work associated with the process of bringing two side chains attached to the protein from "infinite" separation to some close distance. By infinite separation we mean that the two side chains are independently solvated.[66] The process is schematically depicted in Fig. 1.22. The formal definition of the conditional $H\phi O$ interaction is[67]

$$\delta G(R/BB) = \Delta G^*(R/BB) - \Delta G^*(R = \infty/BB), \qquad (1.7.2)$$

where $\Delta G^*(R/BB)$ is the conditional solvation Gibbs energy of the two side chains being at a distance R, given the presence of the BB. Unfortunately, these quantities are not available. Instead, we use small model compounds to mimic the effect of the protein on the environment around the side chains. The requirements of the model compound are the same as those discussed in Sec. 1.5 and Appendix D.

Even with small model compounds, we do not have the required data on *real* side chains. Instead, we have some limited data on the conditional $H\phi O$ interactions between small groups such as methyl or ethyl attached to a small model compound (M) such as cyclohexane or benzene. Instead of the required measure

---

[65]See also Sec. 1.7.3 on the pairwise interactions of side chains.

[66]This is defined more precisely in Appendix F. Here, it is sufficient to say that if the two side chains are far apart, we expect that the "solvation spheres" around the two side chains do not overlap.

[67]Note again that R here is the *distance* between the two groups.

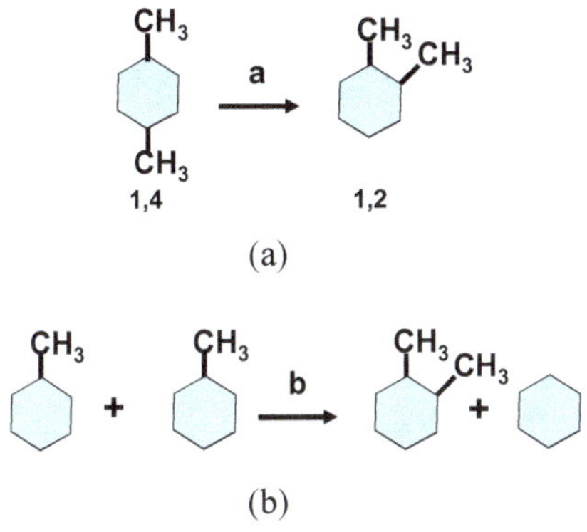

(a)

(b)

Fig. 1.23   Model compounds used to measure the indirect part of the work associated with the process of bringing two $H\phi O$ groups to a close distance.

(1.7.2) we have the quantity

$$\delta G(R/M) = \Delta G^*(R/M) - \Delta G^*(R = \infty/M). \qquad (1.7.3)$$

Figure 1.23 shows an example of such a model compound. Here, M is a benzene ring, and the process is the transfer of a methyl group from position 4 to position 2. The corresponding $H\phi O$ interaction is

$$\delta G(1,2/M) = \Delta G^*(1,4/M) - \Delta G^*(1,2/M), \qquad (1.7.4)$$

where $\Delta G^*(\alpha/M)$ is the conditional solvation Gibbs energy of the isomer α given M, where M is the model compound.

Note carefully that δG does not include the *direct* interaction between the two methyl groups — it is only the indirect, or the solvent-induced, part of the work required to transfer one methyl group from position 4 to position 2, given that a methyl group is at position 1 of the benzyl ring.[68] Table 1.10 shows the values of $\delta G(1,2/M)$ for two methyl and two ethyl groups.

---

[68]The theoretical justification for this interpretation of $\delta G(1,2/M)$ is discussed in Chapter 4 and Appendices D and F.

Table 1.10.   Values of $\delta G(a)$ and $\delta G(b)$ for Methyl and Ethyl Group on a Benzyl Ring at Two Temperatures

Alkyl Group	t(°C)	$\delta G(a)/\text{kJ mol}^{-1}$	$\delta G(b)/\text{kJ mol}^{-1}$
Methyl	10	−0.99	−0.86
	20	−1.25	−1.15
Ethyl	10	−2.52	−2.59
	20	−2.58	−2.55

The connection between $\delta G(1,2/M)$ and the experimental quantities is simply

$$\delta G(1,2/M) = \Delta G^*_{(1,2)} - \Delta G^*_{(1,4)}. \qquad (1.7.5)$$

Thus, to estimate the conditional $H\phi O$ interaction, all we need is the solvation Gibbs energies of the two isomers 1,2-dialkylbenzene and 1,4-dialkylbenzene. Note that in the 1,4-isomer, the distance between the two alkyl groups is assumed to be such that the two groups are independently solvated. How do we know that this assumption is justified?

To examine this question, consider a different process depicted in Fig. 1.23b. Here, we start with two alkylbenzene molecules at infinite separation in the liquid. Thus, the two alkyl groups are by definition at infinite separation and therefore independently solvated.

We next transfer one alkyl group (methyl or ethyl) from one molecule to another to form a 1,2-dialkylbenzene and a benzene ring. The solvent-induced part of the Gibbs energy change for this process is

$$\delta G(b) = \Delta G^*_{(1,2)} + \Delta G^*_{\text{Benzene}} - 2\Delta G^*_{\text{Toluene}}. \qquad (1.7.6)$$

Table 1.10 shows some values of $\delta G(b)$ for this reaction. A comparison of the values of δG for reactions (a) and (b) shows that these are very similar. Since the end product (the

1,2-dialkylbenzene) is the same in the two reactions, we can conclude that in both reactions, the two alkyl groups in the 1,4-isomer are independently solvated.[69]

Note that all values of δG in Table 1.10 are negative. For a methyl group, δG is about -1 kJ/mol and about -2.5 kJ/mol for an ethyl group. In most cases, we see that the absolute values of δG *increase* with the increase in the temperature (except for $\delta G(b)$ for ethyl, where there is a slight decrease). Finally, we note that all the values of δG for the same process are larger in H_2O compared to D_2O (see Part I, Sec. 4.5 for details).

We can conclude from the data that the conditional $H\phi O$ interaction is always negative. It is larger the larger the alkyl group, and it seems that it is also larger at 20 °C than at 10 °C.

It would have been both interesting and useful to have more data of this sort for all the side chains of amino acids attached to backbones that better represent the environment around the peptide bond of the protein. Until such a time when such information becomes available, we shall use these data for comparison with the corresponding data on $H\phi I$ groups.

1.7.2. *Pairwise hydrophilic (HϕI) interaction*

In our discussion of conditional pairwise $H\phi O$ interaction, we relied on information derived from experimental data on the solvation Gibbs energies of model compounds. From what we know on the PMF of two $H\phi O$ solutes in water, there are at least two minima in the Gibbs energy landscape for such a system: One at contact, and a second at a configuration where a water molecule can form a bridge between the two solutes. We use the term "bridge" here only to indicate that the solvent-induced

---

[69]More details in Chapter 4 of Part I and Appendix F.

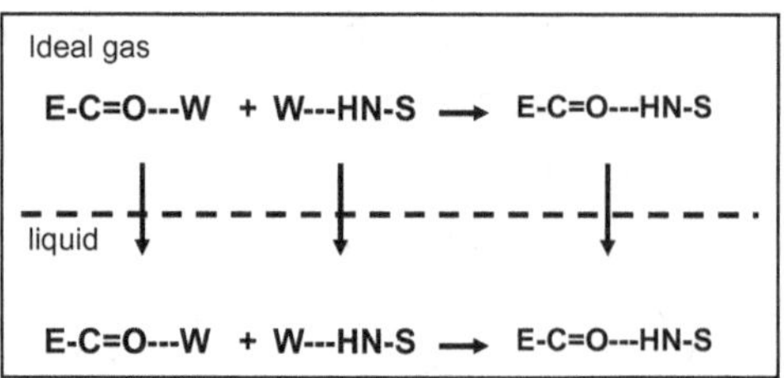

Fig. 1.24 A method of calculating the $H\phi I$ interaction at a distance of $R = 2.8$ Å using the solvation Gibbs energies of the solutes.

interaction is mediated through a single water molecule. This is sometime referred to as water-separated $H\phi O$ interaction.[70]

In the case of $H\phi I$ interactions, we have at least two large solvent-induced effects, one at a distance of about 2.8 Å, and the second at about 4.5 Å.

Two $H\phi I$ groups approaching each other to a distance of about 2.8 Å can form a direct HB provided their orientations are favorable and one group is a donor and the second an acceptor. It is relatively easy to calculate the solvent-induced effect for such a process (Fig. 1.24) if only one arm on each $H\phi I$ group is involved in the process. As we have calculated in Sec. 1.5, the loss of the solvation Gibbs energy of one arm is about 9.42 kJ/mol. Therefore, for two arms, the loss is about 18.84 kJ/mol. This positive change in Gibbs energy is more than compensated for by the energy of formation of the direct HB, which amounts to about -27.2 kJ/mol. Therefore, the net change in Gibbs energy in the formation of a direct HB is about -8.4 kJ/mol. It should be noted that this calculation is very different from the HB inventory argument, where one writes the "reaction" of the HB formation in water as

$$\text{E--C=O}\cdots\text{w} + \text{w}\cdots\text{HO--S} \rightarrow \text{E--C=O}\cdots\text{HO--S} + \text{w}\cdots\text{w}$$

---

[70]See also Part I, Sec. 4.5.

Such a stoichiometric reaction has led to the dismissal of the contribution of hydrogen bonding to protein folding. The flaws involved in this HB inventory argument have been discussed in Part I, Sec. 4.12, and will be further discussed in Appendix I.

The next $H\phi I$ effect which was not recognized until the late 1980s[71] occurs when two $H\phi I$ groups are brought from infinite separation to a distance of about 4.5 Å. If the two $H\phi I$ groups are correctly oriented, they could form HB bridges mediated by a water molecule (Figs. 1.25 and 1.26). It is believed that this $H\phi I$ effect is the strongest solvent-induced interaction between two arms of two $H\phi I$ groups. Therefore, we shall refer to this effect simply as $H\phi I$ interaction. We shall discuss other possible $H\phi I$ effects in Chapter 3 and Appendix G.

There exists no experimental data on the pairwise $H\phi I$ interaction between two $H\phi I$ solutes. The simplest examples of $H\phi I$ "solutes" are water molecules. For a pair of water molecules, we have some information on the PMF as well as on the average

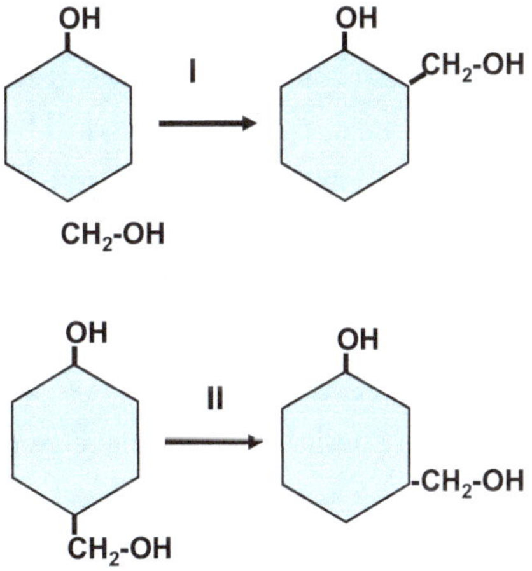

Fig. 1.25   Model compounds used to measure the indirect part of the work associated with the process of bringing two $H\phi I$ groups to a close distance.

---

[71] Ben-Naim *et al.* (1989), Ben-Naim (1990) and Ben-Naim (1992).

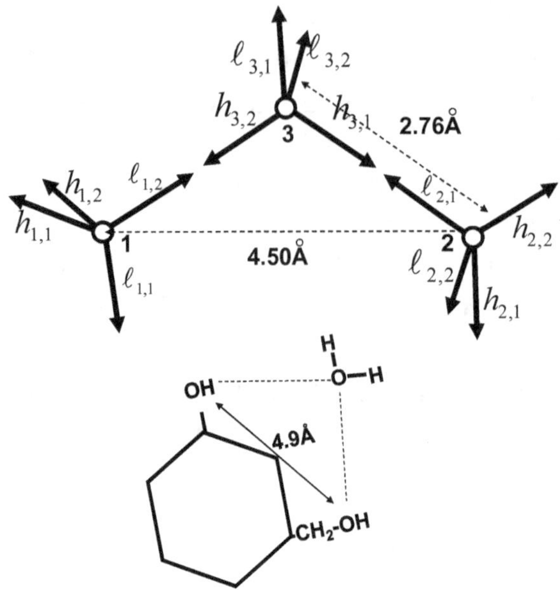

Fig. 1.26   Distances of first and second neighbors in ordinary ice (I_h) and the distance between two $H\phi I$ in a model compound.

force between two water molecules in liquid water. There are also some theoretical estimates of the $H\phi I$ interaction between two water molecules in water.[72] Here, we shall focus on the conditional $H\phi I$ interaction between two $H\phi I$ groups attached to a BB. This is the more relevant quantity in the context of solvent-induced effects in biochemical processes. The available data on the conditional pairwise $H\phi I$ interaction is scarce, inaccurate and not exactly the required data.

Figure 1.25 shows the model compounds used for the extraction of information on the conditional $H\phi I$ interaction. The required quantity is, as before

$$\delta G(R/M) = \Delta G^*(R/M) - \Delta G^*(R = \infty/M), \qquad (1.7.7)$$

---

[72]See Part I, Sec. 4.8.

where ΔG^* in (1.7.7) is the conditional solvation Gibbs energy of the relevant solute in water. What is available is the *difference* between the solvation quantities in water, and in toluene.[73] Denoting by $\Delta\delta G(process)$ the difference between $\delta G(process)$ in water and in toluene, we find the following values based on experimental data on solvation Gibbs energies of the compounds shown in Fig. 1.25.

$$\Delta\delta G[(1,4) \to (1,3)] \approx -13.2\,\text{kJ/mol}, \qquad (1.7.8)$$

$$\Delta\delta G[(1,4) \to (1,2)] \approx -6.5\,\text{kJ/mol}. \qquad (1.7.9)$$

Thus, for the transfer of a $H\phi I$ group from position 4 to position 3 (process II in Fig. 1.25), the difference in δG is about $-13\,$kJ/mol, whereas the difference in δG for the processes in Fig. 1.23 is much smaller.

We attribute the large value of $\Delta\delta G$ for process II to the $H\phi I$ interaction in water. This is supported by the fact that similar values of $\Delta\delta G$ are obtained for the system of water and cyclohexane, and water and hexane.[74] Furthermore, a theoretical estimate of the $H\phi I$ between two water molecules in liquid water at a distance of 4.5 Å is about $-12\,$kJ/mol.[75]

In spite of the approximate nature of the data, it is safe to conclude that the $H\phi I$ interaction between the two $H\phi I$ groups at positions 1 and 3 is about $-13\,$kJ/mol, whereas at positions 1 and 2, it is only about $-6.5\,$kJ/mol. The rationale for these values is discussed in great detail in Part I, Chapter 4. Here, we present some qualitative reasons for expecting a strong $H\phi I$ interaction at a distance of about 4.5 Å. First, note that the distance between the two oxygen atoms in the 1,3-isomer is about 4.9 Å, very close to the distance of 4.5 Å, which is the

---

[73] See Part I, Sec. 4.8 for more details.
[74] Wilf and Ben-Naim (1994).
[75] See Part I, Chapter 4 and Appendix G.

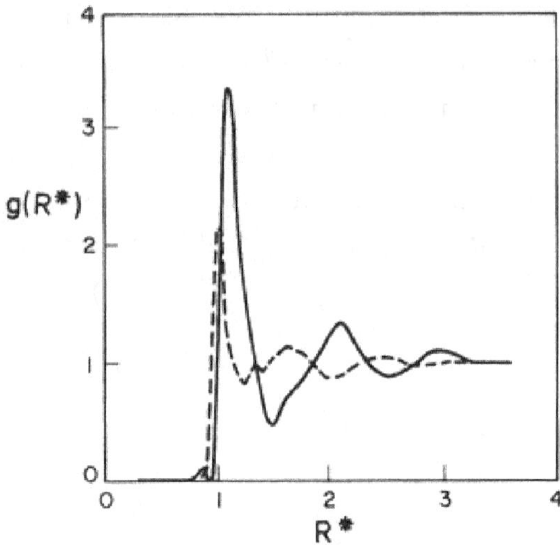

Fig. 1.27 Radial distribution function for water (dashed line) and argon (full line). R^* is the reduced distance $R^* = R/\sigma$, where $\sigma = 2.8$ Å for water. Hence $R^* \approx 1.6$ corresponds to $R^* \approx 4.5$ Å.

distance of the second-nearest neighbors of water molecules in ice (Fig. 1.26). As is clear from the figure, at this specific distance, one water molecule bridges two other water molecules using HBs. This "ideal" structure of ice is only indicative evidence. In liquid water, such a perfect structure does not exist. Nevertheless, we know that the radial distribution function of liquid water has a second peak at about 4.5 Å (Fig. 1.27). This means that the distance of 4.5 Å has a relatively high probability. Furthermore, the second peak of the radial distribution function is an average over all possible orientations of the two water molecules. It is expected that at some specific orientations of the water molecules, the pair correlation function will have a stronger maximum.[76] Indeed, this is confirmed by both simulations of the PMF for specific orientations of the two water

---

[76]For more details, see Part I, Sec. 4.8.

molecules, and the *average force* between two water molecules in liquid water.[77,78]

Additional support for the existence of strong $H\phi I$ interactions comes from analysis of experimental data, as discussed in this section, and some theoretical arguments.[79]

There is one simulated result that seems to be inconsistent with our estimates of the strength of $H\phi I$ interaction at $R = 4.5$ Å. Sun and Kollman[80] calculated the solvation Gibbs energies of the three isomers of hydroxy methylphenol and found that there was no significant difference for the three isomers. They concluded that their results were inconsistent with the water-bridge hypothesis. They also suspected that toluene is at least partially responsible for the difference in the solvation Gibbs energies of the three isomers.

I believe that the conclusions reached by Sun and Kollman were premature and unjustified. First, because there are similar results in cyclohexane and hexane that support our conclusion. Second, simulation of both the $H\phi I$ interaction and the $H\phi I$ force supports our conclusion. And finally, the theoretical arguments discussed in Part I, Chapter 4, and in Appendix G also support our conclusion.

In addition, some details of their calculations are not clear. What are the orientations of the $H\phi I$ groups; are these fixed or free? What is the solute–solvent interaction used in the calculation? To what extent is dipole–dipole interaction between the water molecules and the $H\phi I$ groups responsible for their results, and not the hydrogen bonding? All these points cast serious doubts on the validity of their conclusions.

---

[77] Mezei and Ben-Naim (1990).
[78] Durell *et al.* (1994).
[79] See Part I, Sec. 4.8 and Appendix H.
[80] Sun and Kollman (1996).

1.7.3. *Potential of average force for pairs of side chains of amino acids*

As we have noted earlier, the required data necessary for the study of the solvent effects in biochemical process is not available. What is available is data on a few representatives of hydrophobic and hydrophilic groups. Even this limited data is for backbones that do not represent the environment around side chains attached to the backbone of a protein.

There have been attempts to extract quantities referred to as the PMF between pairs of side chains. Unfortunately, this is not the required data. We shall briefly discuss these quantities in Appendix J.

1.8. Dissection of the Solvation Gibbs Energy of a Globular Protein

As we shall see throughout the rest of the book, the study of solvent-induced effects is equivalent to the study of the solvation Gibbs energy of biomolecules. For small molecules, one can get information on the solvation Gibbs energy from the distribution of the molecule between the liquid and the gaseous phase. Such measurements are not feasible for macromolecules. This fact forces us to study ΔG_α^* using theoretical methods. Calculation of ΔG_α^* using purely theoretical methods is also a formidable task. Nevertheless, because of its importance, one can try either to estimate an approximate value of ΔG_α^* or to get some ideas about the relative magnitude of the various factors that determine ΔG_α^*. The most convenient and efficient way of studying ΔG_α^* is by splitting ΔG_α^* into components, each of which may be studied separately using either theoretical or experimental methods.

Traditionally, solvation Gibbs energies of moderately large molecules were studied by assuming *group additivity* for ΔG^*_α. For example, if α is *n*-hexane, one could write

$$\Delta G^*_{\text{hexane}} = 2\Delta G^*_{CH_3} + 4\Delta G^*_{CH_2}. \qquad (1.8.1)$$

Similarly, for 1-hexanol one can write

$$\Delta G^*_{\text{hexanol}} = \Delta G^*_{CH_3} + 5\Delta G^*_{CH_2} + \Delta G^*_{OH}. \qquad (1.8.2)$$

Empirically, such a group-additivity assumption was found to be quite useful for predicting the solvation Gibbs energies of molecules for which direct measurements are not feasible. However, this empirical recipe for calculating solvation Gibbs energies cannot be founded on solid theoretical grounds. In fact, theoretical consideration shows that such an assumption cannot be justified even when the total interaction energy in the system is pairwise additive. The reader is referred to Appendix E for more details on this question. Here we note that within the simple group-additivity assumptions, the solvation Gibbs energies of the three isomers shown in Fig. 1.28 should be equal to

$$\Delta G^*_\alpha = \Delta G^*_{BB} + \Delta G^*_{R_1} + \Delta G^*_{R_2}, \qquad (1.8.3)$$

where ΔG^*_{BB} is the solvation Gibbs energy of the backbone (BB), here a benzyl ring, and $\Delta G^*_{R_i}$ is the solvation Gibbs energy of the radical R_i. Clearly, for any two such isomers having the same

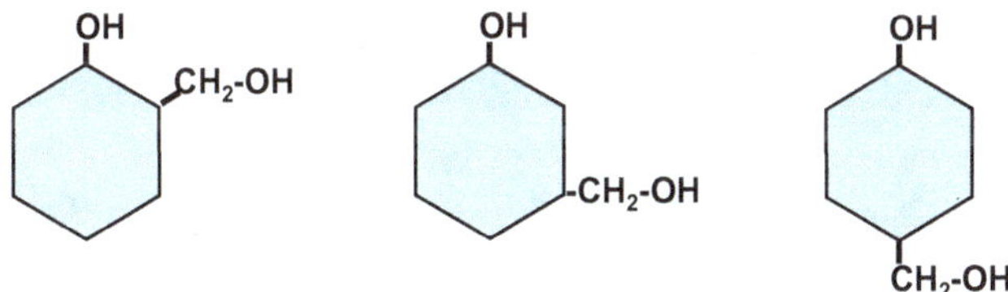

Fig. 1.28   Three isomers.

groups, one would have predicted the same solvation Gibbs energies based on the group-additivity assumption. The experimental data shows that this is not the case, especially when there are strong correlations between the groups constituting the molecule, as we have seen in Sec. 1.7.

We now proceed to split the solvation Gibbs energy of a globular macromolecule into several more manageable components. For simplicity, we assume that the protein is roughly spherical, that there are no solvent molecules within the interior of the protein, and that there are some functional groups (FGs) on its surface, as depicted in Fig. 1.29.

We assume that the interaction energy between the protein and a water molecule can be written as a sum of the form

$$U(\alpha, \mathbf{X}_i) = \sum_k U(\mathbf{R}_k, \mathbf{X}_i), \qquad (1.8.4)$$

where $\mathbf{R}_k$ is the location of the kth atom, or a group in the protein, and $\mathbf{X}_i$ is the configuration (location and orientation) of the ith water molecule.

Since we have assumed that the globular protein is *compact*, in the sense that there are no water molecules in its interior, and

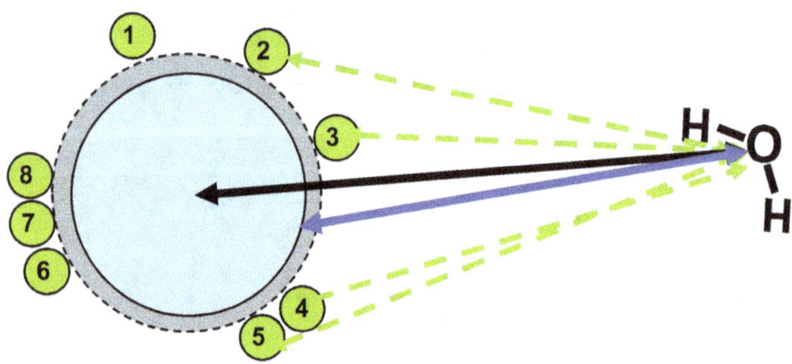

Fig. 1.29 The total interaction energy between the protein α and a water molecule is split into a hard part (black line), a soft part (blue line) and interactions between the FGs and the water (green dashed lines).

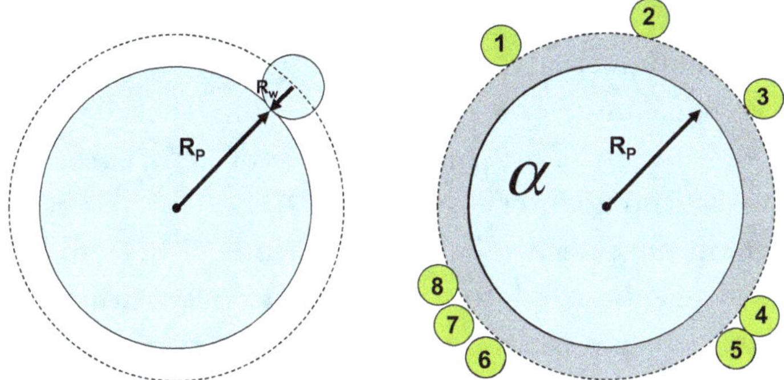

Fig. 1.30   Globular protein with radius R_p and a water molecule with radius R_w. The excluded volume is within the dashed line.

that the groups of the protein are tightly packed, we can replace the detailed sum in (1.8.4) by a simpler, less detailed sum as follows. We assume that we can draw some sphere of radius R_P, such that all the groups within this sphere do not "see" the solvent, which means that the groups in the interior of the sphere do not interact directly with the solvent molecules. These groups contribute to the total interaction energy between α and the water molecule simply by excluding the water molecule from some sphere of radius $R_P + R_W$, where R_W is an effective radius of a solvent molecule (Fig. 1.30). Thus, we can rewrite (1.8.4) as

$$U(\alpha, \mathbf{X}_i) = U^H(\alpha, \mathbf{X}_i) + U^S(\alpha, \mathbf{X}_i) + \sum_k U^{FG}(k, \mathbf{X}_i). \quad (1.8.5)$$

Here, U^H is the *hard* part of the interaction, which is essentially the interaction potential between a hard sphere of radius R_P and a hard sphere of radius R_W, i.e.

$$U^H(\alpha, \mathbf{X}_i) = \begin{cases} \infty & R_{\alpha,i} < R_P + R_W \\ 0 & R_{\alpha,i} \geq R_P + R_W. \end{cases} \quad (1.8.6)$$

The second part of the interaction in (1.8.5) is the *soft* interaction. This can be taken as the van der Waals interaction between a spherical shell of non-polar groups on the surface of the protein and a water molecule.

Finally, the sum over k in (1.8.5) signifies the total interaction energy between all *functional groups* (FGs) that are exposed to the solvent and a water molecule at X_i.

The main simplification in going from (1.8.4) to (1.8.5) is that we do not need to take into account explicitly the interactions between specific groups in the *interior* of the protein (i.e. those groups within the sphere of radius R_P) and the water molecules. It does not matter what kind of groups are within this sphere. Since the water molecules cannot reach into the interior of the protein, they are indifferent to the type of groups buried inside the interior of the protein. These groups collectively combine to produce a region from which a water molecule is excluded. The sum over k, on the other hand, is a sum over all the FGs that are exposed to the solvent. These groups interact with water in different ways according to the type of the FG.

Before elaborating on the details of the expression for the solvation Gibbs energy of α, it is important to stress the fact that group additivity ΔG_α^* does not follow from the assumption of pairwise additivity of the interaction as written in (1.8.5).[81]

We demonstrate this point for a simple solute, an ethane molecule consisting of two methyl groups. The solvation Gibbs energy of ethane may be written as

$$\Delta G_{Et}^* = -k_B T \ln\langle \exp[-\beta B_{Et}]\rangle_0, \qquad (1.8.7)$$

where B_{Et} is the binding energy of ethane. Assuming *pairwise additivity* of the interaction between the two methyl groups and

---

[81] See also Appendix E.

water, we can write

$$B_{Et} = \sum_{i=1}^{N} U(1, \mathbf{X}_i) + U(2, \mathbf{X}_i)$$

$$= B_1 + B_2, \tag{1.8.8}$$

where $U(1, \mathbf{X}_i)$ is the pair potential between one methyl group and a water molecule at configuration $\mathbf{X}_i$. Thus, the total binding energy of ethane is written in (1.8.8) as the sum of the *binding energies* of the two methyl groups. Substituting (1.8.8) into (1.8.7), we get

$$\Delta G_{Et}^* = -k_B T \ln \langle \exp[-\beta B_1] \exp[-\beta B_2] \rangle_0. \tag{1.8.9}$$

Clearly, the average in (1.8.9) cannot be factorized into a product of two averages, each of which corresponds to one methyl group. Instead, one can rewrite (1.8.9) in one of the following forms

$$\Delta G_{Et}^* = -k_B T \ln \langle \exp[-\beta B_1] \rangle_0 - k_B T \ln \langle \exp[-\beta B_2] \rangle_1$$

$$= -2k_B T \ln \langle \exp[-\beta B_{CH_3}] \rangle_0 + \delta G(CH_3 - CH_3)$$

$$= 2\Delta G_{Me}^* + \delta G(CH_3 - CH_3). \tag{1.8.10}$$

In the first expression on the right-hand side of (1.8.10), we carry out the process of solvation of ethane in two steps. First, solvate one methyl group in water. Here, the first average on the right-hand side of (1.8.10) is over all configurations of the solvent molecules in the *absence* of the methyl group. Second, we *conditionally* solvate the second methyl group to a fixed point near the first methyl group. The average in this case is over all the configurations of the water molecules using the *conditional* distribution function, i.e. the distribution of water molecules in

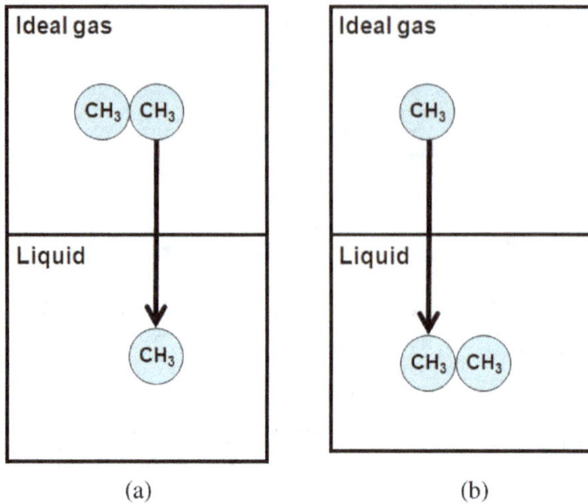

Fig. 1.31   The process of solvation of ethane in the two steps. (a) Transfer one methyl group. (b) Transfer the second methyl group.

the *presence* of the first methyl group.[82] This process is shown schematically in Fig. 1.31.

In the last equality on the right-hand side of (1.8.10), we carry out the solvation of ethane in yet another way. First, we solvate *two* methyl groups at infinite separation (this can be approximated by the solvation Gibbs energy of two methane molecules). Then, we bring the two methyl groups from infinite separation to the final configuration denoted by CH_3–CH_3 and add the solvent-induced part of the work for this process, which we denoted by $\delta G(CH_3$–$CH_3)$.

Clearly, group additivity of ΔG^*_{Et} is valid only when δG is zero, i.e. when the two methyl groups are *independently* solvated. In general, this independence of the solvation cannot be assumed. Therefore, the assumption of group additivity is not supported by theory. It should be noted, however, that practical group additivity uses effective values for the solvation Gibbs

---

[82] For more details, see Ben-Naim (1992, 2006).

energy of a methyl group, which is different from the real solvation Gibbs energy of a methyl group. For instance, one can *define* the solvation Gibbs energy of "methyl" simply by dividing the solvation Gibbs energy of ethane into two equal parts.[83]

Having the simple example of ethane in mind, we proceed to the solvation Gibbs energy of the entire protein. We use the assumption of additivity of the interactions as in (1.8.5) and write the solvation Gibbs energy of α as

$$\Delta G_\alpha^* = -k_B T \ln\langle \exp[-\beta B_\alpha]\rangle_0$$

$$= -k_B T \ln\left\langle \exp\left[-\beta B_\alpha^H - \beta B_\alpha^S - \beta \sum_k B_k^{\mathrm{FG}}\right]\right\rangle_0 . \quad (1.8.11)$$

Here, we use a shorthand notation for the total binding energy of the hard part B_α^H, the soft part B_α^S, and the binding energies of all the FGs that are exposed to the solvent.

Clearly, there is no reason for justifying the factorization of the average in (1.8.11) into three average quantities. Instead, we proceed by stepwise solvating each type of interaction. In the first step we write

$$\Delta G_\alpha^* = \Delta G_\alpha^{*H} + \Delta G_\alpha^{*S/H} + \Delta G_\alpha^{*\mathrm{FGs}/H,S}. \quad (1.8.12)$$

This is a shorthand notation for solvating the protein α in three steps. First, solvate the hard part. The resulting change in Gibbs energy is the first term on the right-hand side of (1.8.12). This is the same as the solvation of a hard sphere of radius R_P. The corresponding work is also equal to the work required to produce a cavity of radius $R_P + R_W$ in the solvent[84] (Fig. 1.32). In the second step, we solvate the soft interaction, given that the hard part of the interaction is already solvated. This part may

---

[83]There are many other types of group additivity. See for example Cabani *et al.* (1981), Eisenberg and McLachlan (1986).
[84]For more details, see Appendix K.

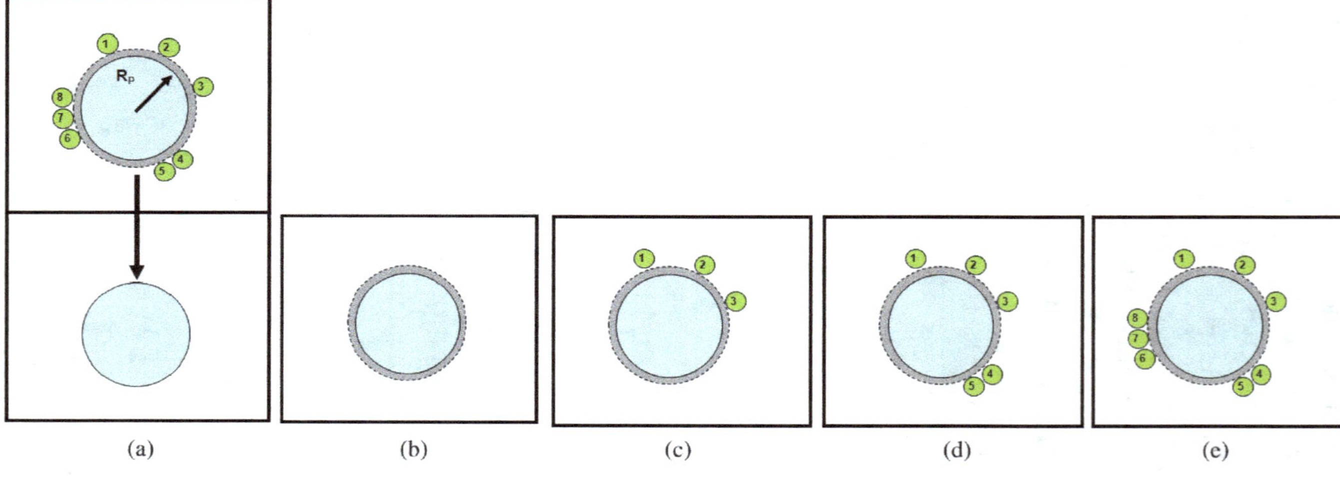

Fig. 1.32   An example of the expansion of the solvation Gibbs energy in steps as in (1.8.14). Solvate (a) the hard (H) part, (b) the soft part (S), (c) the independent FGs (d), the pair-correlated groups, and (e) the triple-correlated groups.

be approximated by van der Waals-type interactions between pairs of non-polar molecules. Finally, we have to *conditionally* solvate *all* the FGs, given that the hard and soft parts of the interaction have already been solvated.

Before grappling with the third term on the right-hand side of (1.8.12), it should be noted that the first two terms do not depend on the specific groups that produce the hard and soft part of the solvation. These two terms are relatively easy to estimate. The more interesting term is the third term on the right-hand side of (1.8.12). This term cannot be split into a sum of the form

$$\Delta G_{\alpha}^{\text{FGs}/H,S} = -k_B T \ln \left\langle \exp\left[-\beta \sum_k B_k^{FG}\right]\right\rangle_{/HS}$$
$$= -k_B T \ln \prod_k \langle \exp[-\beta B_k^{FG}]\rangle_{HS}$$
$$= \sum_k \Delta G_k^{*\text{FGs}/H,S}. \tag{1.8.13}$$

A split of the form (1.8.13) would have been correct if all the FGs were independently solvated.[85] Unfortunately, this is not the case in general. Therefore, we need to proceed more cautiously with the third term on the right-hand side of (1.8.12). We first transfer all the independently solvated FGs. For instance, in Fig. 1.30, the FGs (1), (2) and (3) are independently solvated. Qualitatively, this means that the conditional solvation of one group does not affect the conditional solvation of another group. On the other hand, the two FGs (4) and (5) are pairwise correlated; the solvation spheres around the two groups overlap, hence their conditional solvation Gibbs energy is not the sum of the conditional solvation Gibbs energies of the groups (4) and (5) separately. Similarly, the three FGs (6), (7), and (8)

---

[85]Formally, this means that the average of a product can be written as a product of averages. See also Appendix F.

are triply correlated. Thus, in general, we can write the third term on the right-hand side of (1.8.12) as

$$\Delta G_\alpha^* = \Delta G_\alpha^{*H} + \Delta G_\alpha^{*S/H} + \sum_{k(\text{independent } t)} \Delta G_\alpha^{*k/H,S}$$

$$+ \sum_{i,j(\text{pair correlated})} \Delta G_\alpha^{*i,j/H,S}$$

$$+ \sum_{l,m,n(\text{triply correlated})} \Delta G_\alpha^{*l,m,n/H,S} + \cdots . \quad (1.8.14)$$

This expansion looks complicated, but its meaning is quite simple. Instead of transferring the entire molecule from the gaseous phase into the liquid, we do the solvation process in several steps. First, we transfer the hard part — the corresponding work is the same as creating a cavity of suitable size to accommodate the hard part of the molecule. Next, we "turn on" the soft part of the interaction. This involves the work $\Delta G_\alpha^{*S/H}$. At this stage, we need to transfer all the FGs, but we have to be careful about the *order* in which we do this transfer. For example, two methyl groups on the surface of α might have a different conditional solvation Gibbs energy, depending on the local environment they are found in. Therefore, in (1.8.14), we first transfer all the *independently solvated* FGs, then all the *pair correlated* FGs, then all the *triply correlated* FGs, etc. In this way, we have accomplished the solvation process in several steps (Fig. 1.32). In practice, we cut off the expansion (1.8.14) after the sum over all *pair correlated* FGs. This is done simply because we do not have any data on higher-order correlated FGs.

The expansion (1.8.14) can be rewritten in a slightly different form, which is more convenient for the study of the different contributions to the solvation Gibbs energy. Taking the expansion (1.8.14) up to the sum of the pair correlated FGs, we can do the solvation process in a different order. We first transfer the

H and the S part as before, but now we *transfer all* the FGs and calculate the conditional solvation of *all* the FGs as if they were independent. In the second step, we bring all the pair correlated FGs to the final positions as in the original α molecule.

The corresponding work in the last step is due to the solvent-induced part of the (conditional) potential of mean force between the different pair correlated FGs. The modified expansion is now

$$\Delta G_{\alpha}^{*} = \Delta G_{\alpha}^{*H} + \Delta G_{\alpha}^{*S/H} + \sum_{k(\text{all FGs})} \Delta G^{*k/H,S}$$

$$+ \sum_{i,j} \delta G(i,j/H,S) + \cdots . \qquad (1.8.15)$$

Note that the third term on the right-hand side of (1.8.15) is over *all* the FGs, treated as being independently solvated. The fourth term on the right-hand side of (1.8.15) corrects for the correlation between the groups, each term $\delta G(i,j)$ corresponds to the indirect part of the work required for bringing the groups i and j from infinite separation to the final configuration as in the original α molecule. Figure 1.33 shows a simple example of the last term in expansion (1.8.15).

We summarize what we have achieved so far as follows: We start with writing the protein–water interaction in the form (1.8.5). This involves the assumption of *pairwise additivity* of the *solute–solvent pair potential*. Using this form of the pair potential, we can write the solvation Gibbs energy of the solute α in the form (1.8.11). Finally, we split the process of solvation into several steps: First, solvate the hard part, then conditionally solvate the soft part, and then conditionally solvate all the FGs. The last step is further broken down into smaller components as in Eq. (1.8.15). Note that all the terms on the right-hand side of (1.8.15) have the conditions H and S, i.e. given that the hard and soft parts have already been solvated. The pairwise, as well

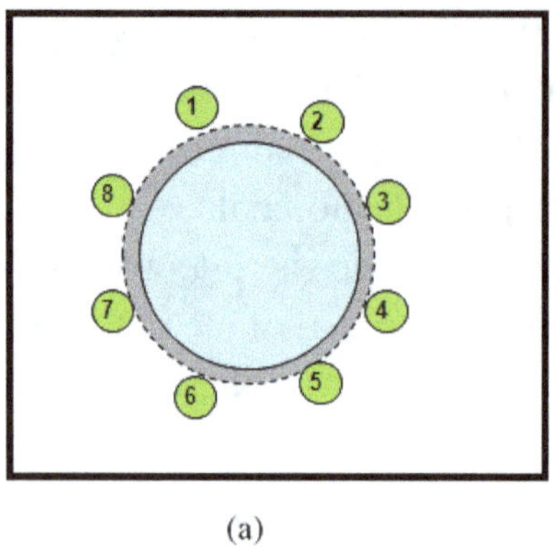
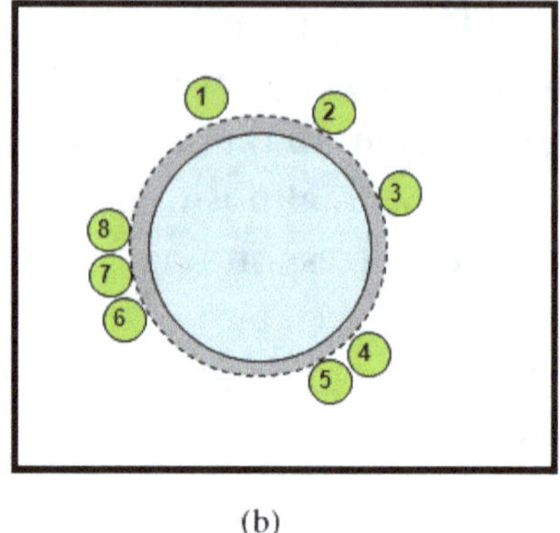

(a)   (b)

Fig. 1.33 A modified stepwise solvation process. Corresponding to Eq. (1.8.15). First, we transform all FGs as if they were independently solvated, then we add the correlations between the FGs.

as higher-order correlations are also conditional quantities, i.e. given that the hard and soft part of the interaction have been solvated.

It should be noted that in this section, we dealt with *one specific conformation* of the globular protein and carried out the dissection of the solvation Gibbs energy into smaller components. Real proteins do not maintain a fixed configuration. What we experimentally conceive as the *natured*, or the stable globular protein consists of a *range* of conformations which are included in the definition of the structure of the protein. Of course, the *values* of the various ingredients of the solvation Gibbs energy of the protein might depend on the specific conformation of the protein. In order to relate the actual solvation Gibbs energy of the *natured* protein to the solvation Gibbs energies of the specific configuration, one must take the appropriate average over all the conformations that are deemed to be "belonging" to the natured protein.[86]

---

[86] See also Appendix B.

In this section we were not interested in the *values* of the solvation Gibbs energies of the specific conformations, but only in the methodology of splitting the ΔG^*_α of the protein into its ingredients. This methodology of splitting is independent of the specific conformation chosen to represent the folded or the unfolded protein.

1.9. Dissection of the Solvation Gibbs Energy of a Denatured Protein

We now turn to the unfolded protein, denoted by **U**. Again, we choose only one conformation of the unfolded, or the denatured form of the protein, bearing in mind that this specific conformation is only one of a huge number of possible conformations that "belong" to the experimentally recognized denatured form. As in the previous section, we are interested here only in dissecting the total solvation Gibbs energy into smaller, more manageable components. Suppose we select the fully extended protein to represent the unfolded protein (Fig. 1.34). We write the total interaction energy of the protein α with a water molecule as

$$U(\alpha, \mathbf{X}_i) = U^{\mathrm{LC}}(\alpha, \mathbf{X}_i) + \sum_k U^{\mathrm{FG}}(k, \mathbf{X}_i), \qquad (1.9.1)$$

where the first term on the right-hand side of (1.9.1) is the interaction energy between the linear chain (LC), or the backbone of the protein, and a water molecule at a configuration $\mathbf{X}_i$. The second term on the right-hand side (1.9.1) is the same as in (1.8.5), i.e. it is the sum of the interactions between all the FGs along the chain with a water molecule. Note that we include in this sum also all the carbonyl (C=O) and the amine (NH) groups that belong to the backbone, and the remaining FGs can be taken as the side chains of the protein α.

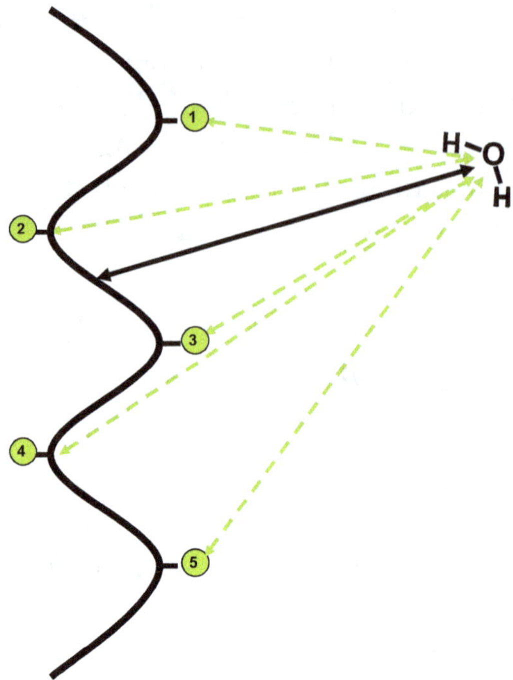

Fig. 1.34 A schematic description of the denatured protein and its interaction with a water molecule corresponding to Eq. (1.9.1) (compare with Fig. 1.29).

The solvation Gibbs energy of the unfolded form can now be written, up to pair correlation terms as

$$\Delta G_\alpha^* = -k_B T \ln \left\langle \exp\left[-\beta \sum_i U(\alpha, \mathbf{X}_i) \right] \right\rangle_0$$

$$= \Delta G_\alpha^{*LC} + \Delta G_\alpha^{FGs/LC} = \Delta G_\alpha^{*LC} + \sum_{k(\text{all FGs})} \Delta G_\alpha^{*k/LC}$$

$$+ \sum_{i,j} \delta G(i,j/LC) + \cdots . \tag{1.9.2}$$

Here, ΔG_α^{*LC} is the solvation Gibbs energy of the linear chain of the protein α, devoid of any functional groups. All other terms on the right-hand side of (1.9.2) have the same significance as the corresponding terms in (1.8.15). The solvation Gibbs energy of the linear chain may be approximated by the solvation Gibbs energy of a long hydrocarbon. It is known that the solvation Gibbs energy of a small linear hydrocarbon is linear in the number of carbon atoms and has the form[87]

$$\Delta G_{LC}^* \approx a + bn, \tag{1.9.3}$$

where n is the number of carbon atoms and a and b are constants; $a \approx 5.94\,\text{kJ/mol}$ and $b \approx 0.75\,\text{kJ/mol}$. The linearity (1.9.3) is valid for hydrocarbons of size $3 \leq n \leq 10$, therefore the extrapolation to a hydrocarbon of a few hundred carbon atoms is not expected to be correct. However, whatever the value we adopt for ΔG_{LC}^*, this will be the relatively uninteresting part of the Gibbs energy of solvation of the unfolded protein. The more important and more interesting contributions are expected to come from the *functional groups*, both of the backbones and the side chains.

1.10. The Relationship between the Standard Gibbs Energy of a Reaction and Solvation Gibbs Energies

In the literature, one usually encounters the *standard* Gibbs energy of a reaction. In Sec. 1.1, we found it more convenient to work with the Gibbs energy of a reaction based on the pseudo-chemical potentials of the various species involved in the reaction. We develop the relationship between the *standard* Gibbs

---

[87]Ben-Naim and Marcus (1984).

energy of a reaction and the corresponding solvation Gibbs energy for the particular case of protein folding.

Consider first the simple process of isomerization

$$U \to F. \tag{1.10.1}$$

U and F will later designate the unfolded and folded forms of a protein, but here the discussion is more general and applies to any conversion between two isomers. Also, for simplicity, we shall assume that both U and F are rigid molecules, i.e. there are no internal rotational degrees of freedom. Later, we shall generalize our discussion to molecules with internal rotations as well as for more general reactions between molecules. In this section, we shall see that the thermodynamic driving force is related to the difference in the solvation Gibbs energies. In the next section, we pose the question: How can we split the total solvent-induced contribution to the driving force of a process into small factors, each of which may be studied in model systems?

We denote by μ_F and μ_U the chemical potentials of the folded (F) and the unfolded (U) species, respectively. These can be written as[88]

$$\mu_F = \mu_F^* + k_B T \ln \rho_F \Lambda_F^3, \tag{1.10.2}$$

$$\mu_U = \mu_U^* + k_B T \ln \rho_U \Lambda_U^3. \tag{1.10.3}$$

where ρ_α is the number density of the species α, and Λ_α^3 is the momentum partition function of the species α. μ_α^* is referred to as the pseudo-chemical potential of the species α. In the specific reaction of isomerization (1.10.1), the momentum partition functions Λ_F^3 and Λ_U^3 have the same value. They depend only on the molecular mass of F and U. At equilibrium we have

$$\mu_F = \mu_U, \tag{1.10.4}$$

---

[88]See Ben-Naim (2006, 2009) and Appendices A and B.

or equivalently

$$\Delta G^*(\mathbf{U} \to \mathbf{F}) = \mu_F^* - \mu_U^* = k_B T \ln \left(\frac{\rho_U}{\rho_F} \right)_{\text{eq}} . \qquad (1.10.5)$$

The quantity $\Delta G^*(\mathbf{U} \to \mathbf{F})$ is equal to the conventional *standard* Gibbs energy of the reaction.[89] Since we are interested only in the solvent-induced effect on the driving force for the reaction (1.10.1), it is easier to use this quantity rather the standard Gibbs energy. The quantity $\Delta G^*(\mathbf{U} \to \mathbf{F})$ has a simple meaning. This is the Gibbs energy change for the process of transferring *one* U molecule being at a fixed position (i.e. a molecule devoid of translational degrees of freedom) to one F molecule being at a fixed position.

From the equality (1.10.5), it follows that when ΔG^* $(\mathbf{U} \to \mathbf{F})$ is large and positive, the ratio $(\rho_U/\rho_F)_{\text{eq}}$ is large and greater than unity. This means that the form U is more favorable than F. On the other hand, when $\Delta G^*(\mathbf{U} \to \mathbf{F})$ is large and negative, the equilibrium ratio $(\rho_U/\rho_F)_{\text{eq}}$ will tend to zero, i.e. the species F will be more favorable.

For this reason, $\Delta G^*(\mathbf{U} \to \mathbf{F})$ is a measure of the *thermodynamic driving force* for the process (1.10.1). The more negative this quantity is, the larger the *driving force* for the reaction, i.e. the process (1.10.1) will proceed in such a way that at equilibrium, $\rho_F \gg \rho_U$.

We stress again that what we have referred to as a thermodynamic *driving force* is not a *force* in the physical sense of the word. The force itself is an important factor in considering the path along which the process U $\to$ F proceeds. We shall discuss the *actual force* operating on the molecule in Chapter 3. Here,

---

[89]Note that $\Delta G^*(\mathbf{U} \to \mathbf{F}) = \Delta G^\circ(\mathbf{U} \to \mathbf{F})$, where ΔG° is the *standard* Gibbs energy of the reaction. However, μ_α^{*l} is not equal to the standard chemical potential of the species α. The relationship between the two is $\mu_\alpha^l = \mu_\alpha^* + k_B T \ln \Lambda_\alpha^3 \rho_s = \mu_\alpha^0 + k_B T \ln \rho_\alpha$. This relationship is valid only for dilute solutions.

the *driving force* is only a measure of the relative stability of the two components at equilibrium, as expressed in the Eq. (1.10.5).

In the process of protein folding, it is helpful to split the driving force into two parts, the enthalpic and the entropic parts. To do this, one simply writes $\Delta G^*(\mathbf{U} \to \mathbf{F}) = \Delta H^*(\mathbf{U} \to \mathbf{F}) - T\Delta S^*(\mathbf{U} \to \mathbf{F})$. Clearly, the relative contributions of ΔH^* and ΔS^* of the reaction (1.10.1) depend on the temperature T. Though both ΔH^* and ΔS^* are dependent on T, one can say that at very low temperatures, the enthalpic term will dominate the driving force, while at very high temperatures, the entropic term will dominate the driving force for the reaction.

We now wish to establish a relationship between the quantity $\Delta G^*(\mathbf{U} \to \mathbf{F})$ and the solvation Gibbs energies of the species $\mathbf{U}$ and $\mathbf{F}$. To do this, we first discuss a simplified version of reaction (1.10.1). Instead of $\mathbf{U}$ and $\mathbf{F}$ being at a fixed position only, we assume that $\mathbf{U}$ and $\mathbf{F}$ also have a *fixed* conformation. We now write the analogue of (1.10.5) for the ideal gas phase

$$\Delta G^{*ig}(\mathbf{U} \to \mathbf{F}) = \mu_F^{*ig} - \mu_U^{*ig}$$
$$= \Delta U(\mathbf{U} \to \mathbf{F}) - k_B T \ln(q_U/q_F). \qquad (1.10.6)$$

Here, we have separated $\Delta G^{*ig}(\mathbf{U} \to \mathbf{F})$ into two contributions. One is due to the change in the internal partition functions of the molecules. q_α includes the rotational, vibrational and electronic partition functions but not the translations (those were already excluded from μ_α^*) or the internal rotations (we have assumed that the molecules are rigid in the sense that there are no internal rotational degrees of freedom). The second part is the change in energy for the process (1.10.1). This is essentially the change in the potential energy of interaction among all the groups or atoms within the molecule.[90]

---

[90]Note that we use U for the potential energy, whereas $\mathbf{U}$ is for the unfolded species.

When the same conversion (1.10.1) takes place in a solvent, (1.10.6) should be modified. We shall always assume that the internal partition functions of $\mathbf{U}$ and $\mathbf{F}$ are not affected by the presence of the solvent. Hence, the analogue of (1.10.6) is

$$\Delta G^{*l}(\mathbf{U} \to \mathbf{F}) = \Delta U(\mathbf{U} \to \mathbf{F}) - k_B T \ln(q_U/q_F)$$

$$+ W(\mathbf{F}|\text{solvent}) - W(\mathbf{U}|\text{solvent}), \qquad (1.10.7)$$

where $W(\alpha|\text{solvent})$ is the coupling work of the solute α to the solvent. The statistical mechanical expression for $W(\alpha|\text{solvent})$ is

$$W(\alpha|\text{solvent}) = -k_B T \ln\langle \exp[-\beta B_\alpha]\rangle_0, \qquad (1.10.8)$$

where $\beta = (k_B T)^{-1}$, B_α is the total interaction energy of α (recall that α is a rigid molecule) with all the solvent molecules being at some specific configuration $\mathbf{X}_1, \ldots, \mathbf{X}_N$ ($\mathbf{X}_i$ is the configuration; location and orientation of the solvent molecule i). This is referred to as the *binding energy* of α to the solvent. The average in (1.10.8) is taken over all configurations of the solvent molecules here in the T, P, N ensemble.

Combining (1.10.6) and (1.10.7), we can write for the solvent-induced part of the driving force;

$$\delta G(\mathbf{U} \to \mathbf{F}) = \Delta G^{*l}(\mathbf{U} \to \mathbf{F}) - \Delta G^{*ig}(\mathbf{U} \to \mathbf{F})$$

$$= W(\mathbf{F}|\text{solvent}) - W(\mathbf{U}|\text{solvent})$$

$$= \Delta G_F^* - \Delta G_U^*, \qquad (1.10.9)$$

where ΔG_α^* is the solvation Gibbs energy of α. Recall that ΔG_α^* is the change in Gibbs energy for the process of transferring a molecule α from a fixed position in an ideal gas phase to a fixed position in the liquid. The quantity $\delta G(\mathbf{U} \to \mathbf{F})$ is referred to as the solvent-induced contribution to the driving force for the process (1.10.1). This quantity will be the central quantity of

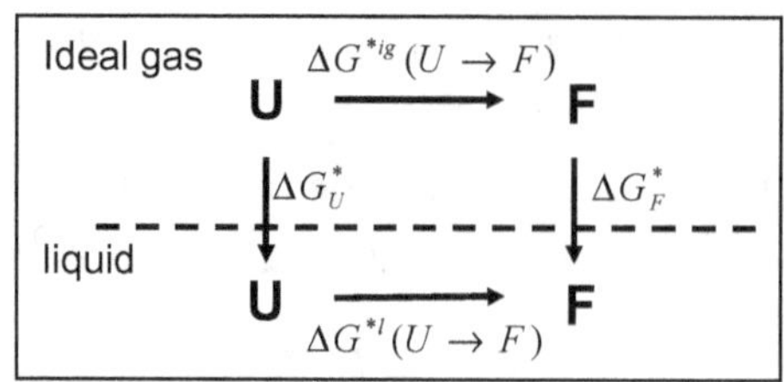

Fig. 1.35 Schematic relationship between the solvent-induced contribution to the driving force for reaction (1.10.1) (see 1.10.9) and solvation Gibbs energies of **U** and **F**.

interest in the entire book. The relation between δG and the solvation Gibbs energies is shown schematically in Fig. 1.35.

Before we proceed with the analysis of the quantity δG in (1.10.9), it is useful to rewrite (1.10.8) in terms of equilibrium constants. Let K^{ig} be the equilibrium constant for the reaction (1.10.1) in an ideal gas phase. We have

$$K^{ig} = \left(\frac{\rho_F}{\rho_U}\right)^{ig}_{eq} = \frac{q_F}{q_U}\exp[-\beta\Delta U(\mathbf{U} \to \mathbf{F})]. \qquad (1.10.10)$$

When the same reaction is carried out in a liquid, the equilibrium constant is modified to take into account solvation effects. Hence

$$K^{l} = \left(\frac{\rho_F}{\rho_U}\right)^{l}_{eq} = \frac{q_F}{q_U}\exp[-\beta\Delta G^{*l}(\mathbf{U} \to \mathbf{F})]$$

$$= \frac{q_F}{q_U}\exp[-\beta\Delta U(\mathbf{U} \to \mathbf{F}) - \beta\delta G(\mathbf{U} \to \mathbf{F})]$$

$$= K^{ig}\exp[-\beta\delta G(\mathbf{U} \to \mathbf{F})], \qquad (1.10.11)$$

or equivalently

$$\frac{K^{l}}{K^{ig}} = \exp[-\beta(\Delta G^{*}_{F} - \Delta G^{*}_{U})]. \qquad (1.10.12)$$

Thus, the solvent-induced part of the driving force is related to the ratio of the equilibrium constants in the liquid and in an ideal gas phase.

Equation (1.10.9) is the required relationship between the solvent-induced contribution to the driving force for the specific reaction (1.10.1) and the solvation Gibbs energies of the species U and F.

We stress again that the relation (1.10.9) was derived for a fixed conformation of U and F. Whenever we deal with the real folded and unfolded forms, we must take the appropriate average over all the configurations of both U and F. This is necessary when applying the relationships to real molecules having internal rotational degrees of freedom (see also Appendix B).

1.11. The Various Solvent-Induced Contributions to the Driving Force for Protein Folding

In Sec. 1.10, we have written the most general expression for the solvent-induced contribution to the driving force for the protein folding process (1.10.9)

$$\delta G(U \to F) = \Delta G_F^* - \Delta G_U^*. \qquad (1.11.1)$$

Recall that this equation was derived for one particular conformation of the unfolded form and one particular conformation of the folded form. Whenever we need to calculate the relevant quantities for a real process of folding, we must take the appropriate averages over all the conformations that we choose to include in the U and F forms. However, at this stage we are not interested in the actual value of δG for a process involving real protein, but only in the factors that contribute to these quantities, and their relative magnitudes.

In Secs. 1.8 and 1.9, we found expressions for the solvation Gibbs energies of the U and F forms. Clearly, having all the contributions listed on the right-hand side of (1.8.15) and (1.9.2), we can estimate the solvent-induced contributions to the driving force for the specific "reaction" in (1.10.1). Such detailed information is not available for any protein. Nevertheless, we can make some order-of-magnitude estimates of the different factors that contribute to $\delta G(\mathbf{U} \to \mathbf{F})$, and of their relative importance.

1.11.1. *The solvation of the hard part*

The first term on the right-hand side of (1.8.14) is ΔG_α^{*H}, which corresponds to solvating the hard part of the protein–solvent interaction potential. This is the same as the free energy of creating a cavity of suitable size at some fixed position in the solvent.[91] We assume that the globular protein is spherical with an effective diameter σ_P. The diameter of the water molecule is taken as $\sigma_W = 2.8\,\text{Å}$. The cavity suitable to accommodate the protein has a radius of

$$R_{CAV} = (\sigma_P + \sigma_W)/2. \qquad (1.11.2)$$

To estimate the work required to create a cavity of radius R_{CAV}, we use the scaled particle theory (SPT).[92] According to the SPT, the Gibbs energy of solvation of a hard sphere is

$$\Delta G_\alpha^{*H} = K_0 + K_1 R_{CAV} + K_2 R_{CAV}^2 + K_3 R_{CAV}^3, \qquad (1.11.3)$$

where the coefficients K_i are given by

$$K_0 = k_B T(-\ln(1-y) + 4.5\,z^2) - \pi P \sigma_W^3/6,$$

$$K_1 = k_B T(6\,z + 18\,z^2)/\sigma_W + \pi P \sigma_W^2,$$

---

[91] See Appendix K.
[92] Reiss (1966), Ben-Naim (2006).

$$K_2 = k_B T (12\,z + 18\,z^2)/\sigma_W^2 - 2\pi P \sigma_W^2,$$

$$K_3 = 4\pi P/3, \tag{1.11.4}$$

and where

$$y = \pi \rho_W \sigma_W^3/6, \quad z = y(1-y). \tag{1.11.5}$$

Here, ρ_W is the solvent density and P is the pressure. Taking the density of water at 298.15 K, $\rho_W = 3.344 \times 10^{22}$ molecules/cm^3 and P as 1 atmosphere, we can calculate the solvation Gibbs energy of the hard part as a function of protein diameter σ_P. This function is a steeply increasing function of σ_P. We shall further discuss this function in Chapter 2.

1.11.2. *The solvation of the soft part*

The second term in (1.8.15) corresponds to the "turning on" of the soft part of the protein–water interaction potential. This is given by

$$\Delta G_\alpha^{*S/H} = -k_B T \ln \langle \exp[-\beta B_\alpha^S] \rangle_H, \tag{1.11.6}$$

where B_α^S is the total van der Waals interaction of the protein with the surrounding water molecules. Note that this is a conditional average quantity, the condition being the solvated hard (H) part of the interaction.

The soft part $\Delta G_\alpha^{*S/H}$ may be estimated by assuming that the surface of the globular protein consists of methane-like molecules, so that the total soft interaction between α and a water molecule is the sum of the interactions between these methane-like molecules and a water molecule. These pair interactions are assumed to be of a Lennard-Jones type. As expected, this part of the solvation Gibbs energy is negative, and about an order of magnitude smaller than the values of the cavity work. Hence, a solute α of the size of a typical globular protein having only hydrophobic groups (methane-like) on its surface will be extremely insoluble in water.[93] This is further discussed in Chapter 2.

1.11.3. *The contribution of the functional groups (FGs) exposed to the solvent*

This step is the most important one in calculating the solvation Gibbs energy of the protein. Assuming that we can use the expansion (1.8.15) up to the pair correlation term, we have at least four terms to consider here. These are:

(i) The conditional solvation Gibbs energy of a hydrophilic ($H\phi I$) group.

(ii) The conditional solvation of a Gibbs energy hydrophobic ($H\phi O$) group.

(iii) The contribution due to the correlation between two $H\phi I$ groups.

(iv) The contribution due to the correlation between a $H\phi O$ and a $H\phi I$ group.

(v) The contribution due to the correlation between two $H\phi O$ groups.

The methodology of obtaining the Gibbs energies of all the ingredients listed above has been discussed in Secs. 1.5–1.7 and at great length in Ben-Naim (1992, 2009). Using all these contributions to the solvation Gibbs energies on the right-hand side of (1.11.1), we can get an estimate of the solvent-induced contribution to the driving force for the process of protein folding. We shall further discuss these quantities in Chapter 3.

---

[93] Here, we refer to solubility from the gaseous phase.

1.12. Concluding Remarks and Some Suggestions for the Future

This chapter dealt with the formal dissection of a very complex process into small ingredients. We shall discuss more specific processes in the following chapters. It is clear at this stage that, based on the very limited amount of data, the various $H\phi I$ effects might be more important than what was thought until quite recently. However, much more data is needed in order to assess the relative importance of the various effects. As we have noted in this chapter, much effort has been expended in the study of the solvation thermodynamics of molecules and the pair correlation between two small solutes in water. There is an urgent need to study also the *conditional* solvation quantities as well as the *conditional* pair correlations. The most important "condition" is of course some model compounds that mimic the effect of proteins on their immediate environment. For instance, molecules such as those shown in Fig. 1.36, or simply di- or tripeptides as model compounds, for the study of conditional solvation quantities near protein. A list of such quantities would serve as an "ideal" hydrophobicity scale. Note that we need only the *conditional* solvation Gibbs energies of the side chains in water, and not in any organic liquid. Similarly, we need to have estimates of the pair correlation contributions to the driving force for all the 20^2 pairs of side chains. This is certainly

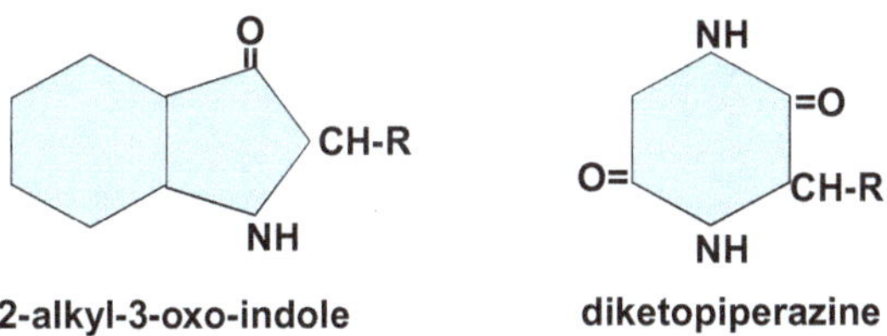

Fig. 1.36 Model compounds for studying more realistic conditional solvation Gibbs energies.

not an easy task. Perhaps only a few representative pairs of side chains would be enough, for instance, those pairs that occur more frequently on the surface of the proteins. Again, the ideal "condition" for the pair correlation would be a small molecule, or a short polypeptide that mimics the environment near the surface of the protein.

CHAPTER 2

Solvation and Solubility of Globular Proteins

The high solubility of proteins in water is well known to any biochemist. Yet, the molecular reason for the solubility of protein is not any less mysterious than the molecular reasons for protein folding and self-assembly of biological macromolecules.

I myself was not aware of the fact that protein solubility was a mystery until the 1980s and the 1990s, when I was visiting NIH and benefited from many conversations with Harry Saroff. I knew, as every chemist and biochemist did, that proteins were highly hydrophilic at their surface. This fact was more than enough to provide a sound explanation for the high solubility of proteins. However, Harry Saroff told me that sometimes replacing a hydrophilic group on the surface of a protein with a more hydrophilic group causes a decrease in the solubility of the protein. An example of such a case is discussed in Sec. 2.5. This fact seems very puzzling. One would have expected that making the surface more hydrophilic should cause an increase in the solubility of the protein, not a decrease.

At that time, I was not sure what the explanation for that phenomenon could be. I suspected that the replacement of one amino acid with another could have a major effect on the structure of the protein — perhaps causing other hydrophobic groups to get exposed to the solvent, causing a decrease in solubility.

For over 20 years, I kept this story at the back of my mind. It was only in the late 1990s, after I discovered the importance of HϕI effects on the stability of a protein, that I attempted to explain this puzzling fact. In 1996, together with Hongwu Wang, I examined the factors involved in the determination of the solubility of proteins. To our surprise, we found that although it is true that the hydrophilic groups on the surface of the protein determine its solubility, it is not only the mere number of such groups, but also the more detailed distribution of hydrophilic groups on the surface of the protein. This chapter tells the story of HϕI interactions and their effect on solubility. It also gives a possible explanation for Harry's puzzling story.

This chapter deals with the simplest process involving proteins — their solubility in water. Although the solubility of proteins is easy to measure, it is extremely difficult to calculate theoretically, and even more difficult to understand. Nevertheless, because of the importance of the solvation Gibbs energy of proteins to the understanding of both protein folding and protein–protein association, we devote this chapter to discussing the factors that contribute to the solvation Gibbs energy of globular proteins.

Knowledge of the solvation of proteins is an essential ingredient in the study of the driving forces for any biochemical process involving proteins, such as protein folding, protein–protein association, and the binding of protein to nucleic acids. The solvation Gibbs energy of a protein is also a major factor in the determination of the solubility of the protein.

The solubility of proteins in water is utilized by the body to transport small molecules from one place to another. This includes the transportation of oxygen by hemoglobin from the lungs to every cell where it is needed, as well as transporting carbon dioxide back from the cells to the lungs, from which it is expelled from the body. The body utilizes the same vehicles to transport drugs from one place to another.

But what makes such a huge organic molecule soluble in water?

Consider a typical protein of say, 150–200 amino acid residues. There are some 20^{150}–20^{200} possible sequences of polypeptides of such length. We know that only a very small fraction of this immense number of polypeptides exist and function in a living cell. These proteins were probably selected by evolution — not randomly chosen from the total number of possible sequences, but rather from the limited number of polypeptides that were soluble in water. We also know that when globular proteins undergo denaturation, i.e. the breakdown of their 3-D structure, they lose their solubility. Therefore, it is clear that the solubility of a protein is intimately related to its 3-D structure. More specifically, it is not the structure of the protein itself, but the distribution of hydrophobic ($H\phi O$) and hydrophilic ($H\phi I$) groups on its surface that determines the high solubility of the protein.

We shall begin this chapter with the general relationship between the solubility and the solvation Gibbs energy of any solute in any solvent. We shall then examine the question of why proteins are soluble in water. We shall see that the conditional solvation of the $H\phi I$ groups on the surface is, as is well known, an important factor. What is less known is that correlations between $H\phi I$ groups are also important, if not the most important, factors in the determination of the high solubility of globular proteins.

2.1. Definition of Solubility and Its Relationship to the Solvation Gibbs Energy

The solubility of any solute α in a liquid w is always defined with respect to some phase of pure, or nearly pure α (Fig. 2.1).

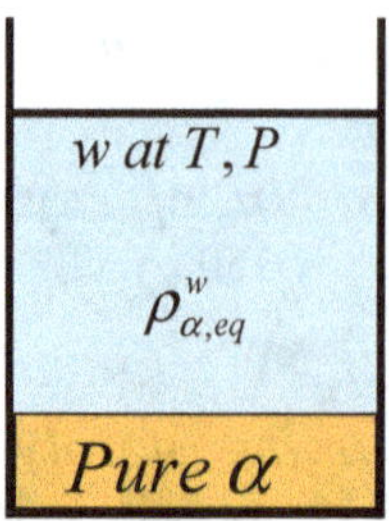

Fig. 2.1   The solubility of α in w is defined as the concentration of α in the liquid phase w at equilibrium with the pure α phase.

Consider, for example, a pure solid α in contact with liquid w. At equilibrium, we have the equality of the chemical potentials of α, in the two phases

$$\mu^{S}_{\alpha} = \mu^{W}_{\alpha}, \qquad (2.1.1)$$

where μ^{S}_{α} and μ^{W}_{α} are the chemical potentials of α in the solid (s) and in the liquid (w) phases, respectively, at equilibrium at some specified temperature T and pressure P.

For simplicity, we assume that the solid phase s is pure α. For simple molecules having a well-defined structure in their solid state, one can estimate the chemical potential μ^{S}_{α} from the inter-action energies between the α-molecules holding lattice points in the solid phase. This is, in principle, also true for proteins. For our purposes, we need not be concerned with the value of the chemical potential in the solid phase, or in any other pure state. We shall be interested mainly in the *solvation* Gibbs energy of the protein.

The solvation Gibbs energy of a protein in water is a well-defined quantity. This quantity is defined as the change in Gibbs energy for transferring one molecule α from a fixed position in an ideal gas phase to a fixed position in the liquid, the process being carried out at a constant T and P (see also Appendices A and B). Note that ΔG^{*}_{α} alone is not enough for calculating the solubility of a protein at equilibrium with its solid phase. However, this is a fundamental quantity that we shall need for calculating the solvent effects on any process involving proteins. It is also a useful quantity for estimating *changes* in solubility caused by the addition of another solute to the system (see Sec. 2.6).

Our central quantity in this chapter is the solvation Gibbs energy of the molecule α, which we write as

$$\Delta G_{\alpha}^{*} = -k_B T \ln\langle \exp[-\beta B_\alpha]\rangle_0, \qquad (2.1.2)$$

where B_α is the total binding energy of α to all the molecules in the "solvent." This includes all solvent molecules, as well as any other solute molecules in the system excluding the specific solute α that we have placed at a specific location in the liquid phase. It should also be noted that, in general, the quantity ΔG_{α}^{*} also includes an average over all the conformations of the protein (see Appendix B). For our purposes in this section, we assume that the globular protein α has a fixed conformation. Hence, the average in (2.1.2) is over all the configurations of the solvent molecules in the system except the specific α molecule that we have placed at a fixed position.

The relationship between the solubility of α and the solvation Gibbs energy is obtained as follows:

We first write the chemical potential of α in the liquid as

$$\mu_{\alpha}^{W} = \mu_{\alpha}^{*W} + k_B T \ln \rho_{\alpha}^{W} \Lambda_{\alpha}^{3}. \qquad (2.1.3)$$

In an ideal gas phase we have the relationship

$$\mu_{\alpha}^{ig} = \mu_{\alpha}^{*ig} + k_B T \ln \rho_{\alpha}^{ig} \Lambda_{\alpha}^{3}. \qquad (2.1.4)$$

For solutes having a measurable vapor pressure, one can use Eqs. (2.1.3) and (2.1.4) together with the condition of equilibrium

$$\mu_{\alpha}^{W} = \mu_{\alpha}^{ig} \qquad (2.1.5)$$

to obtain the solvation Gibbs energy of α in water[97]

$$\Delta G_\alpha^{*W} = \mu_\alpha^{*W} - \mu_\alpha^{*ig} = k_B T \ln (\rho_\alpha^{ig}/\rho_\alpha^{W})_{eq}. \qquad (2.1.6)$$

Thus, from the ratio of the densities of α in the two phases at equilibrium, we can obtain the solvation Gibbs energy of the solute α. Clearly, Eq. (2.1.6) is not practical for proteins. However, to obtain the relationship between the solubility and the solvation Gibbs energy, we assume that the internal degrees of freedom of α are not affected by the presence of the solvent. In this case, we write[98]

$$\mu_\alpha^{*ig} = -k_B T \ln q_\alpha \qquad (2.1.7)$$

and

$$\mu_\alpha^{*W} = W(\alpha|\text{solvent}) - k_B T \ln q_\alpha, \qquad (2.1.8)$$

where q_α is the internal partition function of α and $W(\alpha|\text{solvent})$ is the coupling work, i.e. the work of "turning on" the interaction between α and the solvent molecules.

Therefore, from (2.1.1), (2.1.3), (2.1.7) and (2.1.8), we get

$$\mu_\alpha^S = W(\alpha|\text{solvent}) + k_B T \ln q_\alpha^{-1} \rho_{\alpha,eq}^{W} \Lambda_\alpha^3$$

$$= \Delta G_\alpha^{*W} \left(\rho_{\alpha,eq}^{W} \right) + k_B T \ln q_\alpha^{-1} \rho_{\alpha,eq}^{W} \Lambda_\alpha^3. \qquad (2.1.9)$$

Note that in (2.1.9), we stressed the dependence of ΔG_α^{*W} on $\rho_{\alpha,eq}^{W}$. Thus, in general, even when we know the value of μ_α^S, Eq. (2.1.9) is an implicit equation for $\rho_{\alpha,eq}^{W}$, which is the solubility of α at equilibrium with the pure solid phase of α. When ρ_α^{W} is very dilute, such that solute–solute interactions

---

[97]For more details, see Ben-Naim (2006).

[98]This is normally a good approximation for the internal degrees of freedom. However, for molecules having internal rotational degrees of freedom, one must take a proper average over all the possible conformations. For details, see Ben-Naim (2006) and Appendix B.

may be neglected, ΔG_α^{*W} becomes independent of ρ_α^W. Hence, Eq. (2.1.9) reduces to

$$\mu_\alpha^S = \mu_\alpha^W = \Delta G_\alpha^{*W} + k_B T \ln q_\alpha^{-1} \rho_{\alpha,\mathrm{eq}}^W \Lambda_\alpha^3. \qquad (2.1.10)$$

This relation is useful even when we do not know the chemical potential of the pure α in the solid state. In many problems, we are interested in the *changes* in the solubility caused by the addition of a solute to the system. In general, the solubility of a solute α in two solvents at equilibrium with the same solid phase of α will be given by the ratio

$$\frac{\rho_{\alpha,\mathrm{eq}}^W}{\rho_{\alpha,\mathrm{eq}}^l} = \exp\left[-\beta\left(\Delta G_\alpha^{*W} - \Delta G_\alpha^{*l}\right)\right]. \qquad (2.1.11)$$

Thus, *changes* in the solubility of α are determined by changes in the solvation Gibbs energy of α (provided that the pure solid phase is unchanged).

To conclude, we reiterate that the *solubility* of α depends on the chemical potential of α in the pure state, as well as on the solvation Gibbs energy of α. However, changes in the solubility at constant T, P and μ_α^S caused by changes of the solvent or by addition of a solute to the liquid phase, are determined by the changes in the solvation Gibbs energy of α.[99]

2.2. Solvation Gibbs Energy of a Model Globular Protein

In this section, we shall discuss a very simple "protein." We model the protein α as a giant $H\phi O$ sphere of diameter σ_p.

---

[99]Note carefully that Eq. (2.1.11) is valid for any densities of α in the two phases. For very dilute solutions of α in a solvent relation, (2.1.11) *determines* the ratio of the densities of α in the two phases. However, in general, the solvation Gibbs energies on the right-hand side of (2.1.11) are functions of ρ_α. Hence, Eq. (2.1.11) is an implicit equation for the densities of α in the two phases. See also Sec. 2.4.

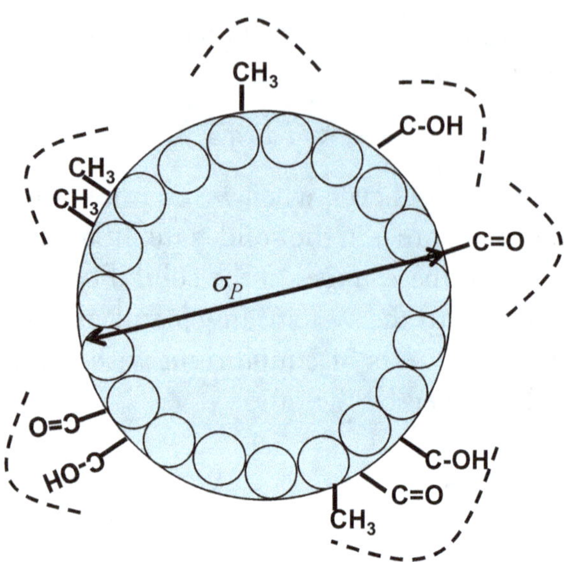

Fig. 2.2 A globular protein is viewed as a sphere of diameter σ_P on the surface of which $H\phi O$ and $H\phi I$ groups are distributed.

It interacts with water molecules through a hard part (H) and a soft (S), van der Waals, part. On its surface, $H\phi O$ and $H\phi I$ groups are distributed (see Fig. 2.2). Also, we use a methyl group to represent a $H\phi O$ group and a hydroxyl or a carbonyl group to represent a $H\phi I$ group. This highly simplified model protein is sufficient for the study of the relative importance of $H\phi O$ and $H\phi I$ groups on the solvation Gibbs energy, hence also on the solubility of protein. As we shall see in Sec. 2.4, the general conclusions of this study are very similar to the ones reached for real proteins.

We shall now solvate this model protein in water. First, we solvate the hard part of the solute–solvent interaction. We have seen in Chapter 1 that this part involves a positive solvation Gibbs energy. We shall now see that this quantity increases sharply with the radius of the hard protein.

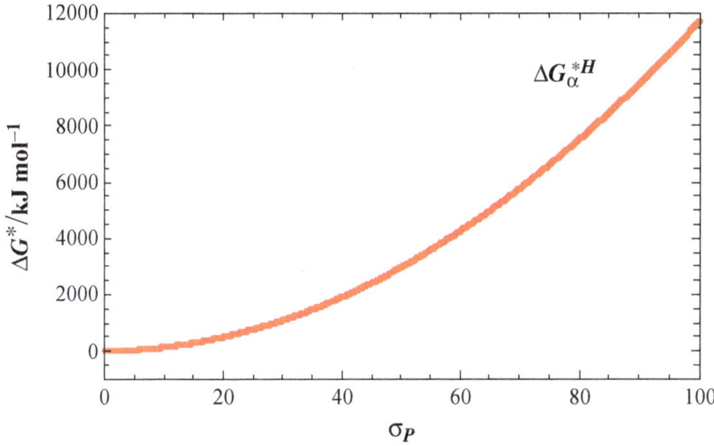

Fig. 2.3 The Gibbs energy of solvation of the hard sphere as a function of its diameter σ_P.

Figure 2.3 shows ΔG_α^{*H} as a function of the diameter of the solute σ_P. The calculations were done according to the scaled particle theory recipe. Define the radius of the cavity formed as α, $R_{CAV} = (\sigma_P + \sigma_w)/2$. The solvation Gibbs energy of α is given by[100]

$$\Delta G_\alpha^{*H} = K_0 + K_1 R_{CAV} + K_2 R_{CAV}^2 + K_3 R_{CAV}^3, \qquad (2.2.1)$$

where the coefficients K_i are given by

$$K_0 = RT(-\ln(1-y) + 4.5z^2) - \frac{\pi P \sigma_W^3}{6}$$

$$K_1 = -RT(6z + 18z^2)/\sigma_W + \pi P \sigma_W^2$$

$$K_2 = RT(12z + 18z^2)/\sigma_W^2 - 2\pi P \sigma_W$$

$$K_3 = 4\pi P/3, \qquad (2.2.2)$$

---

[100] Reiss (1966), Ben-Naim (2006).

where R is the gas constant, here taken as $R = 8.314\,\mathrm{Jmol}^{-1}$ K^{-1}, $T = 298\,\mathrm{K}$, $P = 1\,\mathrm{atm}$[101] and $\sigma_W = 2.8\,\text{Å}^{-3}$, and

$$y = \frac{\pi \rho_W \sigma_W^3}{6}, \quad z = \frac{y}{1-y}, \tag{2.2.3}$$

while $\rho_W = 3.344 \times 10^{-2}$ molecules Å^{-3} is the number density of water at $298\,\mathrm{K}$.

Next, we add the soft part of the solute–solvent interactions. As we noted in Chapter 1, we may assume that these interactions are between $H\phi O$ groups on the surface of the protein α and the water molecules. Figure 2.4 shows a schematic drawing of a globular protein, the surface of which is covered with methane-like molecules.

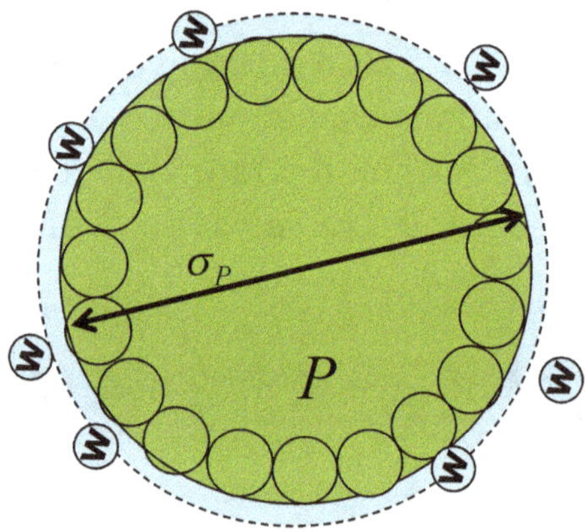

Fig. 2.4 The hard (H) and soft (S) interactions of a globular protein are schematically shown as a hard sphere of a diameter σ_P (green) and a soft, van-der-Waals interaction between methane-like molecules and water molecules (blue).

---

[101]Note that in calculating ΔG^*, we take RT in $\mathrm{kJ\,mol}^{-1}$. Since σ_P and R_{CAV} are taken in Å, we need to convert the pressure P at 1 atm to $\mathrm{kJ\,Å}^{-3}\,\mathrm{mol}^{-1}$ by

$$1\,\mathrm{atm} = 1.3807 \times 10^{-23} \times 273.15 \times 6.022 \times 10^{23}/22.414 \times 10^3$$

$$= 0.1033\,\mathrm{J\,Å}^{-3}\,\mathrm{mol}^{-1}.$$

Following this assumption, we can write the quantity $\Delta G_\alpha^{*S/H}$ as[102]

$$\Delta G_\alpha^{*S/H} \approx -\varepsilon_{\mathrm{Ne,CH_4}} \frac{4\pi}{3} \left[\left(\frac{\sigma_P}{2} + \sigma_w \right)^3 - \left(\frac{\sigma_P}{2} \right)^3 \right] \rho_W. \quad (2.2.4)$$

Basically, the approximation in (2.2.4) means that we view the surface of the proteins as consisting of methane-like spheres. The spherical shell of radius $\sigma_P/2$ and width σ_W is assumed to be filled by neon-like atoms, with the density of water at, say 25°C, $\rho_W = 3.344 \times 10^{22}$ molecules cm^{-3}. Each of these molecules contributes a Lennard-Jones energy parameter $\varepsilon_{\mathrm{Ne,CH_4}}$, calculated from

$$\varepsilon_{\mathrm{Ne,CH_4}} = \left(\varepsilon_{\mathrm{Ne,Ne}} \varepsilon_{\mathrm{CH_4,CH_4}} \right)^{1/2}. \qquad (2.2.5)$$

We use the values of $\varepsilon_{\mathrm{Ne,Ne}} = 34.9/k_B K$ and $\varepsilon_{\mathrm{CH_4,CH_4}} = 148.2/k_B K$ (to convert to J mol^{-1}, we divide by 1.3807×6.022).

Clearly, the soft interaction contributes a negative value to the solvation Gibbs energy. Figure 2.5 shows the sum of ΔG_α^{*H} and $\Delta G_\alpha^{*S/H}$ as a function of the protein diameter σ_P. As can be seen, the addition of the soft part of the potential does not change much the values of the Gibbs energies. One can conclude that the soft part makes ΔG_α^* less positive. However, this effect is proportional to σ_P^2, and as σ_P increases, the dominant contribution to ΔG_α^* is due to the work of creating a cavity, which increases as σ_P^3.

It should be noted that up to this point, the identities of the amino acid side chains in the interior of the protein are of no importance. These side chains contribute to the *volume* of the protein, but not to any specific interaction with the solvent molecules.

---

[102]Ben-Naim *et al.* (1989), Wang and Ben-Naim (1997).

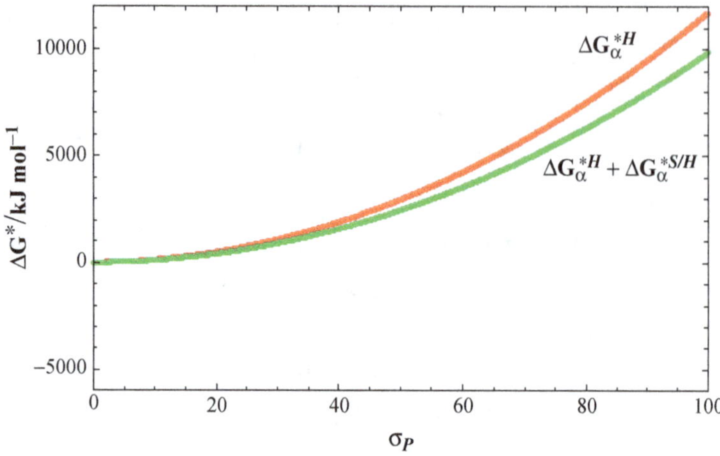

Fig. 2.5   The Gibbs energy of solvation of the hard (H) and the soft (S) parts of the protein α.

Next, we add the FGs that are on the surface of the proteins, i.e. those FGs that are exposed to the solvent. We do that in two steps:

First, we add *all* the FGs that are exposed to the solvent as if they were independently solvated.[103] From what we have learned in Chapter 1, the conditional solvation Gibbs energy of a methyl group is about $2.1\,\mathrm{kJ\,mol^{-1}}$. Clearly, adding hydrophobic groups to the surface of the protein will only make ΔG_{α}^{*W} more positive. On the other hand, adding hydrophilic groups such as OH, C=O or NH is expected to cause a decrease in the value of ΔG_{α}^{*W}. How much of a decrease depends on the average number of "arms" along which the FG can form hydrogen bonds. Remember, we are talking about the solvation of a $H\phi I$ group on the surface of the protein. The number of HBs that such a group can form is the maximum possible number of HBs; an NH group can form one HB, a C=O group can form at most

---

[103]See Appendix F.

two, and an OH group can form at most three.[104] It has been estimated, both theoretically and using experimental data, that the solvation Gibbs energy per arm is about[105] $9.5\,\text{kJ}\,\text{mol}^{-1}$. We use this number to estimate the contribution of $H\phi I$ groups to the solvation Gibbs energy of a globular protein.

We next add the solvation Gibbs energy of the $H\phi I$ groups on the surface of the protein. We start with the curve for the sum of $\Delta G_\alpha^{*H} + \Delta G_\alpha^{*S/H}$ as a function of the protein diameter σ_P, and add the $H\phi I$ groups. The area of the surface of α is $4\pi(\sigma_P/2)^2$. We also assume that the $H\phi I$ groups are uniformly distributed on this surface in such a way that they are *independently* solvated.[106] Thus, the number of $H\phi I$ arms on the surface of a protein of diameter $\sigma_P/2$ will be $4\pi(\sigma_P/2)^2\rho_{\text{arms}}$, where ρ_{arms} is the number of arms per unit area of the surface.

We shall skip the details of the calculations here. These can be found elsewhere.[107] First, we estimated the diameter and the surface area of about 20 proteins. Then, we estimated the average number of $H\phi I$ groups that are exposed to the solvent for each of these proteins. We assumed that each $H\phi I$ group contributes one "arm" that can form HBs with a water molecule.

In Fig. 2.6, we plot the number of $H\phi I$ groups as a function of the surface area of the protein. The average slope of this curve is about $\rho_{\text{arms}} = 0.02$ number of $H\phi I$ groups per Å^2.

From this surface density, we can estimate the average distance between two $H\phi I$ groups to be

$$\bar{d} \approx \sqrt{\frac{\text{surface area}}{\text{number of arms}}} = \sqrt{\frac{1}{\rho_{\text{arms}}}}. \qquad (2.2.6)$$

---

[104] See also Baker and Hubbard (1984) and Thanki *et al.* (1988).
[105] See Sec. 1.5 and Ben-Naim (2009).
[106] See Appendix F.
[107] See Wang and Ben-Naim (1997).

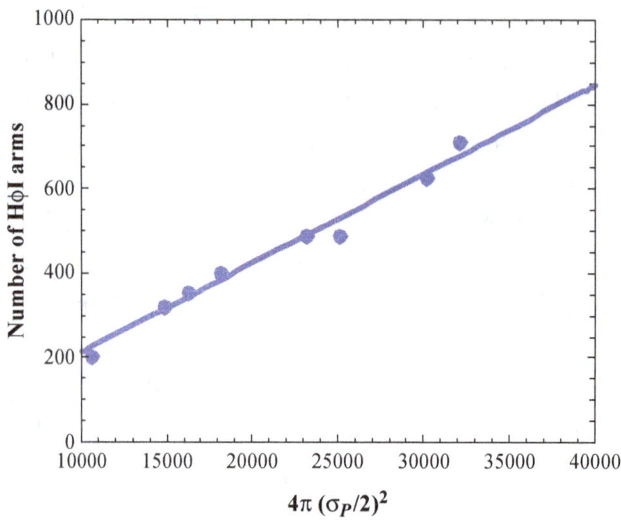

Fig. 2.6 The number of *HφI* arms on the surface of globular proteins as a function of the surface area [from Wang and Ben-Naim (1997)].

Thus, the average distance between the *HφI* groups is about 7 Å.

In Fig. 2.7, we show the effect of adding *HφI* groups on ΔG^*_α for three surface densities ρ_{arms} (0.012, 0.015 and 0.02).

As anyone who has given a thought to the solubility of proteins would have expected, the addition of *HφI* groups to the surface of the protein dramatically reduces the solvation Gibbs energy of the protein. It is also commonly believed that these *HφI* groups are *responsible* for the solubility of proteins in water.

Indeed they are! However, there is also an unexpected addition to the common belief about the contribution of the *HφI* groups to the solubility of proteins. As can be seen from Fig. 2.5, adding all the *HφI* groups as being *independently* solvated does lower the Gibbs energy of solvation of the protein. The question is whether this lowering of the Gibbs energy is sufficient to turn an extremely insoluble solute into a highly soluble protein. It is difficult to give an exact answer to this question. However,

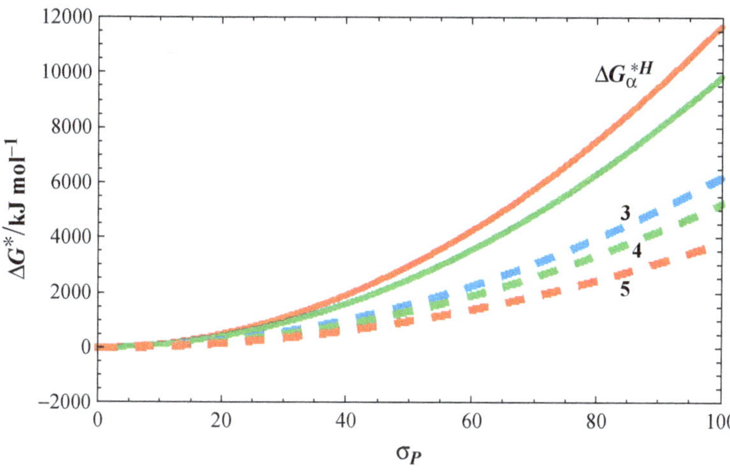

Fig. 2.7   Addition of the conditional solvation of the $H\phi I$ groups for three surface densities. The numbers 3, 4, 5 correspond to $\rho_{\text{arms}} \approx 0.012, 0.015$ and 0.02, respectively. The last density is approximately equal to the one computed from Fig. 2.6.

from a preliminary study of a few real proteins (see Sec. 2.3), it seems that the mere addition of the *sum* of the contributions of all the $H\phi I$ groups is probably not sufficient to make the protein soluble. Remember that we are estimating the *solvation* Gibbs energy of a model of a globular protein. The solubility of such a model compound depends on both the chemical potential of the pure protein and the solvation Gibbs energy of the protein. Thus, whenever we change any attribute of the protein, the chemical potentials of the protein in the two phases will change. Therefore, our inference regarding the effect of changing an attribute of the protein on the solubility is only tentative and approximate.

The probable molecular reason for the high solubility of the protein is the further lowering of the solvation Gibbs energy of protein by pair, and perhaps higher order, $H\phi I - H\phi I$ correlations.

In Sec. 1.7, we have discussed the hydrophilic pair correlation at the most favorable configuration such that the two $H\phi I$ groups can be bridged by a water molecule (See illustration in Fig. 1.26).

It was estimated that each pair-correlated $H\phi I$ group contributes between -10.5 to $-12.5\,\mathrm{kJ\,mol^{-1}}$ to the total Gibbs energy of solvation.[108] When this is translated into a change in solubility, one can estimate an increase in solubility by a factor of 120 to 150 *per pair* of correlated $H\phi I$ groups. Note again that this is true provided that the chemical potential of the pure protein does not change.

In Fig. 2.8, we add the contribution of the pair correlations to the total Gibbs energy. For this estimate, we start with an average surface density of $\rho_{\mathrm{arms}} = 0.02$ and add the pair correlations between these groups.[109] Here, we ignore triplet and

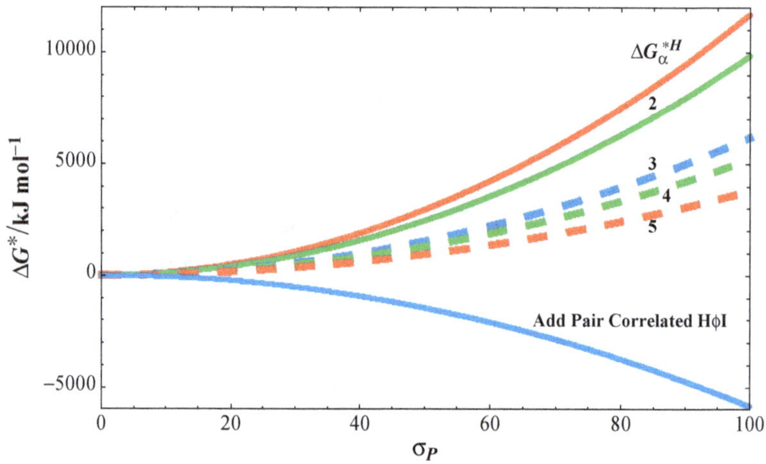

Fig. 2.8   Addition of the pair correlation between $H\phi I$ groups to the total solvation Gibbs energy of a globular protein as a function of the diameter of the protein.

---

[108] For more details, see Wang and Ben-Naim (1997).
[109] See Chapter 1 and Appendix G.

higher order correlations between $H\phi I$ groups. We also ignore pair correlations between $H\phi O$ groups. These also contribute to the solvation Gibbs energy of the protein, but it was estimated that this contribution is at least an order of magnitude smaller than the corresponding correlation between $H\phi I$ groups.

As can be seen from Fig. 2.8, the addition of pair correlations between $H\phi I$ groups dramatically lowers the solvation Gibbs energy of the model protein. We believe that this further reduction in the Gibbs energy of solvation is also responsible for the high solubility of the globular protein.

We can conclude this section as follows: The high solubility of globular proteins is certainly due to the $H\phi I$ groups distributed on the surface of the protein. Furthermore, it is very likely that pair and higher order correlations between $H\phi I$ groups on the surface of the protein contribute significantly, if not decisively, in making the proteins highly soluble.

2.3. Estimation of the Solvation Gibbs Energy of Real Proteins

In the previous sections, we discussed the solvation Gibbs energy of a hypothetical, highly simplified model of a globular protein. We next turn to discuss the solvation Gibbs energy of real proteins. This is an extremely complicated subject. No one has ever measured or calculated the solvation Gibbs energy of any real protein.[110] However, recently, Wang and Ben-Naim (1997) attempted to estimate the various ingredients that contribute to the total solvation Gibbs energies of some specific proteins. The details of the calculations are highly technical, and there is no need to discuss them here. Instead we present

---

[110]Hydration of proteins in the process of protein folding has been studied extensively, e.g. Privalov and Makhatadze (1992, 1993), Makhatadze and Privalov (1993).

here only an outline of the procedure of computation and the final conclusions.

About 20 proteins were selected from the protein data bank (PDB). For each of the proteins, one can easily calculate its "diameter."[111]

From the diameter of the protein σ_P, one can estimate the solvation Gibbs energy of creating a cavity of radius R_{CAV}, given by

$$R_{CAV} = (\sigma_P + \sigma_W)/2. \tag{2.3.1}$$

This procedure leads to the plot of ΔG_α^{*H} as shown in Fig. 2.3. The next step is to estimate the contribution of the part of the solute–solvent interaction to the solvation Gibbs energy. This part was calculated by Eq. (2.2.4) in Sec. 2.2.

The more difficult task is to determine the number of $H\phi O$ and $H\phi I$ groups that are exposed to the solvent. We used two different criteria to decide which of the $H\phi I$ groups on the surface of the protein are actually solvated, and how many "arms" for each of these groups are potentially available to form hydrogen bonds with the solvent.

Assuming that all the "arms" on the surface of the protein which are exposed to the solvent are independently solvated, and assigning the value of $-9.5\,\text{kJ}\,\text{mol}^{-1}$ to the conditional solvation Gibbs energy of each arm, we can calculate the total contribution of the $H\phi I$ group to the solvation Gibbs energy of the protein. As expected, the addition of all the independently solvated $H\phi I$ groups dramatically lowers the curve of ΔG_α^{*W} versus σ_P.

The last step is to include the correlation between the $H\phi I$ groups on the surface of the protein. In contrast to the model protein discussed in Sec. 2.2, where we assumed a uniform distribution of $H\phi I$ groups on the surface of the protein, here we

---

[111]Since the molecules are not spherical, one can only estimate an average diameter from the solvent accessible surface area (SASA).

take the *actual* distribution of the *Hφl* groups. From the actual distribution of the *Hφl* group, one can calculate the distribution of the distances between the *Hφl* groups. This distance distribution between pairs will determine the contribution of each pair to the solvation Gibbs energy.

When two *Hφl* groups are far apart, say $d \geq 5$ Å, they may be considered to be independently solvated. This means that each arm contributes about -9.5 kJ mol^{-1} to the total solvation Gibbs energy. When the distance is around 4.5 Å, the two groups can be bridged by a water molecule. At this distance, there is an enhancement of the solvation energy of a pair of *Hφl* groups by an amount of about -12 kJ mol^{-1}. We refer to this case as a positive correlation, or *Hφl* pair correlation.[112] Finally, at very short distances, say $d \leq 4$ Å, the solvation of one *Hφl* group can interfere with the solvation of the second group. We refer to this case as a negative correlation, i.e. the contribution of the two *Hφl* groups at such distance to ΔG^{*W} is less than (in absolute magnitude) the sum of their contributions when they are independently solvated.

Figure 2.9 shows the distance distribution of all the oxygen and nitrogen atoms on the surface of nine proteins.[113] It is clearly seen that there are many pairs of *Hφl* groups at a distance of between 4 and 5 Å. These groups can potentially form hydrogen bond bridges.

Taking all these contributions into account leads to a solvation Gibbs energy curve, which is very similar to the curves shown in Fig. 2.8.

The conclusion is also the same as the one reached in Sec. 2.2. The solvation Gibbs energy of the hard and soft parts of the protein is large and positive. One can expect that such a "giant"

---

[112]One should also note that higher order correlations can also contribute to the solvation Gibbs energy. These should be accounted for in a more accurate calculation of the solvation Gibbs energies of proteins.

[113]Ben-Naim *et al.* (1989a, 1989b).

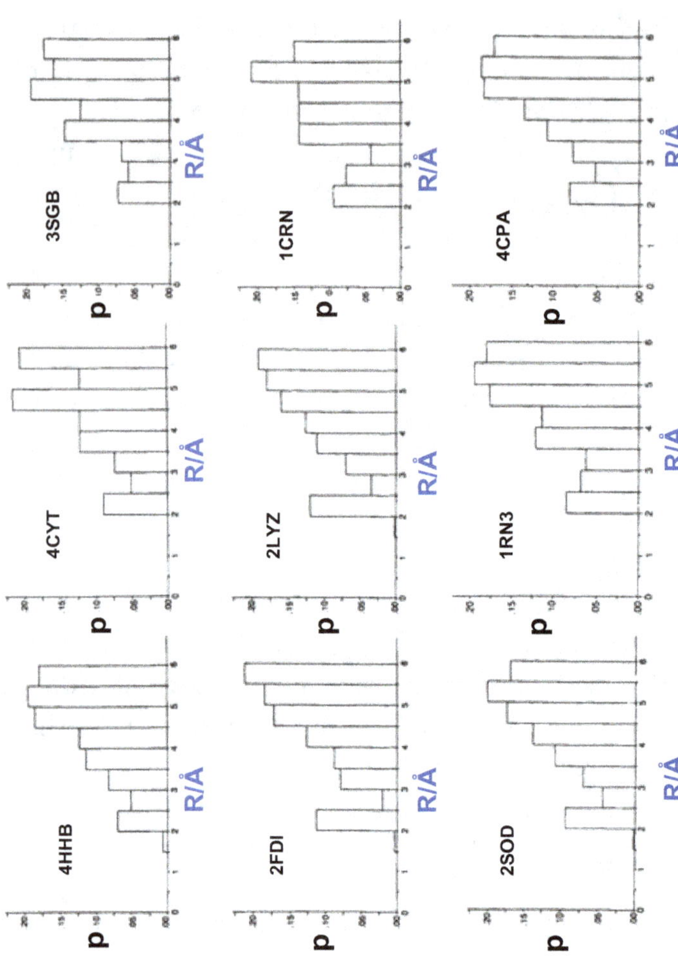

Fig. 2.9   Distance distributions of nitrogen and oxygen atoms on the surface of proteins. The ordinate gives the fraction of pairs found at each distance $(R/\text{Å})$ interval. The symbols for the proteins are 4HHB, hemoglobin; 4CYT, cytochrome C; 3SGB, proteinaise B; 2FDI, ferredoxin; 2LYZ, lysozome; 1CRN, crambin; 2SOD, superoxide dismutase; $1RN_3$, ribonuclease A; 4CPA, carbozypeptidase A [From Ben-Naim *et al.* (1989b)].

HφO molecule would be extremely insoluble in water. Adding all the *HφI* groups to the surface of the protein makes ΔG^{*w} either much smaller and positive, or even negative. This addition of *HφI* groups will increase the solubility of the protein from extremely low to either low or moderate. What makes the actual globular protein highly soluble is probably the contribution of the *pair*, and *higher order correlations* between the *HφI* groups.

In this section, we have discussed only the solvation of proteins. However, it is clear that the arguments presented here are relevant to the solvation of any macromolecule having *HφI* groups on its surface. For instance, segments of DNA are highly hydrated by water molecules.[114] It is also known that the water molecules in the minor groove are highly ordered.[113] It is very plausible to assume that these water molecules form water bridges connecting *HφI* groups on the surface of the DNA. This in turn will affect the stability of a particular conformation of the DNA.

The above conclusion is only tentative. One should always bear in mind that when the solubility is large enough such that solute–solute interactions cannot be ignored, there is no simple relationship between solubility and solvation Gibbs energy. We discuss a simple example in the next section.

2.4. The Relation between Solubility and Solvation Gibbs Energy for Moderately Soluble Proteins

We discuss in this section a simple example where a solute α is not very dilute in a solvent w. It can be shown[115] that

---

[114]Dragan *et al.* (2008), Chalikan *et al.* (1994, 1999), Chalikan and Breslaur (1998), Saenger *et al.* (1986), Arai *et al.* (2005), Chiu *et al.* (1999), Drew and Dickerson (1981), Kopka *et al.* (1983), Shui *et al.* (1998), Minasov *et al.* (1999), Lewin (1967).
[115]For details, see Ben-Naim (2006).

the solvation Gibbs energy of α can be expanded in the solute density ρ_α as

$$\Delta G_\alpha^{*W} = \Delta G_\alpha^{*W(0)} + C\rho_\alpha + C\rho_\alpha^2 + \cdots \qquad (2.4.1)$$

$\Delta G_\alpha^{*W(0)}$ is the solvation Gibbs energy of α in the limit of very dilute solution. This is the ideal dilute limit, where one can neglect all solute–solute interaction, i.e. $\Delta G_\alpha^{*W(0)}$ depends only on solute–solvent and solvent–solvent interactions. When the density of α increases, the solvation Gibbs energy also includes contributions due to solute–solute interactions. The precise expression of the coefficients in the expansion (2.4.1) depends on how we add the solute to the system (keeping T, P, T, μ_w or T, ρ_W constant). We shall assume that the solute α is moderately soluble in the system and take into account only pair correlations between the solutes, i.e. we take into account only the linear term in (2.4.1).

At equilibrium with an ideal gas phase, we have the condition

$$\mu_\alpha^W = \Delta G_\alpha^{*W(0)} + C\rho_{\alpha,\text{eq}}^W + k_B T \ln q_\alpha^{-1} \Lambda_\alpha^3 \rho_{\alpha,\text{eq}}^W$$

$$= \mu_\alpha^{ig} = k_B T \ln q_\alpha^{-1} \Lambda_\alpha^3 \rho_{\alpha,\text{eq}}^{ig}. \qquad (2.4.2)$$

Hence, an analogue of Eq. (2.1.11) (when l is replaced by an ideal gas phase) is

$$\left(\frac{\rho_\alpha^W}{\rho_\alpha^{ig}} \right)_{\text{eq}} = \exp\left[-\beta \Delta G_\alpha^{*W(0)} - \beta C \rho_{\alpha,\text{eq}}^W \right]. \qquad (2.4.3)$$

We see that in this case, there is no simple relation between the equilibrium density of α in w and the solvation Gibbs energy of α at infinite dilution. In the more general case, one must take higher order terms in the expansion (2.4.1) to account for triplet, quadruplet, etc., interactions.

It should be noted that in Eq. (2.4.3), both ρ_α^W and ρ_α^{ig} are determined by the solvation Gibbs energy of α in the liquid phase. If a solid phase of pure α is present in the system, then the equilibrium concentration $\rho_{\alpha,\mathrm{eq}}^{ig}$ is determined by

$$\mu_\alpha^S = \mu_\alpha^{ig} = \mu_\alpha^{*ig} + k_B T \ln q_\alpha^{-1} \Lambda_\alpha^3 \rho_{\alpha,\mathrm{eq}}^{ig}. \tag{2.4.4}$$

In this case, $\rho_{\alpha,\mathrm{eq}}^{ig}$ is fixed in (2.4.3), and this equation becomes an implicit equation for the *solubility* of α at equilibrium with the pure solid phase α.

2.5. A Possible Explanation for an Apparently Paradoxical Experimental Finding[116]

The following is an interesting example of how a puzzling experimental finding may be explained by the formation of pair correlation.

Hughes *et al.*[117] found that replacing an amino group on the surface of a protein with a guanidino group (Fig. 2.10) causes a marked *decrease* in the solubility of the protein. This finding is quite surprising. Assuming that the 3-D structure of the protein does not change upon this substitution (presumably because the amino group is on its surface), and since the guanidino group is

Fig. 2.10   Replacing an amine group with a guanidino group on the surface of a protein.

---

[116]This example was brought to the author's attention by Harry Saroff.
[117]Hughes *et al.* (1949). See also Wang and Ben-Naim (1997).

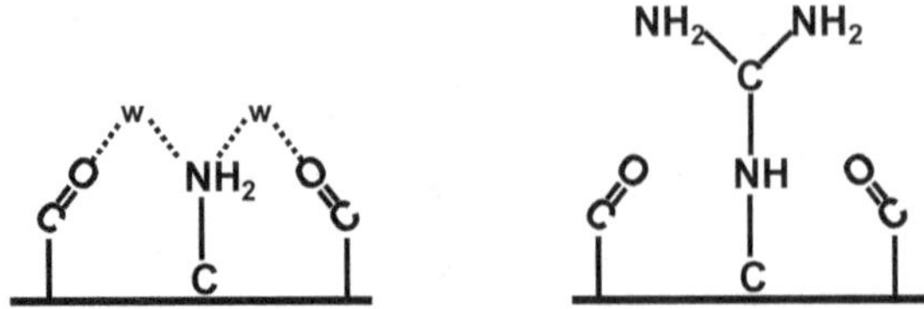

Fig. 2.11   Possible explanation for the decrease in solubility upon replacing an amine group with a guanidino group (see Sec. 2.5).

more polar than the amino group, in the sense that it can form more hydrogen bonds with solvent molecules, one would have expected such a replacement to cause an *increase* in solubility rather than a *decrease*.

One possible explanation is shown in Fig. 2.11. Initially, the amino group is located on the surface of the protein in an environment where it can form one or two HB bridges with two carboxyl groups. Upon replacing the amino group with a guanadino group, the possibility of forming a positive correlation is precluded. Thus, although we have added one polar group which is expected to *increase* the solubility, the elimination of pair correlation would cause a net decrease in the solubility. It would be interesting to find other experimental cases where the addition of a $H\phi O$ group can cause an *increase* in solubility. A possible hypothetical example is shown in Fig. 2.12, where the addition of a $H\phi O$ group might cause an *increase* in solubility, although a *decrease* in solubility is expected.

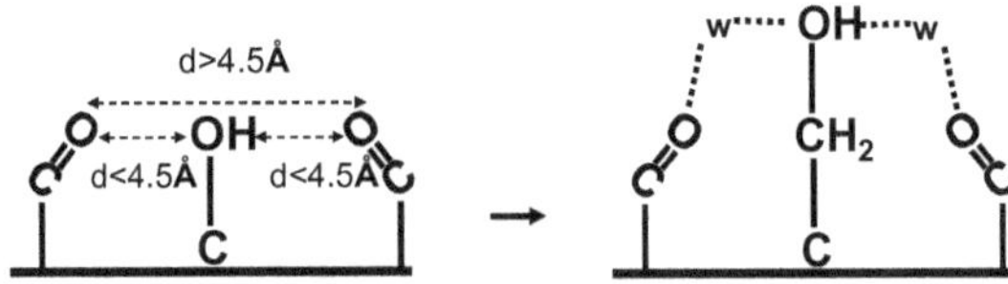

Fig. 2.12   A hypothetical example, where the addition of a $H\phi O$ group (here, methylene) can cause an increase in solubility.

2.6. The Effect of the Addition of a Solute on the Solvation Gibbs Energy

Suppose we have a protein α, diluted in water. We add a solute s and ask how the solubility of the protein will change.

A more general problem is the following. We have again a protein α, diluted in mixed solvents A and B at some composition x_A. We are interested in the response of the solubility to changes in the composition of the solvent. This question is important in connection with studies of denaturation of proteins under changes in the composition of the solvent.[118]

An exact relation between the derivative of the pseudo-chemical potential of α with respect to x_A, in terms of the Kirkwood–Buff integral, is available[119]:

$$\left(\frac{\partial \mu_\alpha^{*l}}{\partial x_A}\right)_{P,T} = \frac{k_B T (\rho_A + \rho_B)^2}{\eta}(G_{\alpha B} - G_{\alpha A}), \qquad (2.6.1)$$

where

$$\eta = \rho_A + \rho_B + \rho_A \rho_B (G_{AA} + G_{BB} - 2G_{AB}) \qquad (2.6.2)$$

and

$$G_{ij} = \int [g_{ij}(\mathbf{R}) - 1]d\mathbf{R} \qquad (2.6.3)$$

are the Kirkwood–Buff integrals (KBI).[120]

The solvation Gibbs energy of α is defined by

$$\Delta G_\alpha^{*l} = \mu_\alpha^{*l} - \mu_\alpha^{*ig}, \qquad (2.6.4)$$

---

[118]Lee and Timasheff (1981), Gekko and Timasheff (1981), Arakawa and Timasheff (1984).
[119]For more details, see Ben-Naim (2006) Chapter 8.
[120]For more details, see Ben-Naim (2006) Chapter 8.

where μ_α^{*ig} is the pseudo-chemical potential of α in an ideal gas phase. Clearly, μ_α^{*ig} is independent of the composition x_A of the liquid phase (it only depends on the internal degrees of freedom of α). Therefore, we can rewrite (2.6.1) as

$$\left(\frac{\partial \Delta G_\alpha^{*l}}{\partial x_A}\right)_{P,T} = \left(\frac{\partial \mu_\alpha^{*l}}{\partial x_A}\right)_{P,T} = \frac{k_B T(\rho_A + \rho_B)^2}{\eta}(G_{\alpha B} - G_{\alpha A}).$$

$$(2.6.5)$$

Note that from the stability condition of the mixture, η must be positive.[121] Therefore, the factor $k_B T(\rho_A + \rho_B)^2/\eta$ is always positive. Furthermore, when the solvent mixture is a symmetrical ideal solution, we have

$$G_{AA} + G_{BB} - 2G_{AB} = 0. \qquad (2.6.6)$$

In which case

$$\left(\frac{\partial \Delta G_\alpha^{*l}}{\partial x_A}\right)_{P,T} = k_B T(\rho_A + \rho_B)(G_{\alpha B} - G_{\alpha A}). \qquad (2.6.7)$$

This is an important result; it states that even when the solvent forms a symmetrical ideal solution, the solvation of α (hence, the solubility of α) may change upon changing the composition of the solvent.

Thus, in any mixed solvent, the sign of the derivative of the solvation Gibbs energy with respect to x_A is determined by the difference in the *affinities* between the pairs αA and αB. In other words, the question of the response of ΔG_α^{*l} to composition changes is equivalent to the question of the preferential solvation of α in the mixed solvents.[122] As we have noted in Sec. 2.4, whenever a solid phase of pure α is present, the equilibrium

---

[121] For more details, see Ben-Naim (2006) Chapter 8.
[122] For more details, see Ben-Naim (2006) Chapter 8.

condition is

$$\mu_\alpha^S = \mu_\alpha^l = \mu_\alpha^{*l} + k_B T \ln \rho_\alpha^l \Lambda_\alpha^3. \tag{2.6.8}$$

Therefore, we can rewrite (2.6.8) as

$$k_B T \ln \rho_\alpha^l \Lambda_\alpha^3 = \mu_\alpha^S - \mu_\alpha^{*l} = (\mu_\alpha^S - \mu_\alpha^{*ig}) - (\mu_\alpha^{*l} - \mu_\alpha^{*ig}). \tag{2.6.9}$$

Clearly, $\mu_\alpha^S - \mu_\alpha^{*ig}$ is independent of the composition of the solvent. Therefore, taking the derivative of (2.6.9) with respect to x_A gives

$$\left(\frac{\partial (k_B T \ln \rho_\alpha^l)}{\partial x_A} \right)_{P,T} = - \left(\frac{\partial \Delta G_\alpha^{*l}}{\partial x_A} \right)_{P,T}. \tag{2.6.10}$$

Furthermore, when α is very dilute in the solvent mixture, ΔG_α^{*l} becomes independent of ρ_α, in which case (2.6.10) determines the change in the solubility of α at equilibrium with a solid phase of pure α. Hence, we have

$$\left(\frac{(\partial \ln \rho_\alpha^l)}{\partial x_A} \right)_{P,T} = \frac{(\rho_A + \rho_B)^2}{\eta} (G_{\alpha A} - G_{\alpha B}). \tag{2.6.11}$$

Thus, the change in the equilibrium concentration of α in the mixture is determined again by the difference in the affinities between the pairs αA and αB.

For simple solutes, it is easy to determine experimentally the solvation Gibbs energy of the solute in mixed solvents. Figure 2.13 shows the solubility (in terms of the Ostwald coefficient) and the solvation Gibbs energy of methane as a function of the mole fraction of ethanol in mixtures of water and ethanol. Thus, the slope of these curves at each point is related to the preferential solvation of the methane with respect to the two solvents, water and ethanol.

For a non-simple solute like protein, relationships (2.6.5) and (2.6.11) are still valid. However, to study the details of the

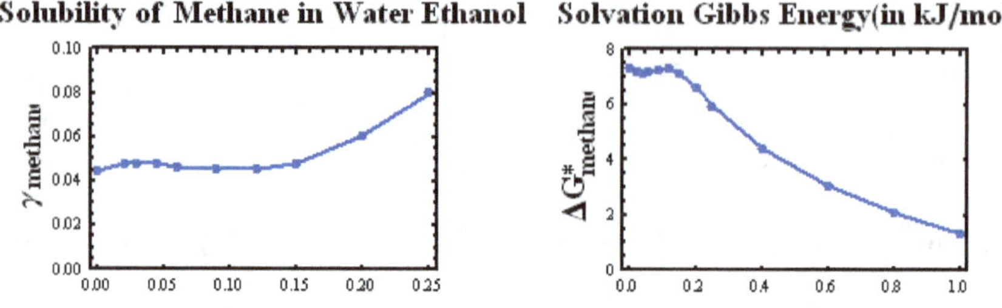

Fig. 2.13 Solubility and Gibbs energy of solvation of methane in mixtures of water and ethanol.

KBIs in this case is not simple. To see that, we write the two KBIs in Eq. (2.6.11) as

$$G_{\alpha A} - G_{\alpha B} = \int_V [g_{\alpha A}(\mathbf{X}_\alpha, \mathbf{X}_A) - 1]d\mathbf{X}_A$$

$$- \int_V [g_{\alpha B}(\mathbf{X}_\alpha, \mathbf{X}_B) - 1]d\mathbf{X}_B. \qquad (2.6.12)$$

In (2.6.12), $\mathbf{X}_A$ and $\mathbf{X}_B$ are the configurational vectors (location and orientation) of an A and B molecule. $\mathbf{X}_\alpha$ is assumed to be fixed in these integrals.

Normally, one averages over all orientations and obtains a one-dimensional integral of the form[123]

$$G_{ij} = \int_0^\infty [\bar{g}_{ij}(R) - 1]4\pi R^2 dR, \qquad (2.6.13)$$

where $\bar{g}_{ij}$ is the angle-averaged pair correlation function, which is a function of the scalar R only. However, we left the KBIs in the form (2.6.12) to stress the fact that the pair correlation function (before averaging over all orientations) is different along different directions. In Fig. 2.14, we show this dependence

---

[123]See Ben-Naim (2006).

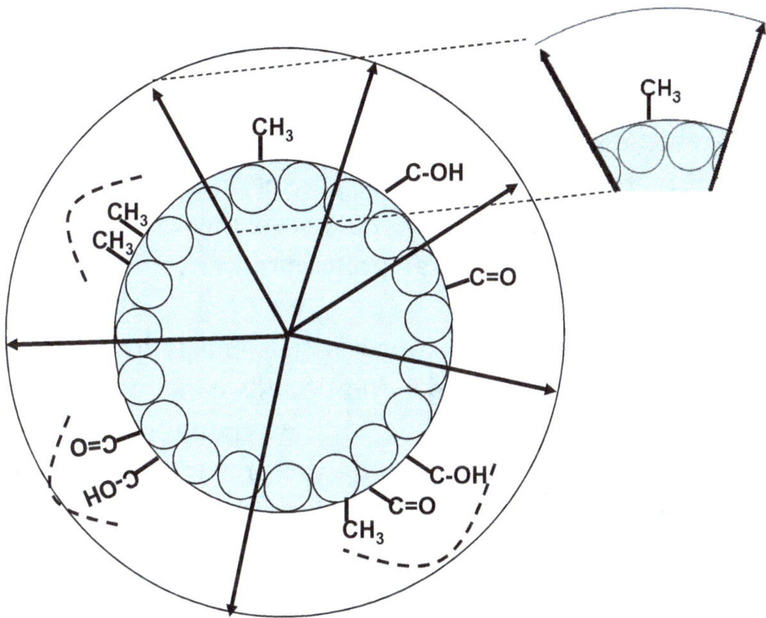

Fig. 2.14 The division of the surface of a protein into segments — each segment may be studied in a model compound. An example of a model compound for studying the conditional solvation of a methyl group is shown on the upper right-hand side of the protein.

schematically in two dimensions. We depict the protein α as a sphere (here, a disk) and a few functional groups on its surface.

Clearly, the overall preferential solvation of the entire protein, encapsulated in $G_{\alpha A} - G_{\alpha B}$, is made up of a sum of preferential solvation contributions due to the different functional groups. In Fig. 2.14, we divide the entire spherical region into segments according to the different functional groups on the surface of the protein.

Thus, in the region around the hydroxyl group, the value of the pair correlation function might be different from the value around a carbonyl group. The same is true for a methyl and an ethyl group. Note, however, that if two FGs are not independently solvated (such as the two methyl groups, or the hydroxyl and carbonyl groups in Fig. 2.14), the value of the pair correlation function might be different from the value around either of the two functional groups.

Thus, although the preferential solvation of the entire protein can be easily determined experimentally through Eq. (2.6.11), understanding the details of the various contributions of the FGs to the overall preferential solvation is not an easy task.

One possible suggestion towards that end is to study the preferential solvation of model compounds containing one or two functional groups only. Thus, for the hypothetical molecule depicted in Fig. 2.14, we can study the preferential solvation around each of the FGs by using small model compound, such as the ones depicted on the right-hand side of Fig. 2.14.

2.7. Concluding Remarks and Suggestions for Future Research

The solvation of proteins is not only important in the study of solubility, but it is also an essential quantity in the study of protein folding and protein–protein association. The study of the solvation thermodynamics of real protein is still in its infancy. However, the methodology described in this chapter might be useful in the study of the solvation thermodynamics of real proteins. The first step towards that end is to find the distribution of the distances between the FGs on the surface of the proteins (i.e. those that are exposed to the solvent). This information will help us to estimate the solvation Gibbs energies of real proteins, as well as to determine the relative importance of $H\phi I$ and $H\phi O$ groups.

It would also be instructive to find small model compounds where increasing the overall *hydrophilicity* of the molecule causes a decrease in the solubility, or, alternatively, increasing the overall *hydrophobicity* of the molecule causes an increase in the solubility.

By solubility here, we mean the Ostwald absorption coefficient, i.e. the solubility from the gaseous phase. As discussed in Sec. 2.5, the interpretation of the solubility of solids in terms of solvation Gibbs energy is more problematic. See also Appendix L.

CHAPTER 3

Protein Folding

My interest in the problem of protein folding started during the year 1986, when I spent a year with Robert Jernigan at NIH. At that time, I had benefitted from interesting discussions with Harry Saroff, who exposed me to many problems associated with proteins, solubilities, cooperativity of binding phenomena, self-assembly, etc.

I started to work on the solvent effects on protein folding only in the late 1980s. It was during those years when I realized that Kauzmann's model, which was supposed to explain the main "driving force" for protein folding, was not suitable for representing the HϕO effect in the protein folding process. At the same time, I also realized that the so-called HB inventory argument was fundamentally wrong. I felt there was an urgent need to construct an inventory of all solvent-induced effects for both protein folding and protein–protein association. This newly found inventory of solvent-induced effects had sharply changed my views regarding the relative importance of HϕO and HϕI effects in protein folding.

For a quick orientation, I shall start with two quotations. One is from Creighton[124]: "An immensely important physical chemical problem is how each protein acquires and maintains its folded conformation."

---

[124]Creighton (1985).

The second is from Rose et al.[125]:

Proteins are linear, un-branched polymers of amino acid residues that can undergo a reversible disorder ⇔ order transition called protein folding. Under suitable conditions, all of the information needed to realize the ordered form of most proteins is encoded in their linear sequence; no auxiliary components are necessary to guide the disordered chain to its unique, biologically relevant three-dimensional structure.

Reading these quotations and many others, one cannot overemphasize the importance of and the challenges involved in solving the so-called protein folding problem.

This chapter is not about the protein folding problem itself — it is only an analysis of the role of water in protein folding. However, analysing the role of water in protein folding might bring us as close as possible to solving the general protein folding problem. There are essentially two problems:

1. *How and why do proteins fold?*
2. *Is there a folding code?*

In an article titled "The Problem of How and Why Proteins Adopt Folded Conformations," Creighton[126] *discusses the two questions, "How" and "Why," as if they were one. Here, we shall make a distinction between the two questions: The "How" question is concerned with the path, or with intermediates along which the protein folds; the "Why" question focuses on the causes or "forces" driving the protein to select a pathway leading from the unfolded to the folded form.*

---

[125]Rose *et al.* (2006).
[126]Creighton (1985).

It seems to me that Creighton discusses only the question of "How," although he repeatedly refers to this question as the "How" and "Why" question.

Is there a universal pathway along which proteins fold? Certainly, the answer to this question is negative. Each protein folds along a particular set of intermediates. In this sense, there exists no common pathway; solving the problem for one protein leaves the same problem for all other proteins unsolved.

Is there a universal cause, *i.e. an answer to the question "Why," that is valid for all proteins? Clearly, the set of forces acting on the protein at each stage of the folding process is specific to each protein. Therefore, it is quite unlikely to find an answer to the general "Why" question. On the other hand, if we ask the "Why" question about a specific protein, we might find the answer to this question by analyzing the actual* forces *exerted on each of the protein's groups at each step of the folding process. Such an answer would be valid only for a specific protein; each protein has a specific set of forces that drives the process of folding along a particular pathway.*

Clearly, if we knew all the forces acting on all the atoms at each intermediate state of the protein, we could answer the "Why" question, and thereby answer the "How" question.

However, if we are only interested in the type of forces *acting on the protein at each stage of the folding process, and if we could find the most important, or the dominant forces acting on the protein, we might consider that to be a* universal answer *to this question.*

The purpose of this chapter is to discuss the "driving forces" as well as the actual forces that are involved in the protein folding process. In a sense, if we find the answer to the question of the "dominant forces" in protein folding, we can claim that we have answered the universal part of the question of "why" proteins fold. In my opinion this will amount, as close as possible to a solution of the *general* problem of protein folding.

In the following sections, we shall first discuss the question of the "driving forces" in the process of protein folding. This is basically an equilibrium problem: What makes the folded 3-D structure stable under certain environments? Next, we shall discuss the actual forces that are exerted on the protein and "force" it to fold along a specific pathway. This is a dynamic problem. We shall see that the answers to both of these questions are intimately related to the presence of the water, and that contrary to the commonly accepted view, the various $H\phi I$ effects are far more important than the corresponding $H\phi O$ effects in determining both the thermodynamic "driving forces" as well as the actual forces in the process of protein folding.

In Sec. 3.10, we shall discuss some aspects of the question regarding the existence of a "folding code" in the sequence of amino acids of proteins. It will be argued that such a code is unlikely to exist.

3.1. The Chemical Equilibrium

Following Anfinsen's classical work,[127] we know that the folding-unfolding of a protein is a reversible process. Here, we use the term "reversible" simply in the sense that the process can be reversed, not implying anything regarding the change in entropy of the process.[128] Under certain optimum physiological conditions, the protein folds spontaneously, and in a relatively short time, into the final 3-D structure. This process does not involve any catalyst. Changing the optimum condition into a denaturating environment, such as one with a high temperature

---

[127]Haber and Anfinsen (1961), Anfinsen (1973).
[128]It seems to me that using this term is superfluous. See also Appendix Q.

or a high concentration of a denaturating agent, will cause unfolding.

The protein folding process may be written as a simple isomerization reaction

$$U \rightleftharpoons F, \qquad (3.1.1)$$

where **U** and **F** will be referred to as the unfolded and folded form of the protein, respectively. Suppose that the species **U** and **F** are well defined and together comprise all the possible accessible states of the protein (**P**) within the experimental time frame.[129] Assuming that the protein is very dilute in the solvent, we can write the chemical potentials of the two species as

$$\mu_U = \mu_U^* + k_B T \ln \rho_U \Lambda_U^3, \qquad (3.1.2)$$

$$\mu_F = \mu_F^* + k_B T \ln \rho_F \Lambda_F^3, \qquad (3.1.3)$$

where ρ_U and ρ_F are the number densities (number of molecules per unit volume) of **U** and **F**, respectively. Since Λ_U^3 and Λ_F^3 depend only on the mass of the molecule (assuming that the conditions of classical statistical mechanics are valid for this system[130]), we can write

$$\Lambda_U^3 = \Lambda_F^3 = \Lambda_P^3 = \left(\frac{h}{\sqrt{2\pi m k_B T}} \right)^3, \qquad (3.1.4)$$

where m is the molecular mass of the protein.

The pseudo-chemical potentials μ_U^* and μ_F^* are defined in terms of the pseudo-chemical potentials of all the conformations that are included in the species **U** and **F**.[131] In an ideal dilute solution, the pseudo-chemical potentials are independent of the protein concentration.

---

[129]We shall further discuss the definition of the two conformers in Sec. 3.2.
[130]The condition is $\rho\Lambda^3 \ll 1$. See, for example, Hill (1956, 1960).
[131]See also Appendices A and B.

The condition of equilibrium is

$$\mu_U = \mu_F. \tag{3.1.5}$$

We define the mole fractions of the two species as $x_U = \rho_U/\rho_P$, $x_F = \rho_F/\rho_P$, where $\rho_P = \rho_U + \rho_F$ is the total concentration of the protein.

From (3.1.2), (3.1.3) and (3.1.5), we can get the equilibrium condition for this reaction:

$$\left(\frac{\rho_F}{\rho_U}\right)_{eq} = \left(\frac{x_F}{x_U}\right)_{eq} = \exp[-\beta(\mu_F^* - \mu_U^*)], \tag{3.1.6}$$

where $\beta = (k_B T)^{-1}$ and $\mu_F^* - \mu_U^*$ are the change in the Gibbs energy of the system for the conversion of one molecule **U** into one **F**. Strictly, this conversion is carried out for molecules devoid of translational degrees of freedom. However, because of the equality (3.1.4), the addition of the translational degrees of freedom to the molecules will not change the meaning of the difference $\mu_F^* - \mu_U^*$. This quantity is also related to the standard Gibbs energy of the reaction (3.1.1).[132]

Defining the equilibrium constant for the reaction (3.1.1) as

$$K = \exp[-\beta(\mu_F^* - \mu_U^*)], \tag{3.1.7}$$

we can rewrite Eq. (3.1.6) as

$$(x_F)_{eq} = \frac{K}{1 + K}. \tag{3.1.8}$$

Experimentally, we know that increasing the temperature usually causes denaturation. The addition of a denaturating agent such as urea or guanidinium chloride also causes denaturation.[133] Examples are shown in Figs. 3.1 and 3.2.

---

[132] See Appendices A, B and Ben-Naim (2006).
[133] See, for example, Schellman (1978).

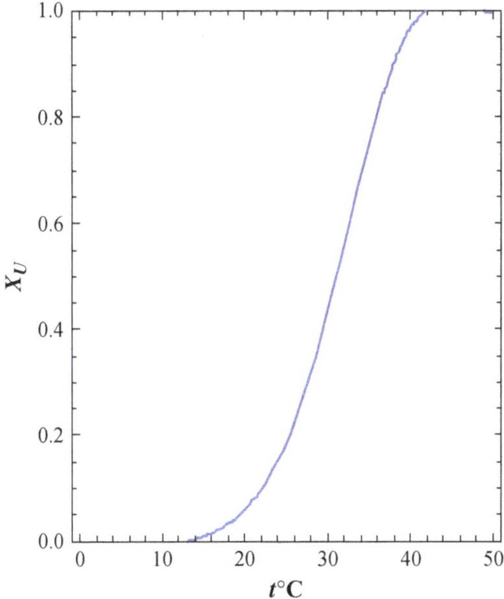

Fig. 3.1 Schematic temperature-denaturation curves of protein (bovine ribonuclease A). The curve is drawn based on an "average" of data from different experimental methods.

Consider the same reaction in (3.1.1) in the gaseous phase. We write the equilibrium constant in the gaseous phase as

$$K^g = \exp[-\beta \Delta \mu^{*g}(\text{U} \to \text{F})]$$

$$= \exp\{-\beta[\Delta H^{*g}(\text{U} \to \text{F}) - T \Delta S^{*g}(\text{U} \to \text{F})]\}. \quad (3.1.9)$$

Qualitatively, we can argue that there are more accessible states (conformations) in the **U** form compared to the **F** form. On the other hand, the compact structure of the **F** form brings more groups close together, hence we expect that $\Delta H^{*g}(\text{U} \to \text{F}) < 0$ and $\Delta S^{*g}(\text{U} \to \text{F}) < 0$. Thus, we see that there are two competing effects that determine the sign of $\Delta \mu^{*g}(\text{U} \to \text{F})$ (and hence the magnitude of K^g). Whatever the values of $\Delta H^{*g}(\text{U} \to \text{F})$ and $\Delta S^{*g}(\text{U} \to \text{F})$ are, we can expect that at very low temperatures ($T \to 0$), the dominating term in the

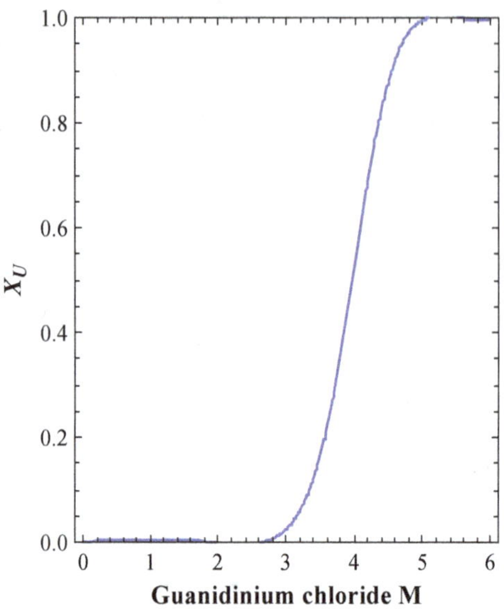

Fig. 3.2 Schematic denaturation curve of chymotripsin induced by guanidinium chloride. The curve is drawn based on an "average" of data from different experimental methods.

exponent of (3.1.9) will be ΔH^{*g}, i.e.

$$K^g \xrightarrow{\text{low }T} \exp[-\beta\Delta H^{*g}(\mathbf{U} \to \mathbf{F})] \gg 1, \qquad (3.1.10)$$

which means that at very low temperatures, K^g is very large, hence $(x_F)_{\text{eq}}$ in (3.1.8) will tend to one.

On the other hand, when the temperature is very high, the limiting value of the equilibrium constant is

$$K^g \xrightarrow{\text{high }T} \exp\left[\frac{\Delta S^{*g}(\mathbf{U} \to \mathbf{F})}{k_B}\right] < 1, \qquad (3.1.11)$$

which means that $(x_F)_{\text{eq}}$ in (3.1.8) will be smaller than unity. The larger the value of $\Delta S^{*g}(\mathbf{U} \to \mathbf{F})$, the more negative the value of the exponent in (3.1.11). K^g becomes smaller, and $(x_F)_{\text{eq}}$ becomes smaller too.

Of course, in real life, we never go to those extremes in temperatures. The analysis done above is only to demonstrate how the competition between the entropy contribution (which favors the unfolded form), and the energy (or enthalpy) contribution (which favors the folded form), is affected by the temperature.

It is easy to show that for a given $\Delta H^{*g}(U \to F)$, and assuming that this enthalpy change is determined only by the difference in the "energy levels" of U and F, the mole fraction of F, x_F will always start at $x_F \approx 1$ at $T \approx 0$, and will drop sharply as the temperature increases. The larger the value of $\Delta S^{*g}(U \to F)$, the sharper the "transition" from almost pure **F** to almost pure **U**. Figure 3.3a demonstrates this behavior for a fixed value of $\Delta H^{*g}(U \to F)$ and different values of $\Delta S^{*g}(U \to F)$. Recall that $\Delta S^{*g}(U \to F)$ is related to the ratio of the number of states of **U** and **F** conformers. Sometimes, the sharp, phase-transition-like behavior is referred to as a *cooperative* process. In fact, this behavior has nothing to do with cooperativity in the sense used in binding phenomena, where a phase-transition-like behavior is observed. Therefore, the usage of this term in the present context seems superfluous.[134]

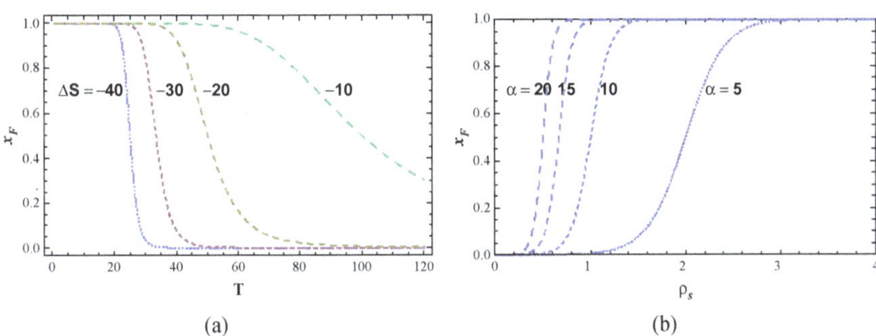

Fig. 3.3 Theoretical denaturation curves. (a) effect of temperature; (b) effect of solute (T and ρ_S in arbitrary units).

---

[134]For more details, see Ben-Naim (1992, 2001). See also Appendix R.

Figure 3.3b shows the effect of added solute on the mole fraction of the "folded" form. These curves were obtained by assuming that the standard Gibbs energy changes linearly with the added solute concentration ρ_s

$$\Delta G^\circ(\rho_s) = \Delta G^\circ(\rho_s = 0) - \alpha\rho_s. \tag{3.1.12}$$

We choose $\Delta G^\circ(\rho_s = 0)$ to be positive [favoring the unfolded form]. As we increase the concentration of the solute, ΔG° becomes negative ($\alpha > 0$). The various plots in Fig. 3.3b are for different values of α.

Note again that for some values of α, the transition from $x_F \approx 0$ to $x_F \approx 1$ is very sharp. There is no cooperativity involved in this process. See also Appendix R.

3.2. Definition of the Folded and Unfolded Forms

In this section, we discuss the question of how to define the two forms which we denoted by **U** and **F** in (3.1.1).

Before we embark on the long and difficult study of the factors that contribute to the "driving force" for protein folding, it is advisable to pause and to ponder the seemingly simple question: What do we mean by the two components **U** and **F** in the reaction (3.1.1)? This sounds like a trivial question because everyone who looks at Figs. 3.1 or 3.2 knows that the **U** component is the one which is the dominant component at high temperatures or at high concentrations of a denaturing agent.

However, for a theoretical examination of the general problem of protein folding, as well as for the examination of the factors involved in the driving force of this process, we need a more precise definition of the two components denoted by **U** and **F**. As we shall soon see, there is some degree of arbitrariness

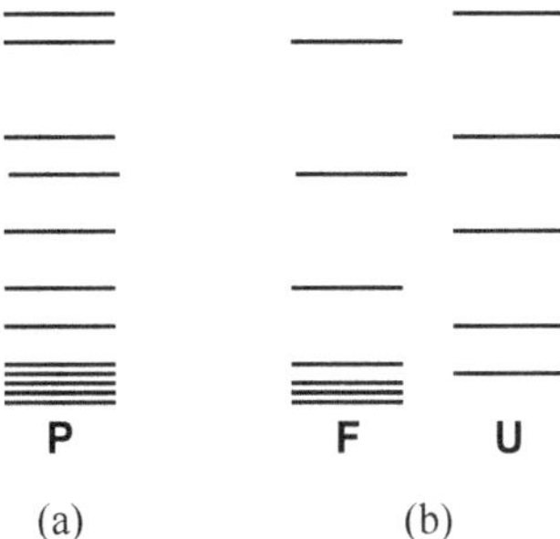

Fig. 3.4 (a) All energy levels of molecules are split into two groups; (b) **F** and **U**.

in defining the two components theoretically, whereas experimentally, the method of measuring the concentrations of **U** and **F** implicitly "defines" these two components.

Consider any molecule **P** having a set of energy levels as depicted in Fig. 3.4. The partition function of a system of N such molecules in an ideal gas phase at temperature T and volume V is

$$Q(T, V, N) = \frac{q_P^N V^N}{N! \Lambda_P^{3N}}, \tag{3.2.1}$$

from which we can derive all the relevant thermodynamic quantities. q_P is the internal partition function of a single molecule.

$$q_P = \sum_i \exp[-\beta \varepsilon_i] = \sum_j w_j \exp[-\beta \varepsilon_j], \tag{3.2.2}$$

where the first sum in (3.2.2) is over all the states of the molecule **P**. The second sum is over all the energy levels; w_j is the degeneracy of the energy level ε_j.

Now, suppose we *choose*, quite arbitrarily, to divide all the states of the system into two groups. There are, of course, many ways of making such a division — one example is shown in Fig. 3.4. Let us denote by u and f the two groups of states.

Define the two quantities:

$$q_U = \sum_{j \in u} w_j \exp[-\beta \varepsilon_j], \qquad (3.2.3)$$

$$q_F = \sum_{j \in f} w_j \exp[-\beta \varepsilon_j]. \qquad (3.2.4)$$

We can refer to a molecule **P** being in one of the energy levels $j \in f$ as an **F** molecule, and as a **U** molecule when being in energy level $j \in f$. Hence, we can write

$$q_P = q_U + q_F, \qquad (3.2.5)$$

and the partition function (3.2.1) is now rewritten as

$$Q(T, V, N) = \frac{(q_U + q_F)^N V^N}{N! \Lambda_P^{3N}}. \qquad (3.2.6)$$

The partition functions in (3.2.1) and (3.2.6) are completely equivalent. The division of all energy levels into two groups is arbitrary — it does not need to be motivated by any theoretical or experimental arguments.

However, if the two species **U** and **F** may be distinguished by any experimental means, then it might be useful to view the system of one component as a mixture of two components at equilibrium, e.g. the two isomers of dichloroethene (Fig. 3.5).

$$\text{Cl}\diagdown \quad \diagup\text{Cl} \qquad \rightleftharpoons \qquad \text{Cl}\diagdown$$
$$\text{C=C} \qquad\qquad \text{C=C}\diagdown$$
$$\text{Cl}$$

Fig. 3.5   The two isomers, cis and trans of dichloroethene.

In this case, we can rewrite (3.2.6) as

$$Q(T,V,N) = \frac{V^N}{N!\Lambda_P^{3N}} \sum_{N_U+N_F=N} \binom{N}{N_U} q_U^{N_U} q_F^{N_F}$$

$$= \sum_{N_U+N_F=N} \left(\frac{q_U^{N_U} V^{N_U}}{N_U!\Lambda_P^{3N_U}} \right) \left(\frac{q_F^{N_F} V^{N_F}}{N_F!\Lambda_P^{3N_F}} \right)$$

$$\times \sum_{N_U+N_F=N} Q(T,V,N_U,N_F). \qquad (3.2.7)$$

In (3.2.7), the sum is over all N_U such that $N_U + N_F = N$. The quantity $Q(T,V,N_U,N_F)$ is the partition function of a *mixture* of N_U molecules of species **U** and N_F molecules of species **F**. It can be shown that one can replace the sum on the right-hand side of (3.2.7) with a single term provided that N_U and N_F are such that they maximize $Q(T,V,N_U,N_F)$.[135]

Denoting by N_U^* and N_F^* the values of N_U and N_F at equilibrium, we can write the partition function (3.2.7) as

$$Q(T,V,N) \approx Q(T,V,N_U^*,N_F^*), \qquad (3.2.8)$$

where on the left-hand side, we have a partition function of a one-component system of N molecules, and on the right-hand side, we have a partition function of a *mixture* of two components with N_U^* of the **U** and N_F^* of the **F** species.

Note carefully that the equalities in (3.2.7) are valid for any division of the entire set of energy levels into two groups. Also, the mixture model view of the one component system is valid for any division into two groups of energy levels. However, such a view will be useful when we can distinguish between the two components by any experimental means. For instance, a system

---

[135]Which is the same as the equilibrium values of N_U and N_F. For details, see Ben-Naim (2008).

of N molecules of dichloroethene may be viewed as a mixture of cis and trans dichloroethene at equilibrium. If we can identify the two species by any experimental means (e.g. the two species have different spectra), then this identification also *induces* the division of all energy levels into two groups — those belonging to the cis isomer and those belonging to the trans isomer.

The situation with proteins is essentially the same, except for an additional complexity resulting from the possibility that different experimental means might induce a different division of the total energy levels (or more generally all the states of the molecules) into two groups. In other words, the two species are not defined by *our choice* of the grouping, but by the choice of the experimental method we employ in determining the concentrations of the two species.

The above paragraph sounds quite abstract. The following simple example should clarify the problem.

Suppose we have a long linear polymer. The rotations about each bond along the polymer produce myriad conformations, or rotational states. We can, if we wish, classify all conformations into two groups. For instance, we may decide to collect all conformations which can be packed in a sphere of radius R as the **F** species, and all other conformations as the **U** species. Figure 3.6 shows such a choice.

However, it is extremely unlikely that we can find experimental means that can distinguish between these two species. There are, of course, properties of the polymer that depend on the average "size," or on the extent of compactness of the conformation (e.g. the viscosity of the solution or self-diffusion of the polymer in a solution). However, these do not discriminate between the two species as we chose to define them. When one measures, say, the viscosity of the solution as a function of temperature, one cannot claim that the transitions that occur in the polymer are those between these two chosen species.

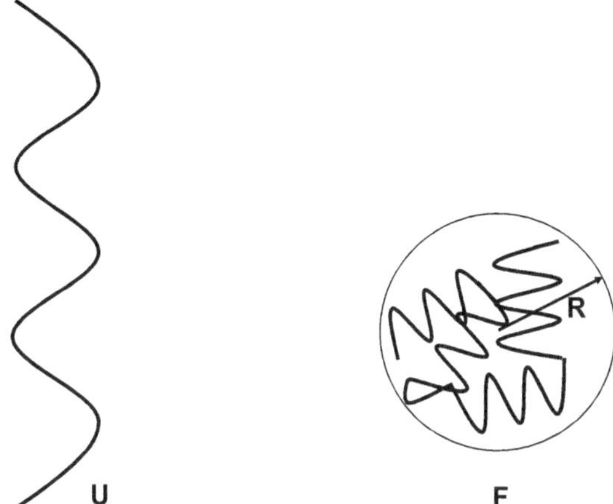

Fig. 3.6 A possible definition of the two species: All conformations that can be packed in a sphere of radius R define the F species, all others, the U species.

As another example, suppose we have again a polymer **P** to which we attach two chromophores, A and B, and we know that these chromophores are attached to different regions of the polymer (Fig. 3.7). Suppose also that each chromophore has two different spectroscopic signals: One when it is exposed to the solvent, and another when it is surrounded by the polymer. Figure 3.7 shows a few possible conformations of such a protein.

We now follow the signal of A as a function of temperature. What we actually follow is the concentration of the conformer for which the chromophore A is surrounded either by water or by the polymer. In this case, the conformers I and II will be considered as **F**, and III and IV as **U**, i.e. $x_F = x_I + x_{II}$, $x_U = x_{III} + x_{IV}$.

On the other hand, had we followed the signal from B, we would have followed different conformations that belong to the species **F** and **U**. In this case, conformers I and III will be considered as **F**, and II and IV as **U**, i.e. $x_F = x_I + x_{III}$ and

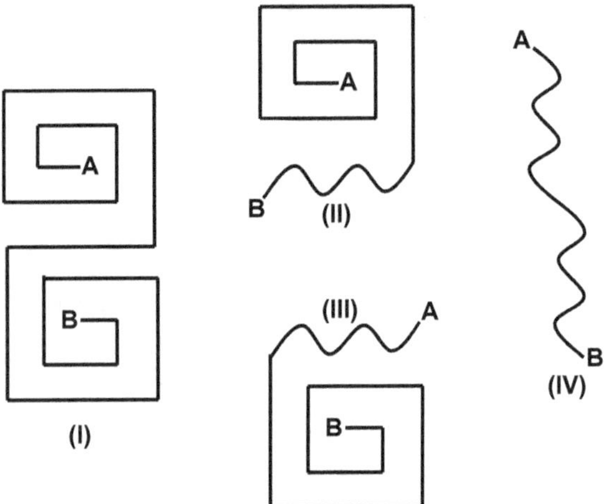

Fig. 3.7   Schematic process of unfolding of a polymer. If we follow the signal of A, then conformers I and II belong to the F form, III and IV to the U form. On the other hand, if we follow the signal of B, then I and III belong to the F form and II and IV to the U form.

$x_F = x_{II} + x_{IV}$. (If we use the same criterion as in Fig. 3.6, perhaps only I will be considered as **F**, and all the other conformations as **U**).

Thus, we see that the very definition of the two groups of states which we call **U** and **F** is determined by the experimental method we use to follow the concentrations of the various species as a function of temperature or concentration of a denaturing agent.

Clearly, in general, different experimental methods will determine different groupings of the state of the proteins into two, or more species.

This observation is important when we try to extract thermodynamic quantities such as Gibbs energies, enthalpies, entropies, etc. for the process of protein folding. Each experimental method *defines* different species **U** and **F** and therefore also different thermodynamic quantities for the process of conversion from **U** to **F**.

3.3. Formal Dissection of the Solvent-Induced Effect on Protein Folding into "Small" Ingredients

In the previous section, we discussed the difficulties and the ambiguity involved in the definition of the folded and unfolded species of the protein. In this section, we focus on one *specific* process of folding. We select a *single* conformer to represent the U species and another *single* conformer to represent the F species. For instance, we select the fully extended all-trans protein, and denote it by U_1, and one specific 3-D structure as determined by X-ray crystallography to represent the F_1 species. We write formally the reaction (Fig. 3.8)

$$U_1 \rightarrow F_1. \tag{3.3.1}$$

This is of course, not a representation of an experimental process of protein folding. However, it is theoretically a well-defined process for which we can write down all the relevant

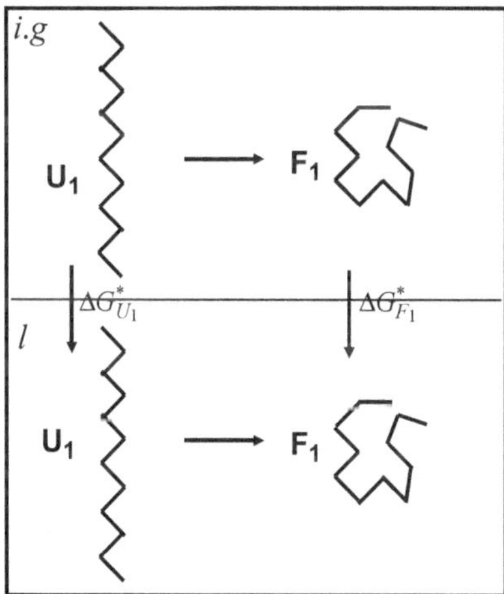

Fig. 3.8   Folding from a specific conformer U_1 to a specific conformer F_1 in the two phases *i.g.* and *l*.

thermodynamic quantities. For instance, we can write the Gibbs energy of this reaction as

$$\Delta G^*(\mathbf{U}_1 \to \mathbf{F}_1) = \Delta U + \delta G. \qquad (3.3.2)$$

Here, ΔU is the change in the potential energy of the protein due to intermolecular interactions *within* the protein molecule. Note that this is not the change in *internal* energy of the system, which is also denoted by ΔU.

Recall that the reaction (3.3.1) is a conversion from one specific conformer, denoted $\mathbf{U}_1$, to one specific conformer, denoted $\mathbf{F}_1$. This is not an experimental process, but only a thought experiment. If we know the potential energy function of the protein, we can, in principle, calculate ΔU for the reaction (3.3.1). This is also the energy change in the reaction (3.3.1) carried out in the gaseous phase. Here, we are interested not in ΔU, but in the *solvent-induced* contribution to the Gibbs energy change, which can be written as

$$\delta G(\mathbf{U}_1 \to \mathbf{F}_1) = \Delta G^{*l}(\mathbf{U}_1 \to \mathbf{F}_1) - \Delta G^{*g}(\mathbf{U}_1 \to \mathbf{F}_1)$$

$$= \Delta G^*(\mathbf{F}_1) - \Delta G^*(\mathbf{U}_1), \qquad (3.3.3)$$

where $\Delta G^*(\mathbf{F}_1)$ and $\Delta G^*(\mathbf{U}_1)$ are the Gibbs energies of solvation of the conformers $\mathbf{F}_1$ and $\mathbf{U}_1$, respectively. The relationship between the Gibbs energies of the reaction in the two phases and the solvation Gibbs energies are shown schematically in Fig. 3.8.

It should be said that the potential function U is a very complicated function, depending on all the locations of the atoms of the protein. However, it is believed that we know its general form and its components, as well as the order of magnitude of these components.[136] The problem of studying the solvent-induced part δG is an order of magnitude more difficult than for studying ΔU. The reason is that both $\Delta G^*(\mathbf{F}_1)$

---

[136]Brooks *et al.* (1988).

and $\Delta G^*(U_1)$ involve averaging over all configurations of the solvent molecules, and this is a formidably difficult problem. Until recently, there was no way of dissecting the quantity δG into "small," manageable quantities, each of which could be studied either theoretically or experimentally using small model compounds.

We shall devote the next few sections of this chapter to develop the methodology of dissecting the quantity δG for the process (3.3.1). The methodology is essentially the same as the one we have presented in Chapter 2 in connection with the solvation Gibbs energy of proteins. Here, however, we have to follow the *changes* in the conditional solvation Gibbs energies of all the FGs of the protein (note that we include the carbonyl and the amine groups of the backbone as FGs) in the folding process.

A glance at Fig. 3.9 shows that we can distinguish between essentially three groups of FGs[137]:

(a) The **E** group (**E** for "external") consists of all the FGs the solvation of which does not change in the reaction (red FGs in Fig. 3.9). Formally, an FG k in the **E** group is defined by the condition

$$\Delta G_F^{*k/H,S} = \Delta G_U^{*k/H,S}. \tag{3.3.4}$$

The notation $k/H, S$ means: The FG k, given that the hard (H) and soft (S) parts of the interaction energy between the protein and a water molecule has already been switched on.

Intuitively, it is clear that an FG k in the group **E** will not contribute anything to δG. This is so because the contribution of this group to $\Delta G^*(F_1)$ and to $\Delta G^*(U_1)$ will be the same,

---

[137]Note the different senses in which the word "group" is used in this section; the group **E** is a set containing functional groups (FGs).

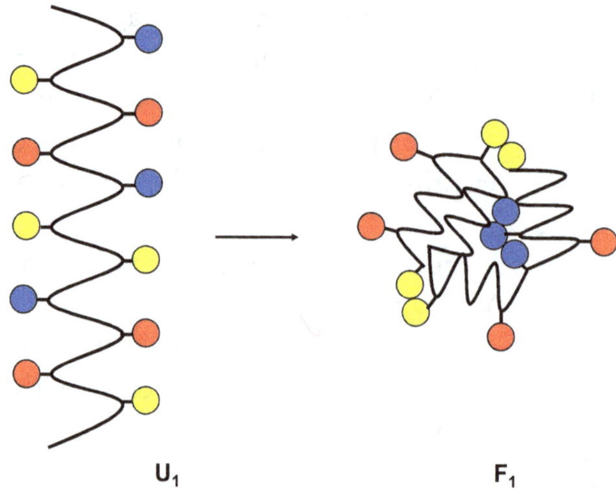

Fig. 3.9 A few examples of FGs belonging to class **E** (red disks), class **I** (blue disks), and class **J** (yellow disks). All FGs in the U_1 conformer are considered to be independently solvated.

and will therefore be canceled out in forming the difference in (3.3.3).

(b) The **I** group (**I** for internal) consists of all the FG's, the solvation of which is completely "lost" in the reaction (3.3.1). Formally, a functional group k belonging to the **I** group is defined by the requirement

$$\Delta G_F^{*k/H,S} = 0, \quad \Delta G_U^{*k/H,S} \neq 0. \qquad (3.3.5)$$

In other words, these are the groups that are initially solvated in the U_1 conformer, but are transformed into the interior of the protein, and hence are not solvated in the **F** conformer (blue FGs in Fig. 3.9).

It should be noted that a group k buried in the interior of the protein is not considered to be "solvated" by the protein. Any interactions between this FG and the rest of the protein are considered to be part of the total interaction energy change ΔU

in (3.3.2). This note is important when considering a *hydropho-bicity scale*. The loss of the conditional solvation Gibbs energy of an FG is the only contribution of that group to the solvent-induced quantity δG, and one does not need to add a conditional solvation Gibbs energy of the same group in any other organic liquid. We shall further discuss this point in Sec. 3.4 and Appendix H. Here, we only note that this process of transferring a *HφO group* from being exposed to the solvent into the interior of the protein (blue FGs in Fig. 3.9) inspired Kauzmann to use the process of transferring a *HφO molecule* from water into an organic liquid. The reader should note here the qualitative difference between these two processes. We shall discuss the quantitative difference in Sec. 3.4.

It is clear from the condition (3.3.5) that the conditional solvation Gibbs energy of the FG k is part of $\Delta G^*(U_1)$ in (3.3.3), but the same FG affects $\Delta G^*(F_1)$ only in determining the *size* of the protein in the F_1 conformer.

(c) The J group (J is used for the joint, or boundary group, see Sec. 4.2) consists of all FGs that do not belong to either the E or the I group. These FGs are solvated in both the F_1 and the U_1 forms, but their conditional solvation Gibbs energies are different in the two conformations of the protein.

Formally, we can write the condition for an FG k belonging to the J group as

$$\Delta G_F^{*k/H,S} \neq \Delta G_U^{*k/H,S}, \tag{3.3.6}$$

and neither of these two quantities is zero. Two examples are shown in Fig. 3.9 (yellow FGs). Intuitively, an FG is deemed to belong to the J group when its solvation changes upon folding. This change can occur either when the FG is partially buried in the interior of the protein or from correlation with other FGs in the F_1 conformer.

Having classified all FGs into the three groups **E**, **I** and **J**, we can write the solvent-induced contribution δG as

$$\delta G(\mathbf{U}_1 \to \mathbf{F}_1) = \delta G^H + \delta G^{S/H} + \delta G^{I/H,S} + \delta G^{J/H,S}. \quad (3.3.7)$$

Here, δG^H is simply the difference in the solvation Gibbs energy of the hard part of the interaction energy in reaction (3.3.1). The term $\delta G^{S/H}$ is the difference in the conditional solvation Gibbs energy of the soft part of the interaction, given that the hard part has already been solvated. $\delta G^{I/H,S}$ includes all the conditional solvation Gibbs energies of the FGs that are in the **I** group, i.e. all the FGs the solvation of which was lost in the reaction (3.3.1). The term $\delta G^{J/H,S}$ includes the contribution of all the FGs belonging to the **J** group. Note that there is no term of the form $\delta G^{E/H,S}$ in (3.3.7). All the FGs belonging to the **E** group do not contribute to the overall solvent-induced quantity δG.

It should be noted that FGs in the **I** group of a real protein are not necessarily independently solvated in the **U** conformer. Specifically, the C=O and the NH of the backbone may be correlated in the **U** form. Thus, when we further examine the various contributions to $\delta G^{I/H,S}$, we have to take into account the burial of independently solvated FGs, as well as pair correlated FGs. The same comment applies to FGs in the **J** group. Thus, not only independently solvated groups in the **U** form change their environments upon reaction (3.3.1) — pair correlated groups in the **U** form can also change their environments in this reaction.

In the next section we shall discuss methods for estimating the various contributions to δG. It is appropriate at this stage to note that none of the terms in (3.3.7) includes the change in the solvation Gibbs energy of a *HϕO molecule*, such as methane or ethane, when it is *transferred* from *water* to an *organic liquid*. The term that comes closest to Kauzmann's model for the transfer of a *HϕO* group from water to an organic liquid is the loss of the *conditional* solvation Gibbs energy of a $H\phi O$ group, which is included in $\delta G^{I/H,S}$. This quantity, as we have seen in Sec. 1.11, is quite different from the solvation Gibbs energy of a $H\phi O$ group in water.

3.4. Methods of Studying and Estimating the Various Contributions to δG

In Sec. 3.3, we have achieved a formal dissection of a very complicated quantity δG into "smaller" ingredients, which in principle can be studied using either theoretical or experimental means. For real proteins, we are still very far from having enough data to study each of these ingredients (note that the terms $\delta G^{I/H,S}$ and $\delta G^{J/H,S}$ include sums of many terms pertaining to single, pair or triplet FGs).

However, because of the importance (and perhaps also the urgency) of the problem of protein folding, we shall study in this section a hypothetical protein: a long linear chain having two kinds of FGs: methyl or ethyl groups representing $H\phi O$ groups, and carbonyl or hydroxyl groups representing $H\phi I$ groups. We shall also assume that all these groups are independently solvated in the unfolded form. We shall use whatever data is available to estimate the relative importance of the various ingredients to the overall solvent-induced contribution to the "driving force" δG.

Again, we discuss one specific reaction such as (3.3.1), i.e. a transformation from a single conformation U_1 to a single conformation F_1. This is done for convenience only. In order to build up a complete *list*, or "inventory," of the ingredients of the solvent-induced effect on the "driving force," it is sufficient to consider one specific reaction. It should be stressed, however, that when we are interested in any *real* process of protein folding, we must take appropriate averages over all conformations that "belong" to the **U** form, and all conformations that "belong" to the **F** form (see also Sec. 3.2). For the present section, we do not need to be concerned with the definition of the two forms **U** and **F**. All we are interested in are the *type* of the ingredients and some estimates of their magnitude.

Thus, for the simple model (Fig. 3.10), and for a specific reaction (3.3.1), we can write the following ingredients of δG:

$$\delta G = \delta G \ (due\ to\ the\ hard\ part\ (H)\ of\ the$$
$$solute\text{--}solvent\ interaction)$$
$$+\ \delta G \ (due\ to\ the\ soft\ part\ (S)\ of\ the\ interaction)$$
$$+\ \delta G \ (due\ to\ the\ loss\ of\ the\ solvation\ of\ H\phi O\ groups)$$
$$+\ \delta G \ (due\ to\ the\ loss\ of\ the\ solvation\ of\ H\phi I\ groups)$$
$$+\ \delta G \ (due\ to\ the\ pair\ correlation\ between$$
$$two\ H\phi O\ groups)$$
$$+\ \delta G \ (due\ to\ the\ pair\ correlation\ between$$
$$two\ H\phi I\ groups)$$
$$+\ \delta G \ (due\ to\ the\ pair\ correlation\ between\ a\ H\phi O$$
$$and\ a\ H\phi I\ group)$$
$$+\ \delta G \ (due\ to\ triplet\ and\ higher\ order\ correlations).$$

$$(3.4.1)$$

As we can see, the list is quite long. For real proteins, it is even longer because we have to account for 20 different side chains in addition to the $C{=}O$ and NH of the backbone. We also have to take into account correlations between the *$H\phi I$* groups in the unfolded protein, which we ignore in the list (3.4.1).

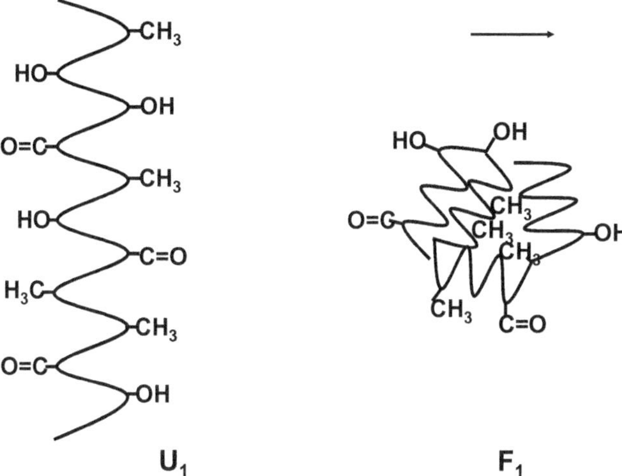

Fig. 3.10   A simple model of a protein having two kinds of FGs; hydrophobic (methyl) and hydrophilic (carbonyl and hydroxyl).

Also, we included in the last term on the right-hand side of (3.4.1) all higher order correlations, on which we currently have almost no information.

We shall now briefly discuss each of the contributions to δG listed in (3.4.1), and how to estimate their magnitudes.

(a) *The hard part*

The term δG^H in (3.3.7) is essentially the difference between the work required to create a *cavity* for the $\mathbf{U}_1$ form, and the cavity for the $\mathbf{F}_1$ form,[138] i.e.

$$\delta G^H = \delta G^{*H}(\mathbf{F}_1) - \Delta G^{*H}(\mathbf{U}_1)$$
$$= -k_B T \ln\langle \exp[-\beta B^H_{F_1}]\rangle_0 + k_B T \ln\langle \exp[-\beta B^H_{U_1}]\rangle_0$$
$$= -k_B T \ln[\Pr(\textit{cavity of } \mathbf{F}_1)] + k_B T \ln[\Pr(\textit{cavity of } \mathbf{U}_1)],$$

$$(3.4.2)$$

---

[138]See Appendix K.

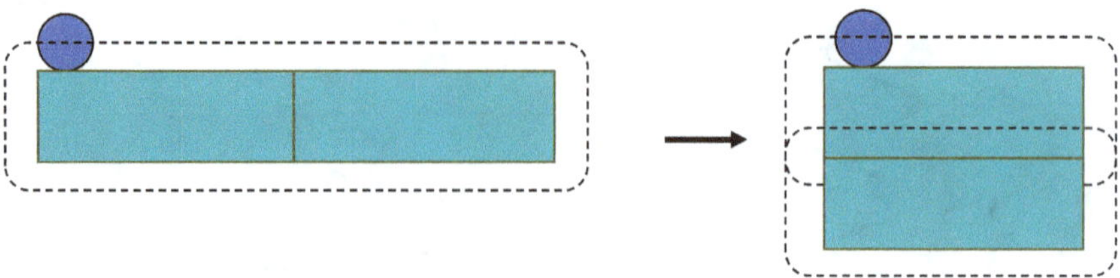

Fig. 3.11 Folding of any hard particle reduces the excluded volume with respect to solvent molecules.

where Pr (*cavity of α*) is the probability of finding a *cavity* for α at some *specific* location in the liquid. The cavities for $\mathbf{F}_1$ and $\mathbf{U}_1$ are simply the regions in the liquid that are excluded from all solvent molecules due to the hard repulsive part of the solute–solvent interaction[139] (Fig. 3.11).

It is intuitively clear that when any hard rod is folded into a compact form, the excluded volume with respect to the solvent molecules is reduced, i.e.

$$V_F^{EX} < V_U^{EX}. \tag{3.4.3}$$

The work required to create a cavity at a fixed point in the liquid can be calculated approximately by the scaled particle theory (SPT).[140] For a large cavity, we can approximate this work as the PV work, i.e.

$$\delta G^H \approx P(\Delta V^{EX}) = P(V_F^{EX} - V_U^{EX}) < 0. \tag{3.4.4}$$

This is only a rough estimation. Since the excluded volume of the protein is reduced upon folding, we can conclude that this process will contribute negatively to δG^H, hence, to the total solvent-induced "driving force" for protein folding. The quantity δG^H is not of much interest in our study. First, because

---

[139] For details, see Appendix K and Ben-Naim (1992) Sec. 5.10. Note that the averages in (3.4.2) are over all configurations of the solvent molecules with the probability distribution of the solvent configurations *before* adding the solute.

[140] For details, see Reiss (1966) and Ben-Naim (1992, 2006).

(3.4.4) is only a rough approximation; second, and more importantly, whatever the magnitude of δG^H is, it is not *specific* to water. Therefore, we cannot expect that this term will determine the specific 3-D structure of the protein apart from the general tendency to fold into a compact structure with the smallest excluded volume.

(b) *The soft part*

The soft part of the solvation Gibbs energy of a protein was discussed in Sec. 2.2. Here, we are interested in the difference between two contributions of the soft part of the interaction *to the solvation Gibbs* energies of the U_1 and F_1 forms, i.e.

$$\delta G^{S/H} = \delta G^{*S/H}(F_1) - \Delta G^{*S/H}(U_1). \tag{3.4.5}$$

Using the assumption regarding the solute–solvent interactions as discussed in Sec. 2.2, we can express $\delta G^{S/H}$ as

$$\delta G^{S/H} = -k_B T \ln\langle\exp[-\beta B^S_{F_1}]\rangle_H + k_B T \ln\langle\exp[-\beta B^S_{U_1}]\rangle_H, \tag{3.4.6}$$

where $B^S_{\alpha_1}$ is the soft part of the solute–solvent binding energy of the solute α (see Sec. 2.2). The averages in (3.4.6) are again over all the configurations of the solvent molecules, as in (3.4.2), but here with the *conditional* distribution, i.e. the distribution of solvent configurations *given* that the *hard* part of α has already been introduced into the solvent.

Because the soft part of the solvation Gibbs energy is determined by the groups on the *surface* of the protein, and since the surface of the protein is *reduced* upon folding, we expect that $\delta G^{S/H}$ will be positive. This is, of course, a qualitative argument. One can make a more precise calculation based on the forms U_1 and F_1, assuming that the soft part of the binding energy is proportional to the number of groups exposed to the solvent, and some energy parameters such as the Lennard-Jones energy parameter between two small non-polar molecules (see Sec. 2.2).

If βB_α^S is very small compared to unity, one can also approximate

$$\langle \exp[-\beta B_\alpha^S] \rangle_H \approx 1 - \beta \langle B_\alpha^S \rangle_H. \tag{3.4.7}$$

Hence,

$$\delta G^{S/H} \approx \langle B_{F_1}^S \rangle_H - \langle B_{U_1}^S \rangle_H \approx -\varepsilon[A_{F_1} - A_{U_1}] > 0. \tag{3.4.8}$$

Here, A_α is the number of non-polar groups exposed to the solvent (which is proportional to the surface area of the solute exposed to the solvent), and $\varepsilon > 0$ could be chosen as the Lennard-Jones parameter for, say, neon-methane interaction (see Sec. 2.2).

What we have found can be summarized as follows: The soft part of the solvation Gibbs energy of a protein is always a negative quantity, and it is proportional to the surface area of the protein. Upon folding, the surface area of the protein (exposed to the solvent) is reduced. Therefore, the contribution of the soft part to the solvation Gibbs energy of the protein is also reduced in absolute magnitude. Hence, $\delta G^{S/H}$ is expected to be positive. We note again, as we have noted in the discussion of δG^H, that whatever the magnitude of $\delta G^{S/H}$ is, it is not specific to water. Therefore, we cannot expect that this term will determine the specific 3-D structure of the folded form of the protein. The more specific solvent-dependent effects are the ones discussed below.

(c) *Loss of the conditional solvation of a $H\phi O$ group*

The third term on the right-hand side of (3.3.7) includes many contributions, one of which is the contribution of the transfer of a $H\phi O$ group from being exposed to water in the U_1 conformer to the interior of the protein. This process is the most closely related to Kauzmann's model using the Gibbs energy change for the transfer of a non-polar molecule, say, methane, from water to an organic liquid (having a similar constituency to the interior of the protein). The implication from Kauzmann's model was that the transfer of the non-polar $(H\phi O)$ groups into the interior of the protein should contribute significantly to the thermodynamic "driving force" for protein folding. This idea has prevailed in the biochemical literature for almost half a century. It is not uncommon to find statements in many books and reviews referring to the $H\phi O$ effect as the *most important* factor in the driving force for protein folding and other biochemical processes. It is also not uncommon to find assertions that the importance of the $H\phi O$ effects is "well established" and "generally accepted."

Although I am aware that the importance of the $H\phi O$ effect to protein folding is "generally accepted," I do not agree that this view is well established, or that the $H\phi O$ effect is the most important effect in protein folding.

I first challenged this generally accepted view in 1980.[141] This was further elaborated upon in 1992,[142] and in Part I.[143] My main disagreement with the accepted view is that Kauzmann's model deals with the Gibbs energy of solvation of a $H\phi O$ *molecule* in water, whereas in the process of protein folding, it is the *conditional* Gibbs energy of solvation of a $H\phi O$ *group* that features in the driving force. In less technical terms, in Kauzmann's model, the $H\phi O$ molecule is initially solvated by water, whereas in the process of protein folding, the $H\phi O$ group is initially solvated by water *perturbed by the presence of the protein*. As we have seen in Chapters 1 and 2, the effect of the backbone (BB) on the conditional solvation Gibbs energy could

---

[141] Ben-Naim (1980).
[142] Ben-Naim (1992).
[143] Ben-Naim (2009).

be one or two orders of magnitude smaller than the solvation Gibbs energy. This fact in itself casts doubts on the adequacy of Kauzmann's model for representing the transfer of a $H\phi O$ group into the interior of the protein, as a result of which the relative importance of the $H\phi O$ effect comes also under a cloud of doubt.

Second, how can one claim that the $H\phi O$ effect (or any other effect for that matter) is the "most important" effect in biochemical processes, when we do not in fact have information on *all* possible factors involved in the biochemical processes? As I have described in Chapter 4 of Part I, the (almost) uncritical acceptance of the $H\phi O$ dogma was partly the result of the dismissal of the contribution of hydrogen bonding based on the so-called HB inventory argument.[144] This argument turned out to be wrong.[145] Furthermore, other $H\phi I$ effects, not recognized until recently, were discovered and should be reckoned with as possible competing factors for the primacy of the solvent-induced effect.

In Sec. 1.5, we have given some representative values of the conditional Gibbs energy of solvation of simple non-polar groups such as methyl and ethyl. We have also seen how strongly these quantities depend on the backbone type. Therefore, based on the present knowledge available, the $H\phi O$ effect, in the sense discussed in this section, seems to be quite small, of the order of about half $k_B T$ at room temperature. Of course, much more has to be done to establish the relative importance of the $H\phi O$ effect. In particular, it would be desirable to have reliable values of the conditional solvation Gibbs energies of all the amino acid side chains attached to a backbone, which mimics the environment around real proteins.

Finally, it should be noted that we are interested in the *solvent-induced* contribution to the overall "driving force" for

---

[144] See Ben-Naim (2009) Sec. 4.12 and Appendix I.
[145] See also Appendix I.

the protein folding process. The relevant quantity that we have discussed in this subsection was the *conditional* Gibbs energy of solvation of a $H\phi O$ group in *water*. There is no need to consider the transfer of a $H\phi O$ group (or molecule) from water to an *organic liquid*, as is used in Kauzmann's model.[146] The reason is that the interaction of the $H\phi O$ group with its surroundings in the interior of the protein, though certainly resembling the surrounding of the same group in organic liquid, is part of the *direct* interaction between the $H\phi O$ group and other groups of the protein. Therefore, these interactions are part of the ΔU in Eq. (3.3.2) and not part of the solvent-induced effect. It might be the case that the interactions of the $H\phi O$ with its surroundings in the interior of the protein have a similar order of magnitude to the solvation of the same $H\phi O$ group in some organic liquid. However, the two quantities are, in principle, different. More on this in Appendix H.

(d) *Loss of the conditional solvation of a $H\phi I$ group*

Another term included in the third term on the right-hand side of (3.3.7) is the contribution due to the loss of the conditional solvation of a $H\phi I$ group in the process. The exact amount of this contribution depends on the number of "arms" along which the $H\phi I$ group can form a HB. As was discussed in Sec. 1.5, the solvation Gibbs energy per arm is about 9.4 kJ/mol. It is also known that most of the backbone polar groups (C=O and NH) that are buried in the interior of the protein are engaged in HBs (e.g. HBs in an α-helix or a β-sheet structure). It should be made clear, however, that these intramolecular HB energies "belong" to the ΔU term in (3.3.2). It is only the loss of the conditional solvation Gibbs energy of the $H\phi I$ group that "belongs" to the solvent-induced part of the driving force δG.

---

[146]See also Appendix H.

It is expected that an amine group which is transferred into the interior of the protein contributes about 9.4 kJ/mol, whereas a carbonyl group is expected to contribute about 18.8 kJ/mol. Each of these two values is smaller than the energy involved in the formation of an intramolecular hydrogen bond.

Thus, one should be very careful in assessing the net contribution of a $H\phi I$ group transferred into the interior of the protein. From the point of view of the solvent-induced effect, the contribution of this $H\phi I$ effect is always positive, and depends on the number of arms of the $H\phi I$ group. It should also be remembered that this effect is much larger in absolute magnitude compared to the corresponding $H\phi O$ effect.

The available information on the solvation of a $H\phi I$ group is quite limited. It would be desirable to have information on the conditional solvation Gibbs energies of all the side chains that are $H\phi I$ and are found in the interior of the protein.

(e) *Pair correlation between two $H\phi O$ groups*

In this and in the next two subsections, we discuss pair correlations between two groups. These contributions belong to the third term on the right-hand side of (3.3.7). Here, we discuss two $H\phi O$ groups that are uncorrelated (i.e. the solvation of which are independent, see Appendix F) in the U_1 form, but which become correlated in the F_1 form (see yellow disks in Fig. 3.9).

It should be noted that originally, the $H\phi O$ bond, or the $H\phi O$ interaction, was identified as the "reversal" of the solvation of the non-polar group. Clearly, two molecules adhering to each other are less exposed to water than the same molecules when they are separated. In this sense, one can view the "dimerization" process as a "partial reversal" of solvation. However, the thermodynamics of the dimerization are quite different from that of solvation.

In the early 1970s, this process of "dimerization" was redefined in terms of the potential of mean force,[147] and as such was extensively investigated using theoretical, experimental and simulation methods. These studies were quite interesting within the context of studying the properties of aqueous solutions. However, it is not clear to what extent these studies are relevant to the correlation between two *HϕO groups* attached to the backbone of a protein.

Figure 3.12 shows how the problem of correlation between two *HϕO* groups was "extracted" from the protein folding process, to be studied in pure water.

The main difference between the two processes is the *presence* of the protein in the original problem. The difference between correlation and conditional correlation is the same as the difference between the solvation process and the *conditional* solvation process. As is shown in Fig. 3.12, the surroundings of

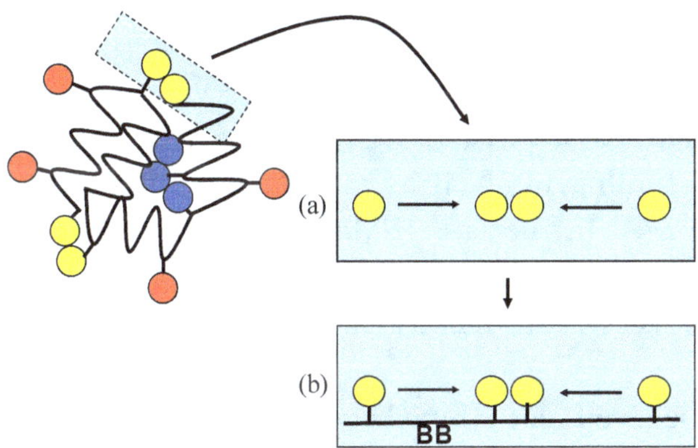

Fig. 3.12 "Extraction" of the problem of *HϕO* pair interaction on the protein for studying (a) the pair correlation between two *HϕO* molecules in water; (b) the conditional pair correlation between two *HϕO* groups attached to the backbone (BB).

---

[147]See Part I and Ben-Naim (1974).

the solvating molecules are the *same* in the two processes (a) and (b), but the distribution of the solvent molecules is not the same. Therefore, the two processes in Fig. 3.12, (a) and (b), measure correlations in two essentially different "solvents."

It would have been desirable to have information on the correlation between the two $H\phi O$ groups attached to a protein, or at least to a small model compound that affects the surrounding water molecules in a similar way to the protein (see discussion in Sec. 1.7). Unfortunately, such information is not available. Instead, we do have some information on the correlation between two $H\phi O$ groups attached to a small model compound. These data were discussed in Sec. 1.7. Whether these results are or are not relevant to the correlation between two $H\phi O$ groups attached to a protein is an open question. Tentatively, we assume that they are, and we accept that the process of bringing two $H\phi O$ groups, such as methyl or ethyl, to a close distance (as determined by the model used) is of the order of about half $k_B T$.

(f) *Pair correlation between two $H\phi I$ groups*

$H\phi I$ interactions of these kinds are relatively new on the scene of protein folding (as well as in protein–protein association, see Chapter 4). In the late 1980s and early 1990s, the study of the factors involved in protein folding was dominated by the $H\phi O$ effect (both solvation and correlations between $H\phi O$ groups).

It was believed that HBing between two $H\phi I$ groups in a solvent (water) had a marginal effect on the "driving force" for protein folding (see Part I Chapter 4). Therefore, no attempts were made to search for other possible $H\phi I$ effects. Most biochemists were satisfied with the $H\phi O$ effect and, to the best of my knowledge, no effort was made to search for any other solvent-induced contributions. A turning point occurred in the late 1980s, when a systematic study of all possible solvent-induced effects was undertaken.[148] The result of this study was the expression for δG in the form (3.3.7). In this expansion, correlation between two $H\phi O$ groups, as well as between two $H\phi I$ groups, appeared. At that time, some data on the correlation between $H\phi O$ groups was available, but none was available for the corresponding correlations between two or more $H\phi I$ groups.

The first indication that such correlation might be important came from some experimental data published by Haberfield *et al.*[149] These data were used to extract the quantity we shall call the correlation between two $H\phi I$ groups.[150]

Soon, more data became available, as well as some simulation of these $H\phi I$ effects,[151] and a theoretical estimate of the strength of these effects.[152] All these data led to the conclusion that correlation between two $H\phi I$ groups (at the correct distance and orientation, see Chapter 4 of Part I) is quite significant, and its role in the process of protein folding should be taken more seriously.

It should be emphasized that the available experimental data are not the data which we would have liked to have, namely data on correlation between $H\phi I$ side chains in a real protein. Therefore, care must be exercised in reaching any final conclusions regarding the relative importance of $H\phi I$ correlation.

However, until more relevant data becomes available, we should use whatever data we have to compare the relative importance of $H\phi O$ correlation and $H\phi I$ correlation. We can also assume that although the environment produced by a

---

[148] Ben-Naim *et al.* (1989), Ben-Naim (1990).

[149] Haberfield *et al.* (1984).

[150] Ben-Naim (1990), Ben-Naim *et al.* (1989a, 1989b), Bahar and Jernigan (1994).

[151] Mezei and Ben-Naim (1990), Durell *et al.* (1994).

[152] Ben-Naim (2009) Chapter 4 and Appendix G.

benzyl or a cyclohexane ring is not the same as the one near a real protein, the *difference* between $H\phi O$ and $H\phi I$ correlations in model compounds might be similar to the difference in real proteins.

Therefore, we can conclude, pending the availability of better data, that $H\phi I$ correlations are much more important than $H\phi O$ correlations. Specifically, the pair correlation between two $H\phi I$ groups such as hydroxyl is about an order of magnitude larger than the corresponding correlation between two $H\phi O$ groups such as methyl or ethyl. It might be true that triplet and quadruplet $H\phi I$ correlations are also important, but at present we have no data on these correlations.

(g) *Pair correlation between $H\phi O$ and $H\phi I$ groups*

For the sake of completeness, we present here one example of the correlation between a $H\phi O$ group, methyl, and a $H\phi I$ group, hydroxyl, on a benzene ring. Consider the formation of 2-methylphenol from toluene and phenol. The δG for this "reaction" is (Fig. 3.13a)

$$\text{phenol} + \text{toluene} \rightarrow \text{2-methylphenol} + \text{benzene}. \qquad (3.4.9)$$

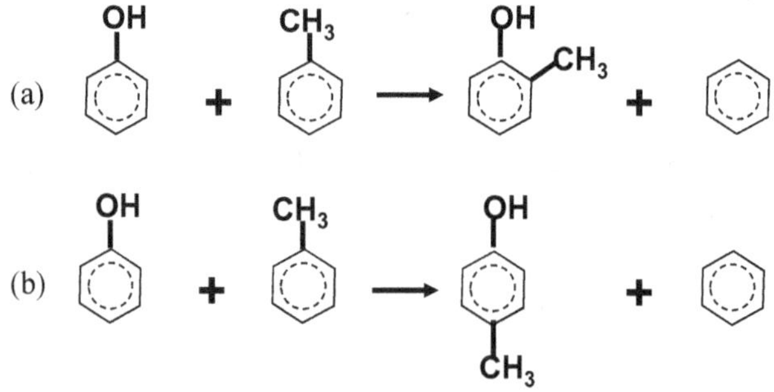

Fig. 3.13   Model compounds for the study of the conditional correlation between a $H\phi O$ and a $H\phi I$ group. Reactions (a) and (b) correspond to Eqs. (3.4.10) and (3.4.11), respectively.

Based on the data by Cabani *et al.*,[153] we have computed

$$\delta G = \Delta G^*_{2\text{-methylphenol}} + \Delta G^*_{\text{benzene}} - \Delta G^*_{\text{toluene}} - \delta G^*_{\text{phenol}}$$
$$= +3.72 \, \text{kJ/mol}. \tag{3.4.10}$$

The fact that δG in this case is positive indicates that the methyl group, brought within a very close proximity to the hydroxyl group, interferes with the solvation of the latter and hence makes this configuration less stable than the corresponding 4-methlyphenol isomer, for which we have (Fig. 3.13b)

$$\delta G = \Delta G^*_{4\text{-methylphenol}} + \Delta G^*_{\text{benzene}} - \Delta G^*_{\text{toluene}} - \Delta G^*_{\text{phenol}}$$
$$= +2.1 \, \text{kJ/mol}, \tag{3.4.11}$$

which is slightly smaller than the previous case. Note also that both values in (3.4.10) and (3.4.11) are positive, whereas the $H\phi O - H\phi O$ and the $H\phi I - H\phi I$ correlations contribute negatively to δG.

Therefore, we cannot expect that these correlations will contribute significantly to the "driving force" for protein folding. Note, however, that although the available data on this kind of correlation are not relevant to the correlation between the same two groups in a protein, the *sign* of these correlations is probably the same as in proteins. The reason is that the theoretical argument provided above is valid for any pair of such groups attached to any backbone.

(h) *Higher order terms in the expansion (3.3.7)*

Up to this point, we have discussed a few $H\phi O$ and $H\phi I$ effects for which experimental data are available, and therefore can say something on the order of magnitudes of the different contribution to the overall solvent-induced effect on protein folding. However, the expansion (3.3.7) contains many more terms

---

[153]Cabani *et al.* (1981).

about which we know very little. We mention here two possible contributions which might be important and deserve further investigation.

The first is higher-order correlations. We note in particular that triplet correlations between three $H\phi I$ groups (at the correct configuration) could contribute significantly to the overall δG in (3.3.7). Some estimates of these effects are given in Appendix G. The second is a longer range $H\phi I$ interactions. We shall discuss these effects in Chapter 4.

3.5. Summary of the Factors Involved in the Stability of the Native Protein

In the previous sections, we have discussed a specific process of protein folding. One specific conformer denoted U_1 was folded into one specific conformer denoted F_1. We did this simplification mainly for one reason: To obtain an inventory of all possible solvent-induced effects on the "driving force" for the process of protein folding. Note carefully that when we say "protein folding," we do not refer to that *specific* process described in (3.3.1). In the real process of protein folding, each of the species denoted by U and F is *defined* by the experimental method we use to measure the concentrations of these two species (see Sec. 3.2). However, we believe that in whatever method we use for defining U and F, each of these species must include conformations of the *type* U_1 and F_1, respectively. Clearly, the *inventory* of all possible solvent-induced effects on protein folding should not be affected by proceeding from the specific reaction (3.3.1) to the more realistic process of protein folding. Of course, the magnitude of each effect is different for each conformation. Therefore, when averaging over all configurations, we must take into account these differences in the magnitude of the effects.

Presently, we do not have enough data to calculate all the possible solvent-induced effects even for folding of the smallest protein. Nevertheless, from the data we already have, it would be safe to conclude that the importance of the $H\phi O$ effects (both the loss of solvation and pair correlation between $H\phi O$ groups) was probably exaggerated. The reasons, as discussed in Part I Chapter 4, are that various $H\phi I$ effects were ignored (mainly due to the HB inventory argument) on the one hand, and the $H\phi O$ effect was overestimated (mainly because of the use of Kauzmann's model for the transfer of a non-polar molecule from water into organic liquid) on the other. Actually, there have been some suggestions in the literature implying that Kauzmann's model *underestimated*, not overestimated the $H\phi O$ effect. These suggestions were found to be wrong and did not live long. We shall discuss one example in Appendix M.

The new picture that emerges is that $H\phi I$ effects are probably more important both for the stability of the native form of the protein, and for the dynamics of protein folding discussed in the next sections.

One could also argue that what we have found so far is based on the comparison of *one $H\phi O$* effect with the corresponding $H\phi I$ effect. Such a comparison led us to the conclusion that $H\phi I$ effects are more important than $H\phi O$ effects. Could it be that the combined effects of *all* the $H\phi O$ groups in a real protein might be larger than the combined effects of *all* the $H\phi I$ ones? In principle, that possibility cannot be ruled out. However, this possibility is quite unlikely. First, because only about one-third of the amino acid side chains are $H\phi O$, and the majority are either neutral or $H\phi I$. Second, the backbone itself contributes two $H\phi I$ groups per amino acid. Therefore, we can safely conclude that it is quite unlikely that the combined effects of all the $H\phi O$ groups would be greater than the combined effects of all the $H\phi I$ groups.

The inevitable conclusion based on evidence from both theory and experiments is that the various $H\phi I$ effects are far more important than the corresponding $H\phi O$ effects.

Yet, in spite of the fact that the evidence in favor of the $H\phi I$ effects was published almost twenty years ago, one can still find statements in the literature about the *dominance* of the $H\phi O$ effects, totally ignoring the evidence in favor of the $H\phi I$ effects. I shall present two quotations from very recent articles.

The first is by Pace[154]:

> After 50 years, Kauzmann's seminal review[155] is still the most important paper published on protein stability. He showed convincingly that the hydrophobic effect is the dominant force in protein folding.

This contention is not true. Kauzmann had a brilliant idea of using data on solvation of non-polar molecules in water to infer the *probable* driving force in protein folding. Kauzmann was cautious enough to say: "The hydrophobic bond is probably one of the more important factors involved in stabilizing the folded configuration ..." This is very far from "convincing" evidence in favor of the "dominant force in protein folding."

In another review, Dill *et al.*[156] summarized the evidence in favor of $H\phi O$ interactions:

> There is considerable evidence that hydrophobic interactions must play a major role in protein folding. (a) Proteins have hydrophobic cores, implying non-polar amino acids are driven to be sequestered from water. (b) Model compound studies show 1–2 kcal/mol for transferring a hydrophobic side chain from water into oil-like media,

---

[154]Pace (2009).
[155]Kauzmann (1955).
[156]Dill *et al.* (2008).

and there are many of them. (c) Proteins are readily denatured in non-polar solvents. (d) Sequences that are jumbled and retain only their correct hydrophobic and polar patterning fold to their native states ..., in the absence of efforts to design packing, charges or hydrogen bonding.

Unfortunately, *none* of these can be used as *evidence* in favor of $H\phi O$ interaction. (a) The fact that $H\phi O$ groups are found in the interior of the protein does not necessarily mean that the interactions are *responsible* for bringing these groups to the interior of the protein (see below). (b) The Gibbs energies of transfer of small model compounds from water to an oil-like media are irrelevant to the driving force in protein folding. This is discussed in many places throughout the book and in Appendix H. (c) The fact that proteins are readily denatured in non-polar solvents means that water is important. It says nothing on the relative importance of $H\phi O$ vis-à-vis $H\phi I$ effects. (d) The last piece of "evidence" says nothing on the role of water, certainly not about the relative importance of the $H\phi O$ vis-à-vis the $H\phi I$ effect.

Such non-evidence appears in the literature almost twenty years after strong evidence in favor of the $H\phi I$ effects was published!

I would like to add one comment on the argument which is often given in the literature, that the mere fact that $H\phi O$ groups are found in the interior of the proteins proves the importance of the $H\phi O$ effect. Such a conclusion is an illusion and cannot be supported by theory. It is similar to the conclusion that the mixing of two ideal gases is the *cause* of the entropy increase upon mixing, or that the "entropy of mixing" is the "driving force" for the process of mixing.[157] Similarly, one cannot say anything about the "driving force" for protein

---

[157]See Ben-Naim (2008).

folding merely by observing the $H\phi O$ groups occupying the interior of the protein.[158] Some specific examples are discussed in Sec. 4.4. Sometimes, one can find in the biochemical literature the "explanation" of the hydrophobic effect in terms of the "iceberg-formation." The fallacy of this "explanation" has been discussed in Chapter 4 of Part I of this series.

The conclusion reached in this chapter regarding the relative importance of $H\phi I$ effects has some overlaps with the conclusion reached by Rose *et al.*[159] Rose *et al.* proposed an inversion of the "side chain/backbone paradigm." We advocate the inversion of the $H\phi O/H\phi I$ paradigm. Clearly, since most of the $H\phi I$ groups are provided by the backbone, it follows that the role of the backbone should be more important than the role of the side chains. The backbone is important not because of its hydrocarbon chain, but because of the $H\phi I$ groups attached to it. Therefore, the two proposals are quite different. The $H\phi O/H\phi I$ inversion is far more general — it provides powerful and highly specific effects.

3.6. The Problem of the Preferential Protein Folding Pathways of Proteins

In the previous sections of this chapter, we discussed the *factors* that contribute to the stability of the 3-D structure of a globular protein.

We now turn to discuss the *factors* that determine the preferential pathways along which a protein folds. We shall discuss only proteins that are soluble in water, and only those that fold spontaneously without the assistance of a chaperon (or a chaperonin).

---

[158]Jesior (2000).
[159]Rose *et al.* (2006). See also Ben-Naim (2009).

We stressed the word *factors* in the two problems because there is some confusion in the literature regarding the nature of these factors. In the case of the stability of proteins discussed in the previous sections, we focused on the Gibbs energy change for the folding process. This can loosely be referred to as a "thermodynamic driving force," but it is not a *force* in the strict physical sense. When "dominant forces in protein folding" are discussed,[160] this almost always means "thermodynamic driving forces," not actual forces!

In this and the following sections, the *factors* involved in the actual process of protein folding are the *real physical forces*. These forces are the ones that guide the protein in folding along a narrow range of pathways, leading to the native form in a relatively short time. Before we continue with the discussion of the forces in protein folding, another caveat: There is also some confusion regarding the questions of "how" and "why" a protein folds in a particular range of pathways. In an excellent review titled "The Problem of How and Why Proteins Adopt Folded Conformations,"[161] Creighton surveys the progress that has been made in "elucidating the *non-random* pathways by which such folded conformations are acquired." In most of this review, Creighton discusses what is known about the *intermediates* in the process of folding. Knowledge of intermediate structures along the folding pathways answers the question of *how* a protein folds, i.e. through which steps the protein moves from the unfolded form to the final native, folded structure. However, this information does not tell us anything about the *forces* that cause the protein to fold along these particular pathways, i.e. it does not answer the question of *why* the protein folds along these particular pathways.

---

[160]Dill (1990). See also Appendix T.
[161]Creighton (1985).

Admittedly, the two questions "Why" and "How" are intimately related. If we knew the (actual) force exerted on each atom of the protein at each moment, we could, in principle, tell in which direction it is most likely to move. Here, the "direction" means the most likely, or the most probable direction (see also Sec. 1.2). If we knew the most probable "direction of motion," then we should also know the pathways along which the protein folds, hence, we should also know all the intermediate conformations.

On the other hand, the knowledge of a few intermediate conformations does not reveal the forces that caused the protein to proceed along these intermediates. The knowledge of many intermediates (in principle infinite) can at best provide only some hints to answering the question "Why."

Why is the question "Why" so important in protein folding?

To answer this question, we start with Anfinsen's classical work on the renaturation of ribonuclease.[162] Anfinsen found that a denatured protein will fold spontaneously when the proper environment for folding is restored (e.g. lowering the temperature or removing a denaturating agent). The folding occurs spontaneously without the need for any additional information *beyond that which is contained in the sequence of amino acids.*

Thus, the *information* required for the folding is somehow inscribed in the sequence of the amino acids. Without elaborating on the nature of the *information* contained in the sequence of amino acids, it is clear that it comes in the form of *instructions.* These instructions cannot be executed unless the proper environment exists.

The situation may be likened to a book containing all the information on how to construct a certain machine. Such a machine will not be constructed spontaneously from the information contained in the book. In order to construct such

---

[162]Anfinsen (1973).

a machine, the information in the book must be *read* and the listed instructions executed step by step. Note that the fact that "information" on the machine is contained in the book does not imply that there exists a "code" that translates a sequence of letters or words into the structure of the machine. The information contained in the book is in the form of a "list of instructions" on how to construct a specific machine.

Likewise, the "information" contained in the sequence of amino acids must be read, and the instructions executed so that the "protein machine," or the native protein, can be constructed. The agent that does that job is the "proper environment," and the most important component in this environment is water.

It is believed that water not only "reads" the information contained in the sequence of amino acids but also translates the instruction into executable orders.[163] These "orders" are the *forces* that are exerted on each of the atoms of the protein that cause the motion of the entire protein towards the end product. Because of the statistical character of these forces, the motion of the protein is not along a unique, deterministic route, but more likely along a narrow range of routes or pathways.

Thus, the information is contained in the sequence, but in order to use this information, water molecules have to *read* the information and execute the instructions. It should be emphasized, however, that the answer to the question of why a protein folds as given by Anfinsen is different from the answers we shall be discussing in the following sections. Anfinsen discussed the overall *thermodynamic* force that drives the protein to fold into the native structure. In the following sections, we shall be interested in the more detailed question of why the protein folds

---

[163]Traditionally, the dynamics of the protein folding process are studied by postulating some kind of kinetic mechanism involving rate constants for the various transformations. Such an approach can describe the overall dynamics of the process, but it does not provide an insight into the microscopic events occurring in the process of protein folding.

in a particular range of pathways. To answer this question, we need to know the actual forces exerted on each atom or group of the protein at each stage of the process of folding.

3.7. Energy Landscapes, Gibbs Energy Landscapes and Forces in Protein Folding

Most reviews on protein folding focus on the thermodynamic "driving forces" rather than the *forces* themselves.[164] This section is devoted to the analysis of the types of forces acting on the protein and, more specifically, on the *solvent-induced forces in protein folding*.

Let $G(\mathbf{R}^M)$ be the Gibbs energy of a system of N water molecules at a given temperature T and pressure P, and a single protein molecule being at a specific conformation denoted by $\mathbf{R}^M = (\mathbf{R}_1, \mathbf{R}_2, \mathbf{R}_3, \ldots, \mathbf{R}_M)$, where $\mathbf{R}_i$ is the locational vector of atom i, or group i of the protein. It is convenient to choose the groups, such as methyl, ethyl, hydroxyl, carbonyl, etc. rather than each atom separately to describe the conformation of the protein.

The statistical mechanical expression for the Gibbs energy of such system is

$$\exp[-\beta G(\mathbf{R}^M)] = C \int dV \exp[-\beta V] \int d\mathbf{X}^N$$

$$\times \exp[-\beta U(\mathbf{R}^M, \mathbf{X}^N)]. \qquad (3.7.1)$$

Here, $\beta = (k_B T)^{-1}$, and C is a constant having the dimensions of length to the power $-3(N + 1)$. This is necessary in order to render the right-hand side of (3.7.1) dimensionless. The integral (without the constant C) has $3N$ dimensions of length (3 per water molecule), and the three dimensions of length due

---

[164]Dill (1990).

to integration over the volume of the system. For all practical purposes, we do not need to specify the constant C since it will always cancel out, either when we take the difference in Gibbs energies (in calculating thermodynamic "driving forces") or when we take the gradient of the Gibbs energy (in calculating the actual forces).

The quantity $U(\mathbf{R}^M, \mathbf{X}^N)$ in (3.7.1) is the total interaction energy between all the molecules in the system being at a fixed configuration. We write this as

$$U(\mathbf{R}^M, \mathbf{X}^N) = U(\mathbf{R}^M) + U(\mathbf{X}^N) + B(\mathbf{R}^M, \mathbf{X}^N), \qquad (3.7.2)$$

where $U(R^M)$ is the internal potential energy of the protein being at a specific conformation $\mathbf{R}^M$. $U(\mathbf{X}^N)$ is the total interaction energy among all water molecules being at a fixed configuration $\mathbf{X}^N = (\mathbf{X}_1, \mathbf{X}_2, \ldots, \mathbf{X}_N)$, and $B(\mathbf{R}^M, \mathbf{X}^N)$ is the binding energy of the protein, i.e. the interaction between the protein at $\mathbf{R}^M$ and the solvent molecules at $\mathbf{X}^N$. These three kinds of interactions are shown schematically in Fig. 3.14.

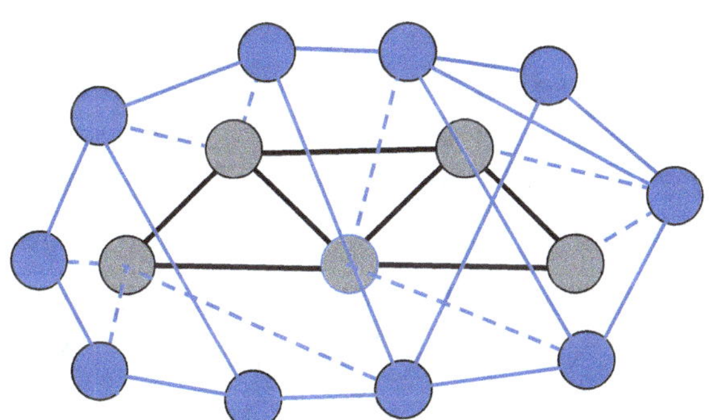

Fig. 3.14 The three types of interactions in a system of five solute molecules (in grey), and ten solvent molecules (in blue). The solute–solute interactions are shown in black lines, the solvent–solvent interactions in blue lines and the solute–solvent interactions in dashed blue lines. Not all interaction lines are shown. See Eq. (3.7.2).

For any change of the conformation of the protein, say, from U_1 to F_1, we have[165]

$$\Delta G(U_1 \rightarrow F_1) = G(F_1) - G(U_1)$$

$$= \Delta U(U_1 \rightarrow F_1) + \delta G(U_1 \rightarrow F_1)$$

$$= \Delta U(U_1 \rightarrow F_1) - k_B T \ln \frac{\langle \exp[-\beta B(F_1)] \rangle_0}{\langle \exp[-\beta B(U_1)] \rangle_0},$$

$$(3.7.3)$$

where on the right-hand side in (3.7.3), we have averages over all the configurations of the solvent molecules in the T, P, N ensemble. Note that by taking the difference in Gibbs energies, the constant C has been canceled out.

From here on, we shall be interested not in a finite change of conformation as in (3.7.3), but in an infinitesimal change in the conformation. The reason for refocusing our interest will be clear from the following considerations: The difference in the Gibbs energy as written in (3.7.3) is sometimes referred to as the "driving force" for the process (here, from U_1 to F_1). This "driving force" gives us the overall thermodynamic force to proceed from U_1 to F_1, but this "driving force" is not a *force*. In other words, knowing for example that $\Delta G(U_1 \rightarrow F_1)$ is negative does not tell us that the system will necessarily *move* from U_1 towards F_1, nor does it tell us in which *direction* the protein will move when it is released at the conformation U_1.[166]

To examine the direction (more precisely, the probable direction, see below) in which the protein will move, we need to

---

[165] Note that the letter U is used here for the potential energy and for the unfolded species.

[166] Here, by "releasing," we refer only to the change in the conformation of the protein. We still assume that the center of mass is at a fixed position. We are not interested in the *motion* of the entire protein but only in the internal changes in the relative locations of the groups comprising the protein. In Appendix N, we also discuss the difference between the "driving force" when acting on a macroscopic system and when acting on a microscopic system.

know the forces acting on each of the M groups of the protein being at the conformation $\mathbf{R}^M$. This force is obtained by taking the gradient of the Gibbs energy with respect to each of the $\mathbf{R}_i$.

In Sec. 1.2, we have seen that the force on, say, group number 1 at $\mathbf{R}_1$, is obtained by taking the gradient of the Gibbs energy with respect to $\mathbf{R}_1$. The same is true for the case of a protein. We take an infinitesimal change in $\mathbf{R}_1$, keeping all other $\mathbf{R}_i(i \neq 1)$ unchanged and averaging over all the configurations of the solvent molecules. Thus,

$$-\mathbf{F}(\mathbf{R}_1)d\mathbf{R}_1 = G(\mathbf{R}_1 + d\mathbf{R}_1) - G(\mathbf{R}_1)$$
$$= [\nabla_1 U(\mathbf{R}^M) + \nabla_1 \delta G(\mathbf{R}^M)]d\mathbf{R}_1. \quad (3.7.4)$$

Note again that here, $\mathbf{R}_1$ is a vector (x_1, y_1, z_1) and $d\mathbf{R}_1$ is an infinitesimal vector (dx_1, dy_1, dz_1). This is different from the notation we use when we integrate over all locations where $d\mathbf{R}$ is an infinitesimal volume $d\mathbf{R} = (dxdydz)$. Note also that in (3.7.4), G is a function of $\mathbf{R}^M$, but we have omitted from the notation all the $\mathbf{R}_i$ which do not change in this process.

Equation (3.7.4) is essentially the same as the equation for the force discussed in Sec. 1.2. The only difference is that we have M particles (or groups) at some fixed configuration $\mathbf{R}^M$, and we are interested in the *force* on one of these groups, given a fixed conformation of the protein but averaged over all configurations of the solvent molecules.

The first gradient on the right-hand side of (3.7.4) is simply the force acting on group 1 being at $\mathbf{R}_1$,[167] which results from all other groups of the protein. In Sec. 1.2, we dealt with the force exerted by one particle on the other. In Chapter 4, we shall also deal with the force exerted by one protein on the other. Here, however, we deal with *intra*-molecular force, i.e.

---

[167] The first term $\nabla_1 U(\mathbf{R}^M)d\mathbf{R}_1$ is the Gibbs energy change associated with the small displacement of group 1 by $d\mathbf{R}_1$.

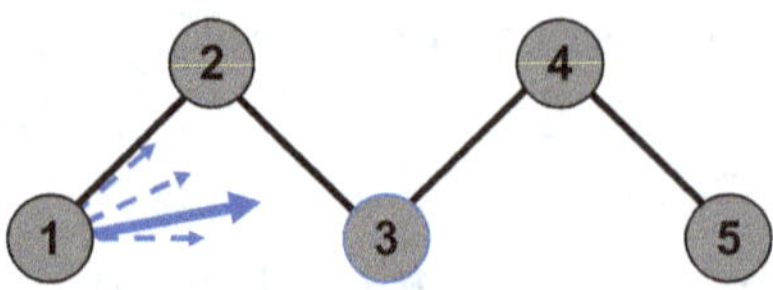

Fig. 3.15 Schematic illustration of the resultant direct force (blue arrow) acting on group 1, by all other groups in the molecule (direct forces exerted by distant groups are weak and may be ignored).

the resultant force exerted by all groups of a single protein on one particular group, say, group 1. This force is schematically shown in Fig. 3.15.

The second gradient on the right-hand side of (3.7.4) is the *solvent-induced force* on group 1, given a fixed conformation of the protein and averaged over all possible configurations of the solvent molecules.

As we have noted in Sec. 1.2, the potential function $U(\mathbf{R}^M)$ is the relatively "easier" part of the thermodynamic "driving force." This is what may be properly described by the term *energy landscape*. It is a multi-dimensional landscape. If we know the landscape, we also know the gradients, and hence the forces on each group at each "point" $\mathbf{R}^M$ on the landscape.

The more difficult part of the landscape is the term δG, which together with $U(\mathbf{R}^M)$ may be referred to as the *free energy landscape*, or better, the Gibbs energy landscape $G(\mathbf{R}^M)$.

For any fixed conformation $\mathbf{R}^M$ of the protein, we can write the equality (see Fig. 3.16)

$$\delta G(\mathbf{R}^M) = \Delta G^*(\mathbf{R}^M) - \Delta G^*(\infty), \qquad (3.7.5)$$

where $\Delta G^*(\mathbf{R}^M)$ is the Gibbs energy of solvation of the protein at the conformation $\mathbf{R}^M$, and $\Delta G^*(\infty)$ denotes the Gibbs energy of solvation of all the M groups when they are at infinite separation from each other. Since the latter is not a function of

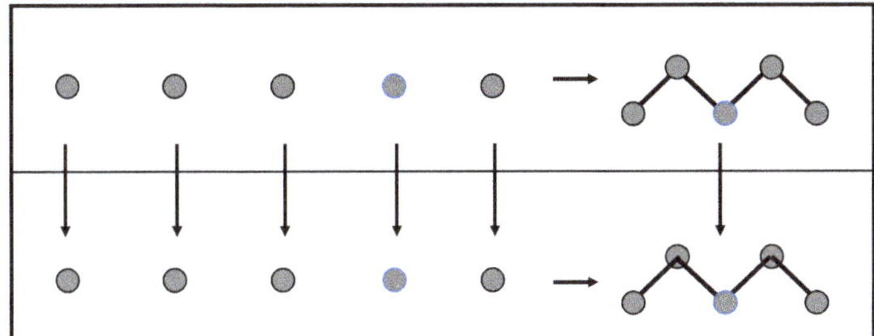

Fig. 3.16   The process of bringing M solute molecules or groups to the final configuration of the protein, in the gaseous and in the liquid phases. The indirect part of the thermodynamic driving force is related to the difference in the solvation Gibbs energies [see Eq. (3.7.5)].

$\mathbf{R}_1$, we can write the solvent-induced force as

$$\mathbf{F}_1^{SI}(\mathbf{R}_1) = -\nabla_1 \delta G(\mathbf{R}^M) = -\nabla_1 \Delta G^*(\mathbf{R}^M). \qquad (3.7.6)$$

Thus, the solvent-induced force is determined by the gradient of the solvation Gibbs energy of the protein at a given conformation $\mathbf{R}^M$.

From (3.7.6), we can make the following general statement: For any given conformation $\mathbf{R}^M$ of the protein, a solvent-induced force will be exerted on group i, if and only if a small change in $\mathbf{R}_i$ causes a change in the solvation Gibbs energy of the protein (keeping T, P, N and $\mathbf{R}_j(j \neq i)$ constant).

We can further simplify (3.7.6) if we assume that the internal degrees of freedom of the protein at $\mathbf{R}^M$ are not affected by the solvation process (see also Appendix B). In such a case, we can write the solvation Gibbs energy as

$$\Delta G^*(\mathbf{R}^M) = \mu^{*l}(\mathbf{R}^M) - \mu^{*ig}(\mathbf{R}^M), \qquad (3.7.7)$$

where μ^{*l} and μ^{*ig} are the pseudo-chemical potentials of the protein at a specific conformation $\mathbf{R}^M$ in the liquid and in an ideal gas phase, respectively. Since both μ^{*l} and μ^{*ig} include the internal partition functions of the protein (rotational, vibrational, electronic, but not the internal rotations), and since these are assumed to be unaffected by the solvation process, they will cancel out in the difference on the right-hand side of (3.7.7). What remains is only the coupling work of the protein to the solvent:

$$\Delta G^*(\mathbf{R}^M) = -k_B T \ln\langle \exp[-\beta B(\mathbf{R}^M)]\rangle_0. \qquad (3.7.8)$$

Hence, the solvent-induced force on group 1 may be rewritten as

$$\mathbf{F}_1^{SI}(\mathbf{R}_1) = -\nabla_1 \Delta G^*(\mathbf{R}^M)$$

$$= k_B T \nabla_1 \ln\langle \exp[-\beta B(\mathbf{R}^M)]\rangle_0. \qquad (3.7.9)$$

From (3.7.9), it is difficult to see the various factors that contribute to the solvent-induced force. A more useful expression which is also easier to analyze and interpret is the following:

$$\mathbf{F}_1^{SI}(\mathbf{R}_1) = \int [-\nabla_1 U(\mathbf{R}_1, \mathbf{X}_W)]\rho(\mathbf{X}_W/\mathbf{R}^M)d\mathbf{X}_W. \qquad (3.7.10)$$

The derivation of this equation is given in Appendix C. The reader should notice that this is essentially the same expression as in Sec. 1.2, only the condition $\mathbf{R}^M$ in (3.7.10) replaces the configuration $\mathbf{R}_1, \mathbf{R}_2$ in Eq. (1.2.17).

Before we analyze the implications of Eq. (3.7.10) on the characterization of the different solvent-induced forces acting on the protein, it is advisable to pause and reflect on the difference between the direct and the indirect forces acting on a simple molecule, such as butane (Fig. 3.17).

Suppose we view a butane molecule as consisting of four spheres linked by chemical bonds. We assume that the bond lengths and the bond angles are fixed. The change in the conformation of this molecule is a result of the rotation about the bond connecting groups 2 and 3. Let us denote by ϕ the dihedral

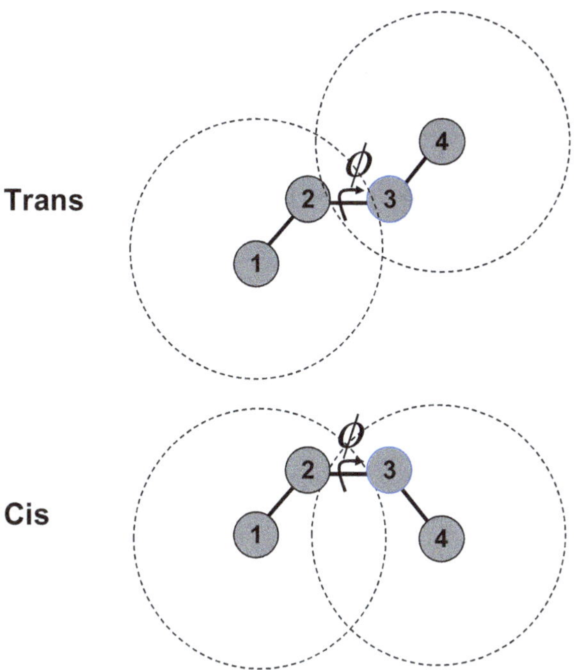

Fig. 3.17 Trans and cis configurations of butane. The energy landscape can be described by either $U(\phi)$ or $U(R_{14})$.

angle of this internal rotation; we choose $\phi = 0$ for the trans configuration and $\phi = \pi$ for the cis configuration (Fig. 3.17).

First, we discuss the force on group 1 for different configurations of the molecule. As we change the dihedral angle ϕ, the only changes in the intramolecular interactions are due to the change in the distance between groups 1 and 4. In the trans configuration ($\phi = 0$), the distance is maximal. As we rotate the molecule, there is a change in $U(\mathbf{R}_{1,4})$. Therefore, there is a direct force acting on group 1 (This is actually torque. Because of the fixed bond lengths and angles, this force will only cause a rotation about the 2–3 bond). Thus, in this example, the potential energy as a function of ϕ is equivalent to the function $U(R_{1,4})$, where $R_{1,4}$ is the distance $|\mathbf{R}_1 - \mathbf{R}_4|$.

The situation is quite different for the solvent-induced force. Here, a solvent-induced force is operative only when the solvation Gibbs energy of the molecule changes. Specifically, we may assume that in the trans configuration, the two groups 1 and 4 are independently solvated. As illustrated in Fig. 3.17, this is equivalent to saying that the *solvation spheres* around groups 1 and 4 do not overlap.[168,169]

It is intuitively clear that as we make a small change in ϕ around $\phi = 0$, the distance $|\mathbf{R}_1 - \mathbf{R}_4|$ changes, but the solvation spheres about 1 and 4 will not change. Therefore, we cannot expect any solvent-induced force at around $\phi = 0$. However, once the angle ϕ becomes larger, say, near the gauche configuration ($\phi \approx \pi/3$ or $\phi \approx 2\pi/3$), the solvation spheres about the two groups 1 and 4 will overlap. Therefore, we should find a solvent-induced force at these configurations.[170] Thus, for butane, we can describe the energy landscape simply as the function $U(\phi)$, or equivalently, $U(R_{14})$. The Gibbs energy landscape can also be described as $G(\phi)$ or $G(R_{14})$.[171]

Describing the energy landscape and the Gibbs energy landscape is much more complicated for pentane molecules. Again, assuming that all the bond lengths and bond angles are fixed, the energy of the molecule will change when we rotate the molecule

---

[168]This is similar to the discussion in Sec. 1.2. As we move the two particles from infinity towards each other, the solvation of each molecule does not change, hence there is no solvent-induced force. Once we reach a distance when the solvation spheres of the two particles overlap, we shall also see a solvent-induced force.

[169]Note also that the relatively small region of the two "solvation spheres" in the trans configuration is too close to 2 and 3, hence, it contains no solvent molecules.

[170]For some calculated changes in the distribution of the different configurations of butane in water, see Ben-Naim (2006).

[171]It should be noted that the *energy* landscape is unique to the molecule, which is butane in this case. On the other hand, the Gibbs energy landscape depends on T and P (and the solvent composition). Therefore, one cannot discuss the "Gibbs energy landscape" without specifying the temperature, the pressure and the composition of the solvent.

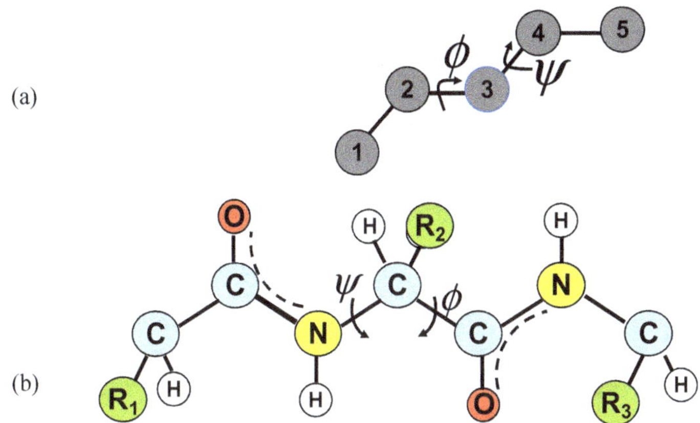

Fig. 3.18 (a) The pentane molecule. Here, the energy landscape may be described by the function $U(\phi, \psi)$; (b) Two rotational angles ϕ and ψ in a peptide bond.

about 2–3 and 3–4 bonds (Fig. 3.18a). In this case, we can describe the potential function either as a function of two angles $U(\phi, \psi)$, or as a function of the two distances $U(R_{14}, R_{15})$, where $R_{14} = |\mathbf{R}_1 - \mathbf{R}_4|$ and $R_{15} = |\mathbf{R}_1 - \mathbf{R}_5|$. This can still be drawn in a 3-D graph. Once we have a larger polymer, it becomes impossible to draw the energy landscape function. However, we can still think of the function $U(\mathbf{R}^M)$, or equivalently the function $U(\phi_1, \psi_1, \phi_2, \psi_2, \ldots, \phi_M, \psi_M)$, in a high dimensional space. In Fig. 3.18b, the two rotational angles in each peptide unit are shown.[172]

Returning to Eq. (3.7.10), we see that the integrand contains two factors [note that the gradient ∇_1 operates only on the potential function $U(\mathbf{R}_1, \mathbf{X}_W)$, not on $\rho(\mathbf{X}_W/\mathbf{R}^M)$]. The first is the force exerted on group 1 by a water molecule at $\mathbf{X}_W = (\mathbf{R}_W, \mathbf{\Omega}_W)$. The second is the conditional density of water molecules at $\mathbf{X}_W$ given the protein at conformation $\mathbf{R}^M$.

---

[172]Note that M in $\mathbf{R}^M$ is the number of atoms or groups, whereas M in $\phi_M . \psi_M$ is the number of amino acids in the protein.

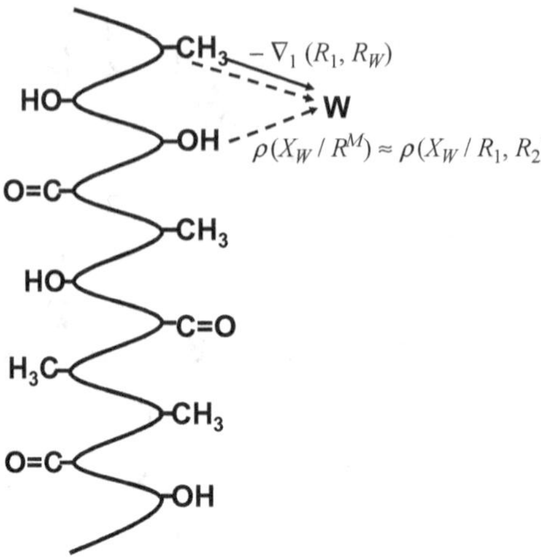

Fig. 3.19 The direct force acting on group 1 (here methyl) by a water molecule at $\mathbf{X}_w$ (full arrow). The conditional density of water at $\mathbf{X}_w$ is mainly due to groups 1 and 2 (dashed arrows).

Figure 3.19 shows schematically the force $-\nabla_1 U(\mathbf{R}_1, \mathbf{X}_W)$ with a full line, and the effect of all the groups of the protein on the density of water molecule at $\mathbf{X}_W$ with a dashed line. The quantity $\rho(\mathbf{X}_W/\mathbf{R}^M)d\mathbf{X}_W$ may also be interpreted as the conditional probability of finding a water molecule within the element of "volume" $d\mathbf{X}_W$ given the conformation $\mathbf{R}^M$.

Had we only a single water molecule at $\mathbf{X}_W$, the force exerted on group 1 would be $-\nabla_1 U(\mathbf{R}_1, \mathbf{X}_W)$. This force would have caused the motion of group 1 in the direction of the vector $-\nabla_1 U(\mathbf{R}_1, \mathbf{X}_W)$. However, we do not have a single water molecule at $\mathbf{X}_W$, but only the probability density of finding a water molecule at $\mathbf{X}_W$, as denoted by $\rho(\mathbf{X}_W/\mathbf{R}^M)$, which when integrated over all possible $\mathbf{X}_W$ gives an average *solvent-induced force*. As we have discussed in Sec. 1.2, this is a genuine physical force. However, since this force is an *average force*, we cannot predict the instantaneous motion of group 1. The only meaning we can ascribe to this average force is probabilistic. In fact, there are two equivalent probabilistic interpretations of this average force.

(a) The first arises from the general relationship between the Gibbs energy $G(\mathbf{R}^M)$ and the probability of finding the conformation $\mathbf{R}^M$ in a system where the protein is free to move and rotate in the solution. This relationship is simply

$$\Pr(\mathbf{R}^M) = C \exp[-\beta G(\mathbf{R}^M)], \qquad (3.7.11)$$

where C is a constant independent of the conformation $\mathbf{R}^M$. This relationship is very general. It implies that the Gibbs energy landscape may be converted to a probability landscape. A maximum in $G(\mathbf{R}^M)$ is translated into a minimum in $\Pr(\mathbf{R}^M)$, and vice versa.[173] The gradient of $G(\mathbf{R}^M)$ is also related to the gradient of the probability function

$$\nabla_1 \ln \Pr(\mathbf{R}^M) = -\beta \nabla_1 G(\mathbf{R}^M). \qquad (3.7.12)$$

Thus, looking at the Gibbs energy landscape and examining the gradients to determine the forces at each "point" $\mathbf{R}^M$ is equivalent to looking at the probability landscape and examining the gradients at each point $\mathbf{R}^M$. Therefore, searching for a dominant *force* in the Gibbs energy landscape is equivalent to searching for the steepest uphill points in the probability landscape.

(b) For the second interpretation of average force, we have to go back to the foundation of statistical mechanics, where we have replaced the time average with an ensemble average. Equation (3.7.10) is already an ensemble average, obtained after the integrations over all the momenta of the solvent molecules have been carried out. In reality, we do not *measure* an ensemble average, but rather, a time average. The real picture is that during

---

[173] One should be careful to specify the parameters with respect to which the Gibbs energy has a minimum. See also Appendix N.

the time one is measuring the force (or the resulting motion) on group 1, the group is bombarded many times by water molecules from different directions. At each instance, there exists a genuine instantaneous force and, by Newton's law, also an instantaneous resulting motion. However, since these bombardments occur on an extremely short time scale, we cannot observe the effect of the instantaneous forces, or the instantaneous motion. All that we observe is the *accumulative* results of the myriad bombardments, and the myriad resulting motions which we observe on a much larger scale of time.

When we follow the change of the conformation of the protein as we change the environment, we actually see the net results of the myriad small forces exerted on all the groups of the protein. This internal motion of the protein molecule is similar to the translational motion of small particles (such as colloids or pollen particles) known as Brownian motion.[174] In both cases, the incessant bombardment of the solvent molecules causes small displacements of the groups of the protein, leading to an overall observable motion or change of conformation.

Returning to the total force in (3.7.4), or to the solvent-induced part (3.7.10), we can say that whenever there exists a strong gradient, there is a strong genuine force. However, a strong force does not imply an instantaneous motion in the direction of the force. All we can say is that there is a certain probability that within the time scale of our measurements, the group or the particle will move in that direction. The stronger the force, the higher the probability of movement in that direction.[175]

So far, we have discussed the force acting on one specific group of the protein. Clearly, at any instance of time, there are

---

[174]Mazo (2002).

[175]Note again that the direct force, i.e. the first term on the right-hand side of (3.7.4), is not an average force. The discussion here applies to either the indirect force, or the total force, which is an average force.

forces exerted on each of the groups of the protein, leading to a concerted motion of the entire molecule from one conformation to another. It is this change in the conformation that we record in the figures such as Figs. 3.1 and 3.2.

3.8. What Kind of Forces are Exerted on the Protein in the Process of Protein Folding?

In this section, we return to the model protein where we have a hydrocarbon chain with two kinds of side chains: Methyl groups representing the $H\phi O$ groups, and hydroxyl groups representing the $H\phi I$ groups. We shall also focus on the solvent-induced part of the force, Eq. (3.7.10), presuming that the direct force is known.

First, note that the integrand in (3.7.10) is a product of two factors: A force and a density. In order to have a significant contribution to the integral, both factors in the integrand must not be too small. That means that the actual force $-\nabla_1 U(\mathbf{R}_1, \mathbf{X}_W)$ must not be too small, and at the same time, the conditional density at $\mathbf{X}_W$ given $\mathbf{R}^M$ must also not be too small. Therefore, the integration in (3.7.10) is effectively only over that region of space where both factors in the integrand are not negligible.

As an example of regions that do not contribute to the integrand, suppose that $\mathbf{R}_W$ is far from $\mathbf{R}_1$, i.e. the distance $|\mathbf{R}_1 - \mathbf{R}_W|$ is such that $U(\mathbf{R}_1, \mathbf{X}_W)$ is almost zero, hence the force is also almost zero (region A in Fig. 3.20). Next, suppose that $\mathbf{R}_W$ is close to $\mathbf{R}_1$, so that the direct force $-\nabla_1 U(\mathbf{R}_1, \mathbf{X}_W)$ is large, but too close to $\mathbf{R}_2$ or $\mathbf{R}_3$, such that the conditional density $\rho(\mathbf{X}_W/\mathbf{R}^M)$ will be very small (region B in Fig. 3.20). Again, this region will make a negligible contribution to the integrand. Another case is when $\mathbf{R}_W$ is very far from the protein, in which case both the force and the conditional density at $\mathbf{X}_W$ will be negligibly small.

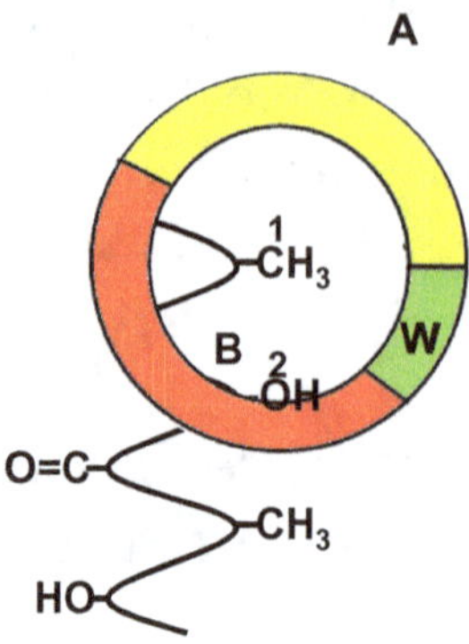

Fig. 3.20 Regions around group 1 (here, methyl) where we expect negligible indirect force (Regions A and B). The colored regions are discussed in Sec. 3.8.

Thus, in order to get a significant average force, we need to have a solvent molecule in a configuration such that the direct force $-\nabla_1 U(\mathbf{R}_1, \mathbf{X}_W)$ is large, and at the same time the local density $\rho(\mathbf{X}_W/\mathbf{R}^M)$ is large too. These requirements for a non-negligible integrand place some constraints on the groups of the proteins near $\mathbf{R}_1$ that can affect the conditional density $\rho(\mathbf{X}_W/\mathbf{R}^M)$. To see this, suppose we start with the fully extended protein. We are interested in the average force on, say, group 1, given the conformation $\mathbf{R}^M$. Clearly, we can only take into account a small region around group 1 (the methyl group in Fig. 3.20).

The first condition for $\mathbf{X}_W$ (or rather for $\mathbf{R}_W$, which is part of the vector $\mathbf{X}_W$), is that it must be in some range of distances from $\mathbf{R}_1$ at which the force $-\nabla_1 U(\mathbf{R}_1, \mathbf{X}_W)$ is large and attractive. We can therefore draw a spherical shell around $\mathbf{R}_1$ in which a water molecule can exert a non-negligible force on group 1[176] (see ring around the methyl group in Fig. 3.20).

---

[176]Note that there is also a region where the water molecule exerts strong repulsive forces on group 1, but this region is of no interest to us because the density of water will be too small at this region.

Thus, a water molecule being in this spherical shell is a *necessary* condition for having a non-negligible *average* force $-\nabla_1 U(\mathbf{R}_1, \mathbf{X}_W)$, but is not a sufficient condition. In order to have a non-negligible average force, the water molecule must be in the spherical shell around $\mathbf{R}_1$ *and* in regions that maximize the conditional density $\rho(\mathbf{X}_W / \mathbf{R}^M)$. In Fig. 3.20, we divide schematically the spherical shell in which $-\nabla_1 U(\mathbf{R}_1, \mathbf{X}_W)$ is large into three regions. The red region is the one from which a water molecule is excluded because of the repulsive forces exerted by the protein (specifically groups 1 and 2). In the yellow region, the density of water is affected by group 1 but not by group 2. In the green region, the density of water is affected both by groups 1 and 2. We therefore expect that this region will contribute the most to the integral (3.7.10).

Note also that groups which are far away from groups 1 and 2 (e.g. groups 3, 4, etc. in Fig. 3.20) do not affect the density of water molecules in the important region which is indicated by green.[177]

The discussion above is, of course, very qualitative. Let us now examine the different kinds of forces for the same model protein and the same conformation as in Fig. 3.20, but more quantitatively. To do this, we first assume that the bulk density of water is very low. In this case, one can write the conditional density in (3.7.10) as[178]

$$\rho(\mathbf{X}_W / \mathbf{R}^M) = \rho_W g(\mathbf{X}_W, \mathbf{R}^M) = \rho_W \exp[-\beta U(\mathbf{X}_W, \mathbf{R}^M)]$$

$$= \rho_W \exp\left[-\beta \sum_{i=1}^{M} U(\mathbf{X}_W, \mathbf{R}_i)\right]. \tag{3.8.1}$$

---

[177]This is true for the specific conformation depicted in Fig. 3.20. For other conformations, many more groups can affect the density around 1. See also Chapter 4 and Appendix S.

[178]For details on the form of the pair correlation at low densities, see Ben-Naim (2006).

In (3.8.1), we have written the conditional density in terms of the bulk density of water ρ_W and the pair correlation function between the protein at $\mathbf{R}^M$ and water. In the limit of very low water density, the pair correlation function is simply related to the pair potential $U(\mathbf{X}_W, \mathbf{R}^M)$. Furthermore, we can assume that the protein–water pair potential is pairwise additive, i.e.

$$U(\mathbf{X}_W, \mathbf{R}^M) = \sum_{i=1}^{M} U(\mathbf{X}_W, \mathbf{R}_i), \qquad (3.8.2)$$

where $U(\mathbf{X}_W, \mathbf{R}_i)$ is the pair interaction between a water molecule at $\mathbf{X}_W$ and a group at $\mathbf{R}_i$. The last form on the right-hand side of (3.8.1) clearly shows that the maximal density of water molecules is expected in regions where water molecules are strongly attracted by the protein. However, since the water molecule must be within the region of the spherical shell around $\mathbf{R}_1$ (to produce a large force), the only groups that will contribute to the sum in (3.8.2) are those that are near group 1. For instance, groups 4 or 5 will contribute a very small interaction energy $U(\mathbf{X}_W, \mathbf{R}_4)$, $U(\mathbf{X}_W, \mathbf{R}_5)$ to the sum (3.8.2), hence also to the local density at $\mathbf{R}_W$.

We conclude that in order to have a large contribution to the integral, we need the two factors in the integrand to be non-negligible. Furthermore, we can focus only on the effects of groups near group 1 and ignore any group of the protein which is too far away from group 1, and therefore does not have any effect on the density of water molecules in the green region of Fig. 3.20. We can therefore reduce the problem of classification of the different forces to the case of a model compound having only two groups 1 and 2 as shown in Fig. 3.21.[179]

---

[179]Note carefully that the choice of groups 1 and 2 is a result of our choice of the fully extended conformation of the protein. For any other conformation, we can repeat the same argument as above, but the groups that will affect the total density at $\mathbf{R}_W$ will be different for different conformations of the protein.

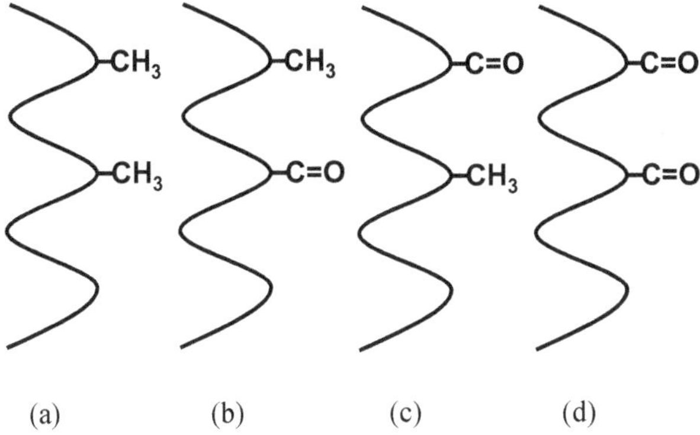

Fig. 3.21   The four possible pairs of functional groups: (a) $H\phi O - H\phi O$, (b) $H\phi O - H\phi I$, (c) $H\phi I - H\phi O$ and (d) $H\phi I - H\phi I$.

Thus, for a model compound having only two groups 1 and 2 and in the limit of low density of the water, the solvent-induced force in (3.7.10) is reduced to

$$\mathbf{F}_1^{SI}(\mathbf{R}_1) = -\rho_W \int [\nabla_1 U(\mathbf{R}_1, \mathbf{X}_W)] \exp[-\beta U(\mathbf{X}_W, \mathbf{R}_1)$$

$$- \beta U(\mathbf{X}_W, \mathbf{R}_2)] d\mathbf{X}_W. \tag{3.8.3}$$

In (3.8.3), we have expressed the average force in terms of the interaction energies between 1 and W, and between 2 and W. In this particular case, it is relatively easy to analyze the regions of integration which contribute largely to the integral.

One more comment before classifying the different kinds of forces. Equations (3.8.1) and (3.8.3) are valid for very low water density (strictly $\rho_W \to 0$). Therefore, the solvent-induced force is also small. However, we shall be interested in the *relative magnitudes* of the four different cases below. As we shall see later on, the *relative magnitudes* of the four cases are maintained even for higher water densities.

Clearly, if we use only two types of groups, say, methyl and carbonyl, to represent $H\phi O$ and $H\phi I$ groups, we have four possible pairs (see Fig. 3.21):

(a) *Group 1, $H\phi O$; Group 2, $H\phi O$*

In this case, the force $-\nabla_1 U(\mathbf{R}_1, \mathbf{X}_W)$ is expected to be weak. Furthermore, since the interaction between a water molecule and the two $H\phi O$ groups is weak even at the most favorable configuration for the triplet at $\mathbf{R}_1, \mathbf{R}_2$ and $\mathbf{R}_W$, we expect that the conditional density at $\mathbf{X}_W$ will be only slightly higher than the bulk density ρ_W.

(b) *Group 1, $H\phi O$; Group 2, $H\phi I$*

In this case, the force $-\nabla_1 U(\mathbf{R}_1, \mathbf{X}_W)$ is expected to be weak, as in case (a). However, because of the presence of the $H\phi I$ group 2, the interaction energy $U(\mathbf{X}_W, \mathbf{R}_2)$ in (3.8.3) will be strong. Therefore, there exists a distance from this $H\phi I$ group where the conditional density might be significantly enhanced due to the term $\exp[-\beta U(\mathbf{X}_W, \mathbf{R}_2)]$. Therefore, we expect in this case to obtain a solvent-induced force larger than in case (a).

(c) *Group 1, $H\phi I$; Group 2, $H\phi O$*

In this case, the force $-\nabla_1 U(\mathbf{R}_1, \mathbf{R}_W)$ is expected to be larger than in cases (a) and (b). Because of the presence of one $H\phi I$ group, the conditional density will be enhanced due to the term $\exp[-\beta U(\mathbf{X}_W, \mathbf{R}_1)]$, which has the same effect as the corresponding term in case (b), presuming the correct orientation of the $H\phi I$ group. Thus, in this case, we expect to get a solvent-induced force larger than in case (b).

(d) *Group 1, $H\phi I$; Group 2, $H\phi I$*

In this case, the force $-\nabla_1 U(\mathbf{R}_1, \mathbf{R}_W)$ will be as large as in case (c). However, because of the presence of two $H\phi I$ groups, we might, under the right conditions, get a higher conditional density. The right conditions are a distance of about $4.5\,\text{Å}$ between the two $H\phi I$ groups, and the correct orientation of the two $H\phi I$ groups. Under these conditions, both of the terms $\exp[-\beta U(\mathbf{X}_W, \mathbf{R}_1)]$ and $\exp[-\beta U(\mathbf{X}_W, \mathbf{R}_2)]$ will be large, hence,

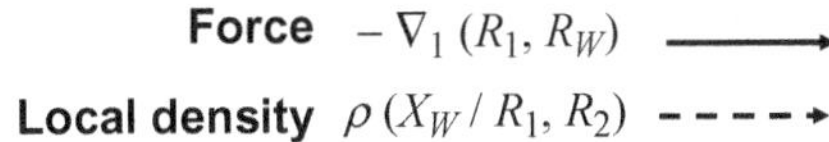

Fig. 3.22 Schematic representation of the change of the force (full arrow) and of the local density (dashed arrow) in going from cases (a) to (d) in Fig. 3.21.

we expect a large conditional density of water molecules. Thus, in this case, both the force and the conditional density will be large and the resulting solvent-induced force is expected to be larger than in case (c).

Figure 3.22 shows schematically the relative order of increasing solvent-induced force from cases (a) to (d). We can conclude that the solvent-induced force between the two $H\phi I$ groups at the correct distance and orientation is the largest of the four cases described above. It should be noted that the conclusion reached above is valid for the case of low density of the water molecules, and where only pair of groups were taken into account. It is qualitatively clear that there might be situations where three, or perhaps four, $H\phi I$ groups might contribute to the conditional density. These are discussed in Appendix S.

For typical densities of water at room temperature and atmospheric pressure [$\rho_W \approx 5.55 \times 10^{-2}$ mol/cm^3], an explicit expression such as in (3.8.3) is not valid. In this case, we must go back to the more general expression (3.7.10) and repeat the same argument as in cases (a) to (d). Now the force term $-\nabla_1 U(\mathbf{R}_1, \mathbf{R}_W)$ is the same as in cases (a) to (d), but the argument regarding the conditional densities is quite different. We present some estimates of local densities near one, two, three and four $H\phi I$ groups in Appendix S.

The general *order* of the solvent-induced forces will be the same as in cases (a) to (d). We expect, however, that now the absolute magnitude of the force will be much larger due to the much higher bulk density of liquid water. We can see that from the following considerations.

We rewrite the general expression in (3.7.10) in the following form:

$$\mathbf{F}_1^{SI}(\mathbf{R}_1) = \rho_W \int -[\nabla_1 U(\mathbf{R}_1, \mathbf{X}_W)]g(\mathbf{X}_W, \mathbf{R}^M)d\mathbf{X}_W$$

$$= \rho_W \int -[\nabla_1 U(\mathbf{R}_1, \mathbf{X}_W)] \exp[-\beta W(\mathbf{X}_W, \mathbf{R}^M)]d\mathbf{X}_W.$$

$$(3.8.4)$$

This looks similar to (3.8.3), except for two differences. Here, ρ_W is the actual density of liquid water, which is much different from the very low density presumed in (3.8.3). This makes the entire solvent-induced force in (3.8.4) much larger than the one in (3.8.2). The second difference is that instead of the potential energy $U(\mathbf{X}_W, \mathbf{R}^M)$ in (3.8.3) (simplified in taking into account only groups 1 and 2), we have the potential of mean force $W(\mathbf{X}_W, \mathbf{R}^M)$ in (3.8.4). The arguments given for cases (a) to (d) above can be repeated for the more general case, and the conclusions are the same, i.e. the local densities increase in going from case (a) to case (d). Thus, the *order* of the strength of the solvent-induced force is as depicted in Fig. 3.22. The absolute magnitude of the force will be very different because of the larger density of water ρ_W in (3.8.4).

We can conclude that for any conformation of the protein, focusing on a $H\phi O$ group, the direct force exerted by a water molecule at the optimum distance and orientation $-\nabla_1 U(\mathbf{R}_1, \mathbf{R}_W)$ is relatively weak. The average solvent-induced force may be enhanced by the presence of $H\phi I$ groups near group 1. On the other hand, focusing on a $H\phi I$ group, the direct force exerted by a water molecule at the optimum distance and orientation (here, the distance and orientation of the water molecule such that it can form an HB with the $H\phi I$ group) is much larger than in the case of a $H\phi O$ group, i.e. $-\nabla_1 U(\mathbf{R}_1, \mathbf{R}_W)$ will be larger in absolute magnitude than the case of a $H\phi O$ group. Again, this direct force will be enhanced by the presence of other $H\phi I$ groups in the neighborhood of group 1.

Thus, the general conclusion is that the strongest solvent-induced force is expected to be exerted on a $H\phi I$ group when this group is also surrounded by other $H\phi I$ groups in its immediate neighborhood. We have discussed in detail only the case of two $H\phi I$ groups, but clearly, the argument may be extended to include the effect of the presence of more $H\phi I$ groups. The force produced by such a one water bridge is operative in the range of distances between the two $H\phi I$ groups of about 4.5 Å. Long range forces are also possible. These will be discussed in Sec. 4.8.

3.9. The Forces in Action

Now that we have some idea of the relative importance of the various solvent-induced forces between different groups along the protein, we turn to the question of how these forces are actually being used in "reading and executing the information" contained in the amino acid sequence of the protein.[180]

The exact mechanism of how these forces operate on each group of each protein is far from known at the moment. However, it is clear that the *most probable* pathways the protein will

---

[180]See Sec. 3.6.

take are downwards in the Gibbs energy landscape or upwards in the equivalent probability landscape.[181]

The eventual motion of the protein towards the folded form is, of course, a result of both the direct and the indirect, solvent-induced, forces. In this book, we are interested in the latter forces, and therefore we shall focus only on the solvation Gibbs energy landscape, i.e. $\Delta G^*(\mathbf{R}^M)$. The gradients on the solvation Gibbs energy landscape determine the solvent-induced forces [see (3.7.6)], and as we have seen in Sec. 3.8, the most important solvent-induced forces are likely to be the $H\phi I - H\phi I$ forces.

Ever since Levinthal attempted to estimate the time that it would take for a protein to fold if it were to move *randomly* from one conformation to another, people have speculated on the "form" of the energy landscape or the Gibbs energy landscape. Levinthal's conclusion is now known as the Levinthal paradox. In my view, there is no paradox in his result,[182] as much as there is no paradox in the so-called Gibbs paradox.[183]

Clearly, a random search of the correct fold of a protein, with an equal transition probability between any two consecutive steps, is not what happens in reality. The fact that there are different kinds of forces acting on the protein at each stage in the folding process rules out the possibility of a completely random search.

On the other hand, it is also clear that a single route from the denatured to the native states is highly unlikely. This would have entailed a transition probability of one to proceed from

---

[181]See Sec. 3.7. Note that even on the probability landscape, we can talk only about the most probable pathways, and not the exact pathways.

[182]Sainsbury (2009) defines a paradox as follows: "A paradox can be defined as an unacceptable conclusion derived by apparently acceptable reasoning from apparently acceptable premises." Clearly, the Levinthal paradox is not a paradox according to this definition simply because it is based on unacceptable reasoning, and on unacceptable premises.

[183]Ben-Naim (2008).

one stage to another along the preferred pathway. The fact that the forces are derived from the PMF (interpreted either as an ensemble average or as a time average force) means that there must be an element of randomness in the choice of pathways for the process of protein folding. The statistical character of the forces precludes also the existence of a single pathway.

There have been several attempts to "resolve" the Levinthal paradox. The most intriguing one was published by Zwanzig *et al.* (1992).[184] Zwanzig *et al.* drew from Dawkins' brilliant ideas for explaining the mechanism of evolution.[185] Briefly, the protein is viewed as sequence of N bonds, and the "connecting bond" between two neighboring amino acids can be characterized as "correct" or "incorrect" (correct means native in biology). Then, they assume some rate constant (k_0) for the transition "correct" $\rightarrow$ "incorrect," and another rate constant (k_1) for the transition "incorrect" $\rightarrow$ "correct." Assuming further that the ratio k_0/k_1 is small, they calculated the mean first-passage time to reach the fully "correct" information.

I believe that the metaphor used by Dawkins is only partially suitable for explaining evolution to the layperson. The mechanism for arriving at the "correct" target as proposed by Dawkins demonstrates the *possibility* of the occurrence of an event which is perceived to be highly improbable. As such, Dawkins' model achieves its goal of taking the mystery out of the evolutionary process. However, even in evolution, there exist no "correct" or "incorrect" results. In fact, Dawkins himself recognized the inaccuracy of his model. However, one can simply *define* a "correct" outcome as one which has some evolutionary advantage. This is not the case for the protein folding process. Therefore, I do not believe that this metaphor is adequate for the process

---

[184]Zwanzig *et al.* (1992).
[185]Dawkins *et al.* (1987). See also Dill (1999).

of protein folding. My main objection to this model is that one cannot justify the preferential transition from an incorrect to a correct bond for any stage of the protein folding process.

In evolution, the transitions from "incorrect" to "correct" are biased according to some selection criterion, i.e. the "correct" result has some advantage, and therefore that result survives. There exists no analog of the selection criterion in the process of protein folding.

It should be noted that what Zwanzig *et al.* refer to as "correct" and "incorrect" bonds are traits that characterize the *entire protein*, not a single peptide bond or a small number of peptide bonds. One can imagine many configurations of the protein for which all the bonds are "correct," but the whole configurations of the protein are not the native one.

Furthermore, starting from the fully unfolded form, i.e. all bonds are "incorrect," it is far from clear that the process of folding will proceed via a series of "corrections." It is also difficult to justify any particular assignment of values for k_0 and k_1. Thus, although the model described by Zwanzig *et al.* might be an interesting one for describing evolution, I do not believe it is relevant to the process of protein folding.

In my view, a more realistic mechanism for folding is one based on the *forces* acting on each group at any given configuration of the protein. In this view, there are no "correct" or "incorrect" bonds or configurations. At each stage of the process of folding, there are many possible transitions, some more probable and some less probable, or even improbable. This view leads to a *range* of *pathways*, which we call the preferential folding pathways (PFPs) along which the protein folds with high probability, and with negligible probability along all other pathways. In other words, this view effectively reduces the immense number of pathways into a narrow "corridor" of PFPs, within which there is some degree of randomness. However, randomness of this kind does not allow the protein to wander in any direction like a "drunken golfer," seeking a single hole in a flat, featureless landscape.

A suitable theoretical framework which is sufficiently general as well as realistic is to view the process of protein folding as a Markov chain.[186] In this view, the protein has a finite number of states. The folded state (or states) may be viewed as an approximate absorbing state (or group of states). Thus, for each state i, there are transition probabilities P_{ij}. These are the conditional probabilities of moving to state j, given that the protein is in state i. Thus, we have for any state i

$$P_{ij} \geq 0, \quad \sum_{j=1}^{n} P_{ij} = 1, \tag{3.9.1}$$

i.e. the transition probabilities are non-negative, and the probability to move from i to any other state j (including the state $j = i$) is one. The magnitude of the transition probabilities are determined by forces acting on the protein, which in turn are determined by the gradients in the Gibbs energy landscape.

For simplicity, we can choose the state n to be an absorbing state, i.e. once the system reaches that state it stays there "forever."[187] The folded state, denoted here by the state n, is characterized by the transition probabilities

$$P_{n,i} \approx 0 \quad \text{for any } i \neq n,$$

$$P_{n,n} \approx 1. \tag{3.9.2}$$

---

[186]Or a Markov process. Here, we assume that the protein has a finite number of states, and in the process of folding there are transitions from one state to another at discrete time intervals.

[187]This is, of course, an oversimplification. The folded state is not a single state, and it is not an absolute absorbing state. However, we can assume that once the folded state is reached, it stays there for a long enough time to function, and for all practical purpose we may assume that it stays in that state "forever."

Within this Markov-chain view of the protein folding process, we can characterize the two extreme cases mentioned above as follows:

The Levinthal model is equivalent to the assumption that at each state i, there is an equal probability to move to each of the states j which are accessible from state i, i.e.

$$P_{ij} \approx p \quad \text{for each } i \text{ and for each } j \text{ accessible from } i. \quad (3.9.3)$$

The other extreme case is that from each i, the protein can move to a single state. Let us denote this state simply as $i + 1$, i.e. at each step we move from i to $i + 1$ until we reach $i = n$, hence

$$\begin{aligned}
P_{i,i+1} &\approx 1 \quad \text{for each } i = 1, \ldots, n - 1, \\
P_{n,i} &\approx 0 \quad \text{for each } i \neq n, \\
P_{n,n} &\approx 1.
\end{aligned} \quad (3.9.4)$$

The more realistic view is that at each state i, there is a group of states which are accessible from i, but have a distribution of probabilities. For instance, starting from $i = 1$, the system can reach states $i = 1, 2, \ldots, n_i$. The transition probabilities to these n_i states is such that some states are reached with relatively higher probability.

It is intuitively clear that if such preferential transition probabilities exist for each intermediate state i, then there will also be preferential pathways to go from state 1 to state n. These will not be the random paths as implied by the Levinthal model, or a single path as implied by the second extreme model. Instead, the transitions will have some preference to go along a specific path, but with some random deviations.

In such a Markov chain, one can also compute the average number of steps to reach the absorbing state.

The important point to stress here is not the specific mechanism of transition from one state to another, and eventually the entire process of folding. Instead, we propose that the highest transition probabilities are determined by the strong $H\phi I$ forces acting on $H\phi I$ groups. If that is the case for most states, then the resulting folding process will proceed along a very narrow "corridor" of pathways, such that the probability of choosing any other pathways will be extremely small and therefore negligible. This is as close as we can get to a solution of the *general* protein folding problem.

Even after we adopt the view that there exist PFPs which effectively reduce the average time required to proceed from the unfolded to the folded form, there is room for speculating about a metaphor for the energy (or the Gibbs energy) landscape.

In 1992,[188] I suggested looking at the solvation Gibbs energy landscape as a multi-dimensional space, each point of which representing one of the conformations of the protein. Gradients on this landscape correspond to lowering the solvation Gibbs energy of the protein. The protein will initially and preferentially move along one of these gradients until it reaches a lower dimensionality space, where again there is a gradient leading to another, yet lower dimensionality space (this was described as a "tunnel within a tunnel"), until the protein is "drained" into the folded form. In a two-dimensional landscape, we can imagine that starting from the top of the hills, the protein flows in the two-dimensional space of x, y towards a "river." Once a "river" is reached, the flow continues along the gradient in the one-dimensional path of the river. In a multi-dimensional case, at any stage, the protein "flows" from a higher dimensional "river" (or tunnel) into a lower dimensional "river" (or tunnel), eventually reaching a one-dimensional river leading to the folded form. This final conformation does not have to be

---

[188] Ben-Naim (1992).

a global minimum in the Gibbs energy landscape as some have speculated following the so-called Anfinsen Dogma.[189]

The rationale underlying this sequential reduction of the dimensionality of the space in which the protein "moves" is the following:

Suppose we start with the fully extended unfolded conformation. Initially, we would expect that the motion of the protein would be random until two or more $H\phi I$ groups are brought to such a distance that they exert a strong $H\phi I$ force on one another. This force reflects the existence of a steep gradient in the original multi-dimensional landscape. Once the two $H\phi I$ groups are brought to a short distance such that they can form a direct HB (as in an α-helix or β-sheet), some of the rotational angles ψ, ϕ will be locked for a short time while the random motion about all other angles continue, but now in a lower dimensional space. Of course, several direct HBs can occur simultaneously, resulting in a further reduction in the dimensionality of the space in which the conformation of the protein moves. Some experimental evidence for the occurrence of water bridges connecting the $H\phi I$ groups of the backbones of proteins has been reported.[190]

This picture is somewhat different from the funnel picture which became popular recently. In the latter, at any initial step, the protein will "drain" though a "funnel" to reach the final native structure. The funnel metaphor is motivated by the "thermodynamic hypothesis" for protein folding. However, it is not clear as to what extent the "thermodynamic hypothesis" is applicable to the process of protein folding.[191]

The Gibbs energy landscape may be described by a function $G(T, P, N; \phi_1, \psi_1, \ldots, \phi_M, \psi_M)$, where $\phi_i \psi_i$ are the internal

---

[189]See, for example, Karshikoff (2006), Plotkin and Onuchic (2002), Dill and Chan (1997).

[190]Sundaralingam and Sekharadu (1989). These authors also discussed the possible role of such water molecules in the folding and the unfolding of the helix.

[191]See also Appendix N.

rotational angles. The thermodynamic hypothesis states that for fixed T, P and N, the configuration $(\phi_1, \psi_1, \ldots, \phi_M, \psi_M)$ will be such that the Gibbs energy will be at a minimum. However, this "thermodynamic hypothesis" does not follow from the second law of thermodynamics. To apply the second law, we assume that the protein can be in one of n discrete number of conformations. Denote by C_i the ith conformation $(\phi_1^{(i)}, \psi_1^{(i)}, \ldots, \phi_M^{(i)}, \psi_M^{(i)})$. Now, suppose we construct a system composed of N proteins, N_1 of which are in conformation C_1, N_2 of which are in conformation C_2, and so on. Denote the mole fractions $x_i = N_i/N$, then for any arbitrary chosen composition $x_1, x_2, \ldots, x_n$, we can define the Gibbs energy of the system $G(T, P, N_w, N; x_1, \ldots, x_n)$, where N and N_w are the number of protein and water molecules, respectively. The vector $\bar{x} = (x_1, x_2, \ldots, x_n)$ may be referred to as the composition vector of the protein. The second law of thermodynamics states that the function $G(T, P, N_w, N; \bar{x})$ has a minimum with respect to *vector* $\bar{x}$, keeping T, P, N_w, N constant. In other words, there exists an equilibrium compositional vector $\bar{x}^{\text{eq}} = (x_1^{\text{eq}}, \ldots, x_n^{\text{eq}})$ which minimizes the Gibbs energy function $G(T, P, N_w, N; \bar{x})$. It is also true that $\bar{x}^{\text{eq}}$ maximizes the probability function $\text{Pr}(T, P, N_w, N; \bar{x})$, and that at $\bar{x}^{\text{eq}}$, there is only one maximum, and at $\bar{x} \approx \bar{x}^{\text{eq}}$ (i.e. at composition near the equilibrium value $\bar{x}^{\text{eq}}$), Pr is nearly one.[192]

Clearly, the vector $\bar{x}$ is itself a probability distribution (x_i is the probability of the protein to be in conformation C_i). However, this probability distribution does not necessarily have a single minimum. It can have no minimum or multiple minima.

Similarly, the Gibbs energy function $G(T, P, N_w, N; x_i)$ viewed as a function of the parameter i (this is the analogue of the PMF viewed as a function of the distance R between two solutes in a solvent, see Sec. 1.2) does not necessarily have

---

[192]See also Ben-Naim (2008) and Appendix N.

a single minimum at some conformation C_i. Rather, the function $G(T, P, N_w, N; \bar{x})$ viewed as a function of the *entire* vector $\bar{x}$ has an absolute minimum at the equilibrium distribution $\bar{x}^{\text{eq}}$. We shall further discuss the thermodynamic hypothesis in Appendix N.

Suppose that we start with any specific initial conformation, say, the unfolded form, denoted C_1. The composition of the system is $G(T, P, N_w, N; (1, 0, \ldots, 0))$. We now release the constraint on the fixed conformation. The Gibbs energy will attain a minimum with respect to the *whole* distribution $\bar{x}$, at a point which we denote by $\bar{x}^{\text{eq}}$. In other words, we do not expect that the initial vector $\bar{x}^{(in)} = (1, 0, \ldots, 0)$ will evolve to a final vector such as $\bar{x}^f = (0, 0, \ldots, 1, 0, \ldots, 0)$, but to a final distribution $\bar{x}^{\text{eq}} = (x_1^{\text{eq}}, x_2^{\text{eq}}, \ldots, x_n^{\text{eq}})$ such that the Gibbs energy has a minimum.[193]

Thus, starting from any initial conformation, the protein does not necessarily "drain" towards a conformation at a minimum in the Gibbs energy landscape. Instead, the distribution of conformations attains an equilibrium distribution for which the Gibbs energy of the system is minimum.

We recall the discussion of the motion of two particles under the potential of mean force (Sec. 1.2). The statistical nature of the solvent-induced force allows motion which is not necessarily in the direction of the average force. Thus, even for a smooth and continuous potential of mean force, there is no guarantee that the particles will move towards each other. This "rugged" motion occurs not because of the ruggedness of the potential function, but because of the statistical nature of the forces.

Likewise, in the multi-dimensional Gibbs energy landscape, the ruggedness of the motion of the protein is basically due to

---

[193]In the continuous case, let ϕ denote the conformation of the protein. The Gibbs energy landscape is $G(T, P, N; \phi)$. This function does not necessarily have a minimum. The second law states that the *functional* $G(T, P, N; \text{Pr}(\phi))$ has a single minimum with respect to all possible distribution functions $\text{Pr}(\phi)$.

the statistical nature of the forces acting on each group of the protein. This will cause a *rugged motion*, whether or not the landscape is rugged.

It is often said that the criteria for detecting an absolute (or global) minimum in the Gibbs energy landscape is that if we start from different initial configurations and end up in the same stable structure, this state must be the absolute minimum. This statement may be true, but it is not necessarily true. The question of local versus global minima is not really relevant to the problem of protein folding. It might be the case that most of the initial states (conformations) of the denatured (or recently synthesized) protein are on "one side of a hill," hence, all the physically realizable initial states will be drained into the same minimum. If the protein spends enough time in that minimum and can perform its biological functions during that period of time, the question of the existence of other minima, lower or higher than the one reached, becomes irrelevant. In any case, arguments based on the "thermodynamic hypothesis" might not apply to the case of protein folding.[194] The most general answer to the question why protein folds is now very simple: The protein folds (if it folds) at a reasonable time, simply because at any stage of the folding pathway, there are strong $H\phi I$ forces acting on the various $H\phi I$ groups, which *force* the protein to fold, with high probability towards the folded form.

This is as close as one can get to answering the question of *Why* protein folds. There exists no general answer to the question of *How* protein folds. However, implementing the answer to the question *Why* for any specific protein, we can also get an answer to the question of *How* a specific protein folds.

---

[194]See also Appendix N.

3.10. Is there a "Folding Code"?

The fact that a denatured protein can refold spontaneously into the native structure once the correct environment (temperature, pressure and composition of the solvent) is restored, without the help of an enzyme or a template, was correctly interpreted to mean that all the *information* required for the folding process is already contained in the sequence of the amino acids. The term "information" is used here in its colloquial meaning — not in the information-theoretical sense.[195]

From the fact that the information for folding is *contained* in the sequence, scientists conjectured the existence of a "code," sometimes referred to as the "second translation of the genetic message,"[196] or the "second half of the genetic code."[197] The elucidation of this code, if it exists, is a major unsolved challenge known as the protein folding problem.[198]

If such a code exists, i.e. a translation from a 1-D sequence of an amino acid into a 3-D structure of a protein, perhaps one could also find a reverse code, i.e. translation from a 3-D structure into a 1-D sequence. Such a reverse code could widely open the door to a vast new field of bioengineering, as Fasman wrote in the preface to the book Prediction of *Protein Structure and the Principles of Protein Conformation*[199]:

> The time is approaching when a new protein will be designed on the drawing board, using predictive algorithms, and its subsequent synthesis, via cloning or peptide coupling, will offer new and interesting challenges for biochemists and molecular biologists.

---

[195]See Ben-Naim (2008).
[196]Goldberg (1985).
[197]Kolata (1986).
[198]Behe *et al.* (1991).
[199]Fasman (1989).

First, it is clear that there exists no code that translates letter by letter, word by word or phrase by phrase. For instance, there is no code that would translate each amino acid or group of amino acids into a series of coordinates. Such a code can easily be ruled out; a given protein might have a repeated phrase (i.e. of a small sequence of amino acids) forming very different structures. Therefore, it is clear that neither a single amino acid nor a small "phrase" of amino acids can determine its own eventual location in the 3-D structure of the native form. It is more likely that either very large phrases or perhaps the entire protein determine the final 3-D structure. This means that if a code exists, it would be in the form of a huge dictionary, translating the whole sequence of amino acids into 3-D structures. Such a dictionary could certainly be created from the known structures of proteins, which are a very tiny fraction of all possible sequences of amino acids.

The question still remains whether such a code or dictionary would be useful for the more important problem of designing new proteins which perform a prescribed function. As Fasman stated, one can in principle *design* a protein that will have a specific function, say, catalyzing a specific reaction. Having designed such a structure on the drawing board, one can go back and synthesize the sequence of amino acids that fits the structure. The main question is whether such a sequence, though fitting the designed structure, would actually fold into that designed structure.[200] In my opinion, it is highly improbable that such a sequence would fold into the required structure, the reason being that one can *design* — on the drawing board — a protein to perform a well-defined function, but it is not enough to

---

[200]This procedure is often described as the "inverse protein folding problem." Gutte *et al.* (1979), Hecht *et al.* (1990), Eisenberg *et al.* (1986), DeGrado (1988), De Grado *et al.* (1989), Mutter and Vuilleumier (1989), Mutter *et al.* (1989), Cohen and Parry (1990), Hill *et al.* (1990), Shaw *et al.* (1990), Hahn *et al.* (1990), Yue and Dill (1992).

design the end product, i.e. the final 3-D structure. One must also design the sequence of amino acids in such a way that the forces acting at *each stage* of the folding process, if it folds at all, will be such that they will lead to the final 3-D structure. This kind of a design is, of course, much more demanding, and it is unlikely to be achieved in the foreseeable future. Thus, designing a protein on the drawing board is one thing, and making it fold into the required structure is another, almost impossible, task.

Notwithstanding the high improbability of success of such a reverse-designed protein, one can think of an alternative way of achieving the goal of synthesizing a new protein. Suppose we had that huge "dictionary" which translates sequences into structures, or some complicated algorithm which calculates the structure for any given sequence. Clearly, there must be a very large number of sequences that would fit the drawing-board-designed structure. One could try one fitted sequence and look it up in the dictionary, or use the algorithm to see which structure it would fold to. If the required structure is obtained, we have succeeded; if not, take a different fitted sequence and look up the corresponding structure. This process can be iterated until one finds a sequence that will produce the required structure. Unfortunately, there is no guarantee that such a procedure would succeed.

A stronger argument against the existence of a unique code which translates from sequence to structure is that such a code, if it exists, would depend on temperature, pressure and composition. We know that many proteins attain different structures in different environments. For instance, hemoglobin has at least two stable conformations — call them Oxy and Deoxy. If a code existed, then the question would arise: To which structure would the sequence of hemoglobin be translated, to Oxy or to Deoxy? It is clear that for the "general" protein, one would need to create an infinite number of codes, i.e. a different code for each set of external variables.

Finally, one wonders: If designing a functioning protein is so difficult, how could Nature have achieved such a task, and on numerous occasions? This question has been asked many times by numerous authors. In a recent review article, Wolynes (2005) writes: "Evolution solved the protein-folding problem. A major goal of biomolecular science has been to understand how this was done."

My opinion is quite different. Evolution does not "solve" any problem. Evolution only *evolves*. The results of this evolution seem to us as if they were reached by "solving a problem." Nature also did not design any protein to perform certain pre-programmed functions. Polypeptides and proteins were synthesized (either randomly in the early stages of life on Earth, or by the more sophisticated machinery that has evolved since then). Only those proteins which happened to be useful or had an advantage survived.

Nature does not *design* the structure that it needs but *exploits* those structures that are available and happen to be useful or confer some advantage.

3.11. Concluding Remarks and Suggestions for Future Research

There are essentially two problems referred to as the problem of protein folding. The first addresses the questions of "how" and "why" a protein folds; the second is whether or not there exists a code which translates from a sequence to a structure. The first problem has different answers for different proteins. Perhaps the only aspect of the first problem which is common to all proteins is the identification of the types of interactions and the types of force that are involved in the process of folding. This part of the problem may be considered solved once we will have a complete inventory of all the factors involved in the process of protein folding and the important factors are identified. In my opinion, we are approaching the solution of this problem with the discovery of the various $H\phi I$ effects.

Regarding the second question, on the existence of the "second genetic code" and the associated problem of reverse engineering, I do not believe that an answer is to be found anywhere in the foreseeable future, if such an answer exists at all.

Thus, the part of the problem that is solvable seems to be almost solved, while the second part is not likely to have a solution.

In spite of this, there is much to be done in the study and understanding of the problem of protein folding. Instead of expending effort in envisaging the shape of the Gibbs energy landscape, I believe that it would be more beneficial to focus on the *type* of *forces*, both direct and indirect, acting on the protein along the folding process. These forces may be studied both theoretically and experimentally with small model compounds. Once we have a complete inventory of all the forces, or at least the most important ones, we can use this knowledge to predict the preferential folding pathways of *specific* proteins. Perhaps some general or universal principles will emerge from such a study. When that time comes, there will be no need to search for a general solution to the "protein folding problem."

CHAPTER 4

Association and Self-Assembly of Biomolecules

My interest in the solvent-induced effects on the association between macromolecules started at the same time I undertook to examine the solvent-induced effects on the process of protein folding.[201] I was surprised to learn that although the two processes are very different, the "inventories" of all possible solvent-induced effects for these processes are the same. It was much later that I realized that my conclusion regarding the relative importance of the HφI effect in protein folding also applied to the process of the association of proteins.[202] Another aspect of the binding process in which HφI interaction can play an important role is its specificity, which is referred to as molecular recognition. It turns out that whenever HφI interaction features, the prevailing lock-and-key model might be rendered irrelevant to the recognition. This could have a profound effect on the field of drug design. Some specific examples were suggested by Wang and Ben-Naim (1996).

Further examination of the HφI effects revealed a rich repertoire of effects; some are much stronger and some have a longer range compared to the simple one-water bridge studied earlier. These forces could play an important role in the dynamics of both protein folding and protein–protein association.

---

[201] Ben-Naim (1990, 1992).
[202] Ben-Naim (1992, 2006).

Binding, association and self-assembly processes abound in biological systems. These processes range from binding small ligands such as drugs to protein or to DNA, to association between proteins, as in hemoglobin, to self-assembly of a large number of sub-units to form macromolecules such as the Tobacco mosaic virus. There are also many studies of binding between non-biological macromolecules.[203]

As in Chapter 3, there are essentially two puzzles associated with these processes. The first is similar to the question of the stability of the folded protein. In the folding process, the large number of conformational states of the unfolded form tends to favor the denatured protein. To understand the "driving force" for folding, we need to find out what factors stabilize the native structure of the protein. Similarly, in any association process, there are far more configurations to the separated units than to the bound aggregates. Again, to understand the "driving force" for the binding process, we must find out which factors are responsible for the stabilization of the dimer, or the oligomer, relative to the separate monomers.

The second puzzle is similar to that regarding the preferential pathways of protein folding. It is concerned with the specificity of the binding mode. There are many ways two globular proteins can bind, yet only one specific binding mode is stable. Specificity of binding is essentially the same as molecular recognition. These phenomena are relevant to diverse biochemical systems ranging from the binding of drugs to protein (hence also to drug design), to the binding of protein to DNA (controlling genetic expression), to the way the immune system works. As in the rest of the book, we shall focus mainly on the solvent-induced part of the driving forces for the binding processes and their specificity. We shall also see that in all these processes, the various

---

[203] Schneider (2009).

*HφI effects are the more important part of the "driving forces."
This conclusion was not accepted initially. Most people identified the HφI effect with HBing in water, and this was deemed
to contribute insignificantly to the "driving force." However,
recently, there are an increasing number of biochemists who
recognize the importance of the HφI effects, which include not
only direct HBing, but also water bridges of different kinds,
which will be discussed in this chapter.*[204]

4.1. Thermodynamics and Statistical Thermodynamics of the Association Process

Consider the process of binding a ligand **L** to a protein **P** to
form the complex **PL**. **P** and **L** could be a protein and a small
ligand, two proteins or a protein binding to DNA. For such
macromolecules, the Gibbs energy of binding depends on many
interactions between groups on one molecule and groups on the
other molecule. In addition, the changes in the solvation Gibbs
energies of all the solutes involved in the process determine the
solvent-induced contribution to the total "driving force" for the
binding (or association) process.

In this section, we shall dissect this very complicated problem
into smaller, more manageable problems. To do so, we shall
discuss the process of binding one specific conformation of the
protein, which we denote by **P**, and one specific conformation
of the ligand, denoted by **L**, to obtain a specific conformation
of the complex **PL**.[205] Thus, we write

$$P + L \rightarrow PL. \tag{4.1.1}$$

---

[204]Xu *et al.* (1997), Jesior (2000).

[205]In the case of protein folding, we have denoted these specific conformations by F_1 and U_1. Here, we do not use an index to denote the specific conformation. It is clear, however, that when we discuss the thermodynamics of binding of real molecules, we must average over all possible conformations.

As in Chapter 3, we write the chemical potential for each species as

$$\mu_\alpha = \mu_\alpha^* + k_B T \ln \rho_\alpha \Lambda_\alpha^3, \qquad (4.1.2)$$

where μ_α^* is the pseudo-chemical potential of the molecule α (see also Appendices A and B), ρ_α is the number density and Λ_α^3 is the momentum partition function of α.

In an ideal gas phase, the chemical potential of α is given by

$$\mu_\alpha^{ig} = U_\alpha + k_B T \ln \rho_\alpha \Lambda_\alpha^3 q_\alpha^{-1}, \qquad (4.1.3)$$

where U_α is the potential energy function of the α molecule, and q_α is the internal partition function of a single α molecule. The equilibrium condition in an ideal gas phase is

$$\mu_P^{ig} + \mu_L^{ig} = \mu_{PL}^{ig}. \qquad (4.1.4)$$

Hence, the equilibrium constant is

$$K^{ig} = \left(\frac{\rho_{PL}}{\rho_P \rho_L}\right)_{eq} = \frac{q_{PL}}{q_P q_L} \frac{\Lambda_L^3 \Lambda_P^3}{\Lambda_{PL}^3} \exp[-\beta \Delta U(P + L \to PL)]. \qquad (4.1.5)$$

Note that unlike the case of protein folding, where the momentum partition functions of **U** and **F** were the same, here, the change in the momentum partition function is a major contribution to the "driving force" for the binding process. $\Delta U(P + L \to PL)$ is the change in the potential energy for the reaction (4.1.1). Since we have assumed that there is no change in the conformations of **P** and **L** upon binding, ΔU in (4.1.5) is simply the sum of all the interactions between groups of **P** and groups of **L** in the final complex **PL**.

As in the case of protein folding, we can define the solvent-induced part of the Gibbs energy change for any process as

$$\delta G = \Delta G^l - \Delta G^{ig}, \qquad (4.1.6)$$

where ΔG^l and ΔG^{ig} are the Gibbs energy changes of the same process in the liquid and ideal gas phases, respectively. Because of the particular reaction (4.1.1), where only one conformation of each solute was chosen, ΔG^{ig} is simply $\Delta U(\mathbf{P}+\mathbf{L} \to \mathbf{PL})$, i.e. the interaction energies between all groups of $\mathbf{L}$ and groups of $\mathbf{P}$ at the final configuration of the complex $\mathbf{PL}$.

It is easy to show, either theoretically or simply from the cyclic process in Fig. 4.1, that δG for the process (4.1.1) is

$$\delta G(\mathbf{P} + \mathbf{L} \to \mathbf{PL}) = \Delta G^*_{PL} - \Delta G^*_P - \Delta G^*_L, \qquad (4.1.7)$$

where ΔG^*_α is the solvation Gibbs energy of the solute α.

We also assume that all the solutes involved in the reaction (4.1.1) are very dilute in the solvent, hence all the solvation quantities on the right-hand side of (4.1.7) do not depend on the concentrations of the solutes.[206]

Thus, with the identification of the solvent-induced effect (4.1.7), we can write the equilibrium constant of the reaction (4.1.1) in the liquid (l) as

$$K^l = K^{ig} \exp[-\beta(\Delta G^*_{PL} - \Delta G^*_P - \Delta G^*_L)]. \qquad (4.1.8)$$

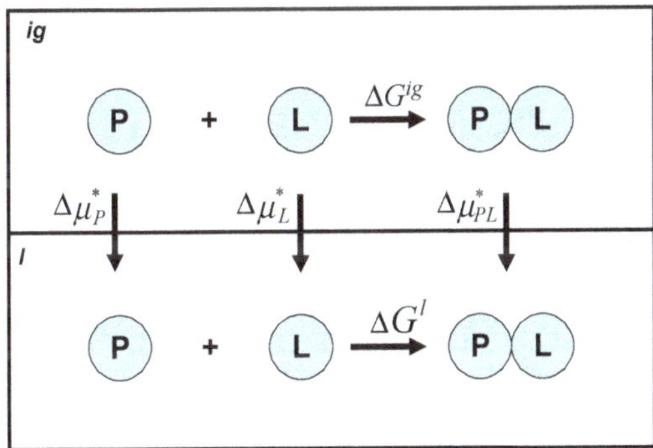

Fig. 4.1   A cyclic process corresponding to Eq. (4.1.7).

---

[206]For more details, see Ben-Naim (2006).

In writing (4.1.8), it is assumed that the internal partition functions of all the solutes involved in the reaction are not affected by the presence of the solvent.

We can now write the standard Gibbs energy change for a unit reaction in the liquid as

$$\Delta G^0(\mathbf{P}+\mathbf{L} \to \mathbf{PL}) = \Delta U(\mathbf{P}+\mathbf{L} \to \mathbf{PL}) + k_B T \ln \left(\frac{\Lambda_{PL}^3}{\Lambda_L^3 \Lambda_P^3} \right)$$

$$+ k_B T \ln \left(\frac{q_L q_P}{q_{PL}} \right) + (\Delta G_{PL}^* - \Delta G_P^* - \Delta G_L^*). \qquad (4.1.9)$$

The standard Gibbs energy is defined as the change in Gibbs energy of a unit reaction when $\rho_L = \rho_P = \rho_{PL} = 1$ (one particle per unit of volume. The units of volume must be chosen to be the same as used for the momentum partition function).

In (4.1.9), we have four terms that contribute to the standard Gibbs energy, or the driving force, for the reaction (4.1.1):

(a) The direct interaction energy between **P** and **L** at the configuration of the complex **PL**. This term includes all the interactions between groups of **P** and groups of **L**.

(b) The change in the momentum partition function. In forming the complex **PL**, there is a loss of the translational Gibbs energies of two molecules and a gain of the translational Gibbs energy of the complex **PL**. This is a major thermodynamic "driving force" favoring *dissociation* (see Sec. 4.3).

(c) The change in the internal partition functions consists of various terms. We assume that the electronic and nuclear partition function do not change in the reaction. We also assume that all the vibrations in **P** and **L** are the same as in **PL**. However, there is a new vibrational degree of freedom of the complex **PL**, which we assume to contribute negligibly to the "driving force." Also, by definition of the species involved in reaction (4.1.1), each of the species has a fixed conformation. Therefore, there are no changes in any internal rotational degrees of freedom. The only important contribution to the third term in (4.1.9) is due to the changes in the rotational partition functions (see example in Sec. 4.3).

(d) The change in the solvation Gibbs energies of the solutes in the reaction (4.1.1). This is the term which may be referred to as the solvent-induced contribution to the "driving force." In Sec. 4.3, we shall describe a quantitative example of dimerization of the two model globular proteins. Here, we can describe qualitatively the main contributions to the "driving force" for the association process.

There are mainly two opposing "forces" which are operative in determining the "driving force" or the equilibrium constant for the association process.

On the one hand, there is the loss of translational Gibbs energy and the gain of a new rotational Gibbs energy. The corresponding net change in Gibbs energy should be positive, i.e. favoring dissociation. This effect is largely an entropic effect, and can easily be understood in terms of changes in the *locational* Shannon's measure of information.[207] Initially, **P** and **L** can access the entire volume of the system V. When a complex **PL** is formed, we can say that one solute, either **L** or **P**, can still access the entire volume V. But since **P** and **L** are now moving together as a new entity, the second solute, either **P** or **L**, is now limited to a small volume around L (or P). Denoting this volume by v, the change in the entropy due to change in the locational

---

[207]Ben-Naim (2008).

Shannon's measure of information is roughly

$$\Delta S \approx k_B \ln \frac{v}{V} < 0. \qquad (4.1.10)$$

Therefore, this negative entropy change contributes positively to the driving force, i.e. favoring the dissociation. Note that since $v \ll V$, this entropy change is quite large.

Now, the question is: From where does one get a negative contribution to the driving force so as to make the association favorable? Since there are no covalent bonds between **P** and **L**, it is expected that ΔU in (4.1.9) is small and therefore cannot provide the necessary negative contribution to the driving force.[208]

We can conclude that the only candidate that can provide a large negative contribution is the solvent-induced quantity δG. Traditionally, it was assumed that the $H\phi O$ effect is the major contribution to δG. We shall see in this chapter that it is the $H\phi I$ rather than the $H\phi O$ effect that is the most important contribution to δG. We shall further analyze the contributions to δG for the association process in the next section.

4.2. The Factors Involved in the Association of Two Biomolecules

In this section, we dissect the "grand problem" of the "driving force" for the association reaction into small ingredients. We shall be brief, since the components that contribute to the ΔG^0 of the association reaction (4.1.1) are much the same as those we have found for the protein folding process in Chapter 3.

---

[208]The simplest example where ΔU is large and negative is the formation of H_2 from two atoms of hydrogen. Here, a chemical bond is more than enough to compensate for the loss of the translation of one hydrogen atom. For details, see Hill (1960).

The solvent-induced contribution to the driving force δG is given by

$$\delta G(\mathbf{P} + \mathbf{L} \to \mathbf{PL}) = \Delta G_{PL}^* - \Delta G_P^* - \Delta G_L^*. \qquad (4.2.1)$$

As we have done in Chapters 2 and 3, we can split each of the solvation Gibbs energies in (4.2.1) into various components

$$\Delta G_\alpha^* = \Delta G_\alpha^{*H} + \Delta G_\alpha^{*S/H} + \sum_i \Delta G^{*i/H,S}$$

$$+ \sum_i \delta G(i,j/H,S) + \cdots , \qquad (4.2.2)$$

where the first term is the solvation of the hard part of the solute–solvent interaction, the second term is the solvation of the soft part of the interaction, the third term includes the conditional solvation Gibbs energies of all the functional groups (FGs) that are exposed to the solvent, and the fourth term includes all the (conditional) pair correlations between FGs on the surface of the protein. There are also higher-order correlations about which we do not know much, and we shall therefore stop after the fourth term on the right-hand side of (4.2.2).

Applying (4.2.2) to δG in (4.2.1), we have the following terms:

$$\delta G(\mathbf{P} + \mathbf{L} \to \mathbf{PL}) = \left[\Delta G_{PL}^{*H} - \Delta G_P^{*H} - \Delta G_L^{*H} \right]$$

$$+ \left[\Delta G_{PL}^{*S/H} - \Delta G_P^{*H} - \Delta G_L^{*/H} \right]$$

$$- \left[\sum_{i \in I} \Delta G_P^{*i/H,S} + \sum_{i \in I} \Delta G_L^{*i/H,S} \right] + \left[\sum_{k,l} \delta G^{k,l} \right]. \quad (4.2.3)$$

We shall now discuss each of the terms in the square brackets in (4.2.3).

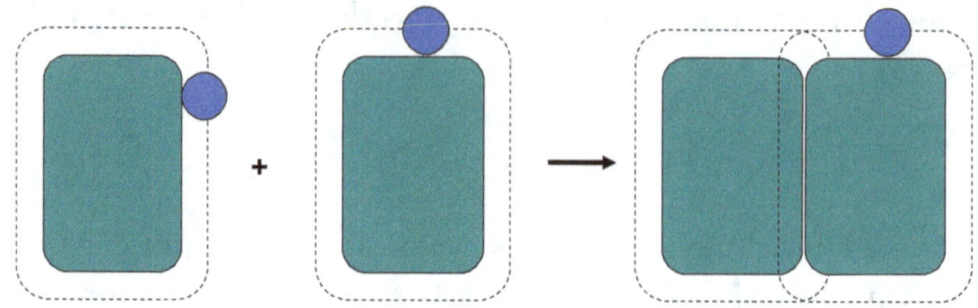

Fig. 4.2　Reduction of the excluded volume with respect to a solvent molecule (blue) upon association.

(a) The contribution of the hard part

$$\delta G^H = \Delta G_{PL}^{*H} - \Delta_P^{*H} - \Delta_L^{*H}. \qquad (4.2.4)$$

Each of the quantities on the right-hand side of (4.2.4) is essentially the work of creating a cavity in the solvent (see Appendix K). Without getting into the details of the calculation of the cavity work, it is clear that for any association reaction, δG^H is negative. This follows from the fact that the excluded volume (with respect to a solvent molecule) of the complex **PL** is always smaller than the excluded volume of the two separate monomers (Fig. 4.2). For very large solutes, we can use the approximate expression for the cavity work for each solute

$$\Delta G_\alpha^{*H} \approx P V_\alpha^{EX}, \qquad (4.2.5)$$

where P is the pressure and V_α^{EX} is the excluded volume of the solute α with respect to the solvent molecule (Fig. 4.2).

Thus, we can estimate δG^H as

$$\delta G^H \approx P\left[V_{PL}^{EX} - V_P^{EX} - V_L^{EX} \right] < 0. \qquad (4.2.6)$$

Since the excluded volume of the complex **PL** is always smaller than the excluded volumes of the two separate components **P** and **L**, we can predict that δG^H is negative. Note, however, that this quantity depends on the sizes of the solutes and of the solvent molecules. It will be negative for any solvent, not necessarily water. Therefore, we cannot expect that this term will dominate δG for the association reaction in water, nor can we expect that this term will determine the specificity of the binding mode.

(b) The second term on the right-hand side of (4.2.3) is

$$\delta G^{S/H} = \Delta G_{PL}^{*S/H} - \Delta G_P^{*S/H} - \Delta G_L^{*S/H}. \tag{4.2.7}$$

This term arises from the van der Waals interactions between non-polar groups, such as methyl groups and water. We can assume that the conditional solvation Gibbs energy of each solute α is proportional to its surface area A_α.[209] Hence, we write

$$\Delta G_\alpha^{*S/H} \approx -\varepsilon A_\alpha, \tag{4.2.8}$$

where ε is an energy parameter of the same order of magnitude as the Lennard-Jones parameter for neon–methane interaction.[210] With this assumption, we can write (4.2.7) as

$$\delta G^{S/H} \approx -\varepsilon[A_{PL} - A_P - A_L] > 0. \tag{4.2.9}$$

Since the total surface area exposed to the solvent decreased in the process, and since $\varepsilon > 0$, we expect that this term will be positive.[211] As in the case of the hard contribution discussed earlier, we note that this contribution depends on the sizes of the solutes (more precisely on the size of their surfaces), but is independent of the specific properties of water and the properties of the FGs on the surfaces of the solutes.

(c) Contributions due to FGs on the surface of the protein

---

[209] For more details, see Ben-Naim (1992).
[210] See Sec. 4.3.
[211] See Sec. 4.3.

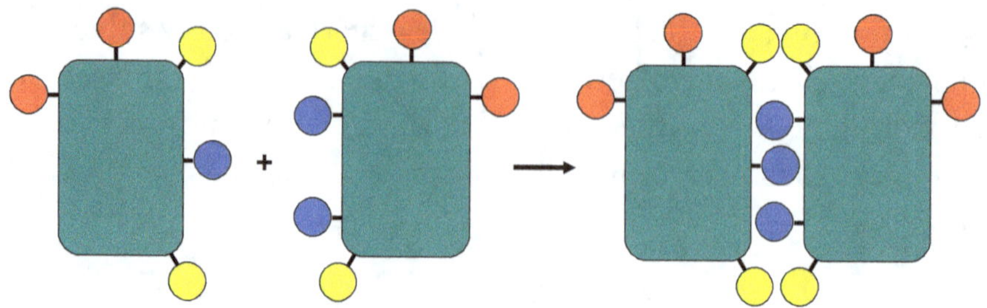

Fig. 4.3   Examples of FGs belonging to group **E** (red disks), group **I** (blue disks) and group **J** (yellow disks).

As we have done in Chapter 3, we can divide all functional groups on the surface of the two solutes **P** and **L** into three groups, **E**, **I** and **J**.

(i) The **E** group (**E** for external) consists of all the FGs on both **P** and **L** the solvation of which is unchanged in the process of association. They are referred to as "external" FGs because they are not involved in the process of binding. Examples of such FGs are shown in Fig. 4.3 (red disks). The formal definition of an FG belonging to the **E** group is that

$$\Delta G_P^{*i/H,S} = \Delta G_{PL}^{*i/H,S}, \tag{4.2.10}$$

$$\Delta G_L^{*i/H,S} = \Delta G_{PL}^{*i/H,S}. \tag{4.2.11}$$

Clearly, a FG the solvation of which does not change in the association process will not contribute to the "driving force" of the process, and will not appear in the third sum in (4.2.3).

(ii) As in the process of protein folding, the **I** group (**I** for internal) consists of all the FGs the solvation of which is completely lost in the process. In protein folding, these were the FGs which were transferred into the interior of the protein. In the binding process, the FGs in group **I** are those that are "buried" in the binding domain, or the interface between

P and L (blue disks in Fig. 4.3). The contribution to δG is the third term on the right-hand side of (4.2.3). Note the minus sign and that we sum over all $i \in I$, i.e. all the FGs in the **I** group in both **P** and **L**. The formal definition of FGs belonging to the **I** group is similar to the definition in the protein folding process.

$$\Delta G_P^{*i/H,S} \neq 0 \quad \text{and} \quad \Delta G_{PL}^{i/H,S} = 0, \quad (4.2.12)$$

$$\Delta G_L^{*k/H,S} \neq 0 \quad \text{and} \quad \Delta G_{PL}^{k/H,S} = 0. \quad (4.2.13)$$

The methods for calculating the various conditional solvation quantities in (4.2.3) are the same as discussed in Chapter 3. We shall use some of these quantities in the example discussed in Sec. 4.3.

(iii) The **J** group consists of all FGs that do not belong to either the **E** or the **I** groups. The **J** group (**J** for joint, or boundary) includes many FGs whose solvation changes to some extent in forming the complex **PL**. There are, of course, many possibilities for changes involving pair and higher correlations between FGs. A few examples are shown in Fig. 4.3. We shall discuss only one kind of FG in this group, namely, those which are independently solvated in **P** and in **L** but became correlated in the complex **PL**. An illustration of such a pair of FGs is shown in Fig. 4.3 (yellow disks).

Thus, we see that the "inventory" of solvent-induced effects for the binding process is essentially the same as for the folding process discussed in Chapter 3. Therefore, the order of magnitude and the relative importance of the various contributions will also be the same. We shall discuss a numerical example for the association of two hypothetical proteins in the next section.

To the best of my knowledge, no specific example of association of two proteins has been studied in detail. Perhaps the first work on the thermodynamics of association of insulin was carried out by Steinberg and Scheraga.[212] Some aspects of the various contributions to the driving forces of protein–protein associations were discussed in the literature[213]; an interesting review was also published by Yu *et al.*[214]

4.3. Association of Two Hypothetical Globular Proteins

In this section, we discuss a dimerization process: Two globular proteins, denoted by M_1 and M_2, forming a dimer, denoted D. We assume that the two monomers are nearly spherical in shape, and that two kinds of FGs are distributed on their surface.

There are essentially two ways of looking at the association process. One is to assume the existence of a dimer as a chemical entity and write a chemical equilibrium constant for the reaction

$$M_1 + M_2 \rightarrow D, \tag{4.3.1}$$

$$K = \frac{\rho_D}{\rho_{M_1} \rho_{M_2}}, \tag{4.3.2}$$

then examine the molecular causes that make K large in aqueous solutions.

The second approach is to not assume the existence of a dimer. Instead, the two monomers M_1 and M_2 are presumed to move freely in the solution. In this view, there is neither sacrifice nor gain of translational–rotational degrees of freedom in the process of formation of the dimer. Instead of the Gibbs energy of dimerization, we examine the potential of mean force (PMF) to find the clues for strong and negative contributions to the "driving force." Instead of an equilibrium constant such as

---

[212] Steinberg and Scheraga (1963).

[213] Steinberg and Scheraga (1963), Makhatadze and Privalov (1993), Amzel (1997), Tamura and Privalov (1997), Baker and Murphy (1997), Sear (2007), Siebert and Amzel (2002), Karplus and Janin (1999), Janin (1995).

[214] Yu *et al.* (2001).

(4.3.2), we examine the probability of, or the average density of pairs of the two monomers at some close separation, and at a specific relative orientation.

In both approaches, we look at the solvent-induced interactions for clues to help understand the thermodynamic *driving forces* for dimerization or higher-order aggregation. In both cases, we can express the solvent-induced effect in terms of the solvation Gibbs energies of the species involved. We shall discuss the two approaches in the next two sections. We shall analyze all possible (direct and indirect) contributions to the driving force. The conclusion reached is that the factors most likely responsible for the formation of stable aggregates are $H\phi I$ interactions. This is in accordance with a conclusion reached regarding the process of protein folding in Chapter 3.

4.3.1. *The "driving force" for dimerization*

We consider two spherical proteins M_1 and M_2 forming a dimer D. In anticipating the application of this methodology to real proteins, we assume that the two proteins are different, but for simplicity, we shall assume that the mass density is constant throughout the interior of the protein. Viewing D as a single molecular entity, we can write the reaction as

$$M_1 + M_2 \rightarrow D, \qquad (4.3.3)$$

and the corresponding equilibrium condition

$$\mu_{M_1} + \mu_{M_2} = \mu_D. \qquad (4.3.4)$$

Each of these chemical potentials can be written in the form

$$\mu_\alpha = U_\alpha + \Delta G_\alpha^* + k_B T \ln \Lambda_\alpha^3 q_\alpha^{-1} \rho_\alpha, \qquad (4.3.5)$$

where U_α is the energy of the particle measured relative to some arbitrary state for which the energy is chosen as zero. ΔG_α^* is the solvation Gibbs energy of the species α. Λ_α^3 is the momentum partition function, q_α is the internal partition function of the species α and ρ_α is the number density.

For the unit reaction (4.3.3), we can write the free energy change as:

$$\Delta G = \mu_D - \mu_{M_1} - \mu_{M_2} = \Delta U + \Delta G_D^* - \Delta G_{M_1}^* - \Delta G_{M_2}^*$$

$$+ k_B T \ln\left[\frac{\Lambda_D^3}{\Lambda_{M_1}^3 \Lambda_{M_2}^3}\frac{q_{M_1} q_{M_2}}{q_D}\frac{\rho_D}{\rho_{M_1}\rho_{M_2}}\right]. \qquad (4.3.6)$$

This is referred to as thermodynamic "driving force" for the reaction (4.3.3).

The equilibrium constant for the reaction (4.3.3) in an ideal gas (or in a vacuum) is

$$K^{ig} = \frac{\rho_D}{\rho_{M_1}\rho_{M_2}} = \frac{q_D \Lambda_{M_1}^3 \Lambda_{M_2}^3}{q_{M_1} q_{M_2} \Lambda_D^3}\exp[-\beta\Delta U], \qquad (4.3.7)$$

where $\beta = (k_B T)^{-1}$, and ΔU is the energy change for bringing the two proteins from infinite separation to the final configuration of the dimer D. The equilibrium constant in the liquid is related to K^{ig} by

$$K^l = K^{ig}\exp[-\beta\delta G]$$

$$= K^{ig}\exp\left[-\beta\left(\Delta G_D^* - \Delta G_{M_1}^* - \Delta G_{M_2}^*\right)\right]. \qquad (4.3.8)$$

Thus, the difference in the equilibrium constant in the two phases is determined by the solvation Gibbs energies of the three species involved in the reaction (4.3.3). This is essentially the same as in Fig. 4.1.

We shall now rewrite the thermodynamic driving force for the reaction (4.3.3) in three terms, corresponding to the three terms in Eq. (4.3.6), i.e.

$$\Delta G = \Delta G_{T-R} + \Delta U + \delta G. \qquad (4.3.9)$$

The first term is the contribution to the driving force due to changes in the translational and the internal degrees of freedom of all the species. For simplicity, we shall assume that q_α is only the rotational partition function of the species α. We neglect the vibrational degree of freedom of the dimer, and we assume that all internal vibrations and rotations within the monomers are unaffected by the process of dimerization. We take $\rho_D = \rho_{M_1} = \rho_{M_2} = 1$ (one particle per cm^3) to obtain a *standard* free energy for the reaction (4.3.3). In this calculation, we use cm^3 for the units of volume in calculating Λ_α^3, which renders the expression under the logarithm dimensionless.

For real proteins, we can calculate all the moments of inertia from the knowledge of the coordinates of each atom, or each group in the protein. In the present calculation, we assume that the protein has a homogenous mass density of $\delta = 0.73\,g/cm^3$.[215] In the general case, the moments of inertia along the three axes are different. For the spherical protein, the moment of inertia is calculated according to the equation

$$I_z = I_y = I_x = \frac{\pi\delta}{2} \int_{-R}^{R} dz(R^2 - z^2)^2 = \frac{8\pi\delta R^5}{15}. \qquad (4.3.10)$$

In the following calculations, we shall assume that two spheres are slightly different, with a radius of R and $0.01\,R$, respectively.

For the dimer, we need the moment of inertia about an axis tangent (tan) to the sphere. This is easily obtained from the

---

[215]J. M. Sanchez-Ruiz, private communication.

moment of inertia about the center of mass (cm) by the transformation (Steiner's theorem)[216]

$$I_{\text{tan}} = I_{\text{cm}} + \frac{4\pi\delta R^5}{3}.\qquad(4.3.11)$$

From (4.3.11), we can calculate the three moments of inertia for the dimer, where the density δ is related to the volume of the sphere by[217]

$$M = \delta\frac{4\pi R^3}{3}.\qquad(4.3.12)$$

With M being the total mass of the sphere of radius R.

The momentum partition function is calculated from

$$\Lambda_\alpha^3 = \left(\frac{h}{\sqrt{2\pi M_\alpha k_B T}}\right)^3.\qquad(4.3.13)$$

With $h = 6.626 \times 10^{-34}\,\text{J}\cdot\text{s}$, $k_B = 1.3807 \times 10^{-23}\,\text{J K}^{-1}$ and $T = 298.15\,\text{K}$.

The rotational partition function for the species α is calculated from

$$q_{\text{rot},\alpha} = \sqrt{\pi}\left(I_{x,\alpha}\,I_{y,\alpha}\,I_{z,\alpha}\right)^{1/2}\left(\frac{8\pi^2 k_B T}{h^2}\right)^{3/2}.\qquad(4.3.14)$$

With these assumptions, we can calculate the first term on the right-hand side of Eq. (4.3.9). This quantity is the Gibbs energy change for the formation of one mole of dimers from one mole of M_1 and one mole of M_2, assuming that the two monomers M_1 and M_2 do not interact either directly or indirectly. Let us refer to this term as the translational–rotational (TR) contribution to the driving force. Figure 4.4 shows the

---

[216]Joos (1958), Goldstein *et al.* (2002).
[217]For more details, see Ben-Naim (2006).

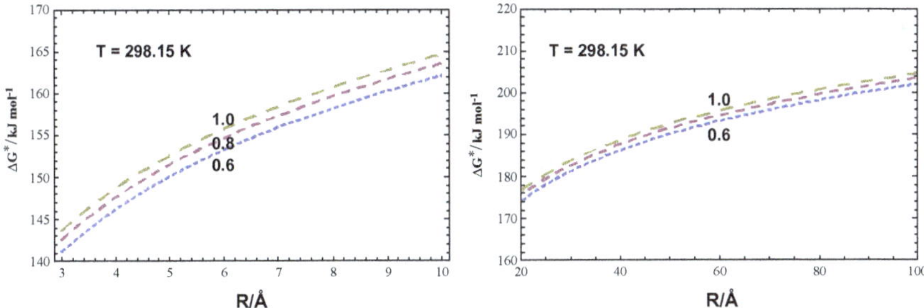

Fig. 4.4   The TR contribution to the "standard Gibbs energy" of the dimerization as a function of the radius of the proteins for three different mass densities $\delta = 0.6, 0.8, 1.0$.

dependence of the TR contribution on the radius of the globular protein for a few mass densities. The Gibbs energy changes are all large and *positive*, which means the TR contribution does not favor dimerization. The larger the spheres, the larger the positive value of the TR contribution. Figure 4.5 shows similar plots for

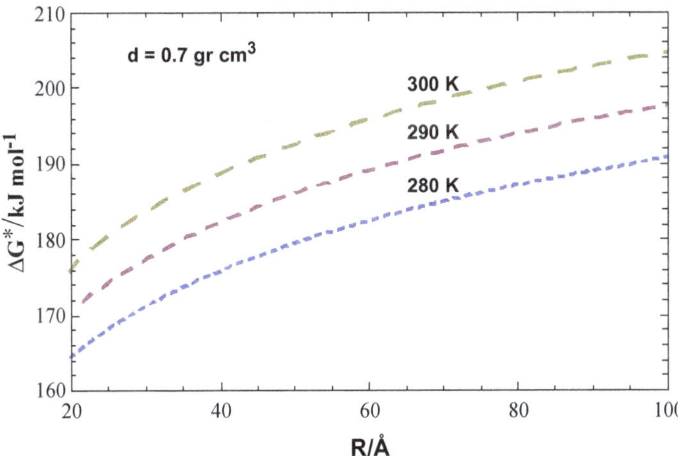

Fig. 4.5   The TR contribution to the "standard Gibbs energy" of the dimerization as a function of the radius of the proteins for three different temperatures $T = 280, 290, 300\,\mathrm{K}$.

different temperatures. As expected, the higher the temperature, the larger the (positive) TR contribution.

These effects are well known and well understood. Since no interaction energy is involved in the dimer formation, the entropy of the two separated monomers is larger than the entropy of the dimer, hence, the formation of a dimer is unfavorable. The main question is: From where does one obtain a *negative* contribution to ΔG that would render the formation of the dimer favorable?

The addition of the direct interaction ΔU cannot be of much help. Suppose that the protein is composed of only methylene groups interacting via van der Waals forces. Clearly, at contact distance between the two spherical proteins, there are only a few groups on one monomer that interact with groups on the second monomer. Since this contribution is about 2.1 kJ/mol, the additional negative contribution to ΔG will be minor, and cannot possibly make ΔG negative.[218]

Direct hydrogen bonds (HBs) between the two proteins could contribute negatively to ΔU. However, this would not change the picture much. First, because there could not be many FGs on the surface of the protein that can form *direct* HBs between one monomer and another. Second, even when direct HBs can be formed, they do not contribute the full HB *energy*, but the HB energy *minus* the loss of the solvation Gibbs energy of the FG participating in such HBing. It was estimated that each direct HB between an FG on M_1 and a group on M_2 would contribute about -6.3 kJ/mol[219] (or about $3k_B T$ at room temperature).

At this stage, it is clear that a large negative contribution must come from the solvent-induced effect. We shall postpone

---

[218]The direct interaction might be much larger for two cubical proteins.
[219]Ben-Naim (1990).

the discussion of this contribution to Sec. 4.3.3, after examining the problem of dimerization from the point of view of probability. Before that, it should be noted that a few authors have suggested "modifying" the translational and rotational partition functions of the monomers and the dimers by using some kind of cell model or free volume model for the liquid state.[220] In this model, the proteins are restricted to wander in a small cell or a small free volume rather than in the entire volume of the liquid. However, I do not think this approach is useful. Since the particles are treated as classical particles, they can access the entire volume of the liquid. The whole idea of a particle moving in a restricted "free volume" to account for the loss of translational entropy is now obsolete.

4.3.2. *Virtual dimers, probabilistic considerations*

An alternative and, in some cases, better way of looking at the problem of association is to view the two proteins M_1 and M_2 as the only chemical entities in the solution. No "dimer" as a chemical entity is recognized to exist in the solution. In this approach, both M_1 and M_2 have their full translational and rotational degrees of freedom, and no permanent dimers are presumed. In this view, there exists no concentration of *real* dimers. Instead, one can define the probability of finding the two monomers at any specific range of distances (or of distances and relative orientations). For convenience, we assume that the two monomers are spherical, and we *define* a *virtual dimer* as a pair of monomers at some interval of distances, say, between σ_p and $\sigma_p + d\sigma_p$, where σ_p is the diameter of the protein. In this view, the monomers do not lose nor gain any translational or rotational degrees of freedom upon "dimerization." The first term of Eq. (4.3.9) does not feature in this view. Instead, only the

---

[220]See, for example, Siegbert and Amzel (2004).

potential of mean force (PMF) is used to determine the relative probability of occurrence of a specific configuration of the two monomers.

For two spherical proteins, we can write the PMF[221] as

$$W(r) = \Delta U(r) + \delta G(r). \qquad (4.3.15)$$

Here, $\Delta U(r)$ and $\delta G(r)$ have the same significance as in (4.3.9), but are now viewed as a function of the distance r between the two proteins. [In (4.3.9), the distance r is fixed and is the distance between the two monomers in the configuration of the dimer D.]

The probability of finding the two monomers at certain range of distances is given by

$$P(X_1, X_2)dX_1dX_2 = \frac{\rho^{(2)}(X_1, X_2)dX_1dX_2}{\iint dX_1 dX_2 \rho^{(2)}(X_1, X_2)}, \qquad (4.3.16)$$

where $\rho^{(2)}(X_1, X_2)$ is the pair distribution function and X_i is the configuration (i.e. location and orientation of the monomer M_i). The denominator in (4.3.16) is, in a closed system, given by

$$\iint dX_1 dX_2 \rho^{(2)}(X_1, X_2) = N_1 N_2, \qquad (4.3.17)$$

which is simply the total number of pairs (consisting of one M_1 and one M_2) in the system.

For spherical proteins, one can always integrate over all the configurations of one monomer and transform to polar coordinates to get the average number of pairs at any specified configuration. If we define a "dimer" to be a pair of M_1 and M_2 at some range of distances $\sigma_p \leq r \leq \sigma_p + d\sigma_p$, then the average

---

[221]See Ben-Naim (2006) and Part I.

number of such pairs is

$$\overline{N}(\sigma_p \le r \le \sigma_p + d\sigma_p) = \iint_{\sigma_p \le r \le \sigma_p + d\sigma_p} dX_1 dX_2 \rho^{(1)}(X_1, X_2)$$

$$= \rho_1 \rho_2 V \int_{\sigma_p}^{\sigma_p + d\sigma_p} g_{12}(r) 4\pi r^2 dr, \tag{4.3.18}$$

and the average density of such pairs will be

$$\rho(\sigma_p \le r \le \sigma_p + d\sigma_p) = \frac{\overline{N}}{V} = \rho_1 \rho_2 \int_{\sigma_p}^{\sigma_p + d\sigma_p} g_{12}(r) 4\pi r^2 dr, \tag{4.3.19}$$

where $g_{12}(r)$ is the pair correlation function for the pair M_1 and M_2 at distance ρ_1, and ρ_1, ρ_2 are the densities of the two monomers. The PMF is related to the pair correlation function as[222]

$$g(r) = \exp[-\beta W(r)] = \exp[-\beta U(r) - \beta \delta G(r)], \tag{4.3.20}$$

where $U(r)$ is the direct pair potential, and $\delta G(r)$ is the solvent-induced contribution to the PMF.

Suppose that the proteins are modeled as hard spheres of diameter σ_p in an ideal gas phase (for the present discussion, assume the two spheres have the same or nearly the same diameter). In which case, $W(r)$ is zero for distances beyond $r \ge \sigma_p$, hence (4.3.19) reduces to

$$\rho(\sigma_p \le r \le \sigma_p + d\sigma_p) \approx \rho_1 \rho_2 \frac{4\pi}{3} \left[(\sigma_p + d\sigma_p)^3 - \sigma_p^3 \right]. \tag{4.3.21}$$

Clearly, for very dilute solutions, ρ_1 and ρ_2 are very small, hence the density of such dimers in (4.3.21) is extremely small.

---

[222]See also Appendix C.

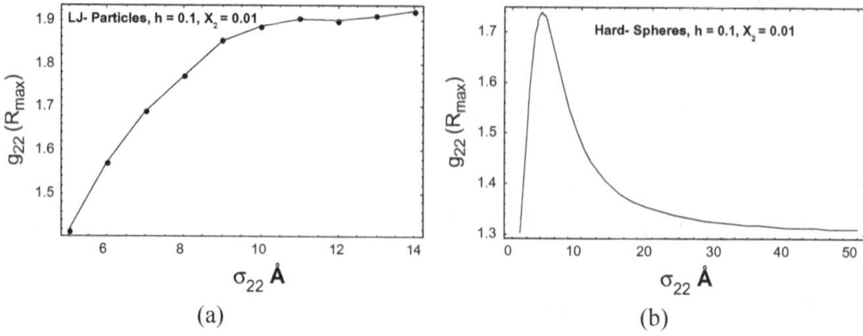

Fig. 4.6   Values of the pair correlation function for two large spheres at the first maximum, $R_{\max}$. The left plot (a) is for Lennard-Jones particles, and the right one (b) for hard spheres. Both as a function of the diameter of the "proteins."

Figure 4.6 shows the value of the first peak of the pair correlation function. These values were obtained for Lennard-Jones particles (left) and for hard particles (right)[223] by solving numerically the Percus–Yevick equations for a mixture of two components. The diameter of the solvent molecules was chosen as 1 Å, and the diameter of the "protein" was varied $4 \leq \sigma_{22} \leq 12$ for the LJ particles, and $1 \leq \sigma_{22} \leq 50$ for the hard spheres. In both cases, we chose the total volume density $\eta = 0.1$ and the mole fraction of the second component $x_2 = 0.01$. Details of the method of solution of the PY equations has been described elsewhere.[224] Clearly, the maximum value of $g(R)$ at the first peak is about 1.9. If we *defined* a "dimer" as a pair of monomers at a distance between $\sigma_{\max} \leq \sigma_{22} \leq \sigma_{\max} + 0.5\,\text{Å}$, the density of such dimers is estimated to be at most 5×10^{-3} particles Å^{-3} (or about $8.3 \times 10^{-3}\,\text{mol}\,\text{cm}^{-3}$). This is a very small concentration reflecting the fact that "association" is very unfavorable.

---

[223] For details see Ben-Naim (2006).
[224] For details see Ben-Naim (2006).

It is clear that in the absence of any kind of strong attractive interaction between the monomers, there will be an extremely low density of "dimers" in a gaseous phase, either in this view or in the previous view. Therefore, we next turn to analyze the contributions of the direct and indirect interactions to the PMF in (4.3.20).

4.3.3. *Some numerical estimates of various contributions to the total PMF*

As we have done in Chapter 3, we split the intermolecular potential energy between any two species as

$$U = U^H + U^S + \sum_{i,j} U^{FG}(i,j). \qquad (4.3.22)$$

The first term is the *hard* part of the potential. This is essentially the repulsive part of the pair potential. The second term is the *soft* part. This is basically a van der Waals interaction between the two interacting molecules. The third term is a sum over all the specific interactions between the FGs on one molecule and the FGs on the other.

Thus, we view the protein as a sphere having a hard core and a weak attractive van der Waals interaction. In addition, there are FGs that are exposed to the solvent. For simplicity, we shall assume that there are only two types of FGs: A methyl group (CH_3) representing a hydrophobic group ($H\phi O$), and a hydroxyl or a carbonyl group (OH or C=O) that represents a hydrophilic group ($H\phi I$). Figure 4.7 shows a schematic split of the interaction potential between the two proteins.

Each of the three components of the pair potential (4.3.22) contributes both to the *direct* interaction $\Delta U(r)$ and to the indirect interaction $\delta G(r)$. We shall discuss each of these separately.

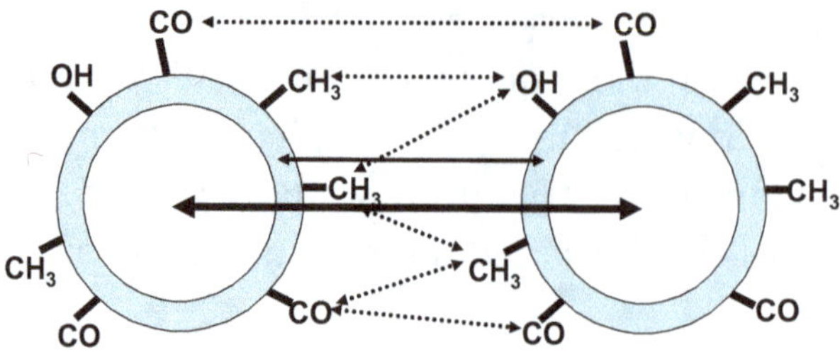

Fig. 4.7   Schematic split of the interaction energy between a protein and a water molecule into three parts, corresponding to Eq. (4.3.22): hard (heavy double arrows), soft (thin double arrows), and specific functional groups (dotted arrows, only a few are shown).

(i) *The direct interactions between the two proteins*

The hard part of the potential function is by definition zero at distances greater than the diameter σ_p. The soft part will contribute negatively to the PMF. However, for two spheres at contact distance, these interactions are weak and one cannot expect much from these interactions.[225] The only significant contribution can be due to FGs on each protein. Specifically, these FGs can form *direct* hydrogen bonds (HBs). This could contribute some $25\,\text{kJ}\,\text{mol}^{-1}$ per HB formed. Again, we note that for two spherical proteins, we cannot expect too many direct hydrogen bonds between FGs on one protein and FGs on the other. Thus, at most, we can expect only very few such FGs to be in the right configuration to form direct HBs. The situation will be different for proteins of different structures, say, two cubes that can form a large interface between the two monomers. Thus, although these *direct* HBs can contribute significantly to the PMF, their contribution in aqueous solutions will be diminished due to the loss of the solvation Gibbs energy of these FGs upon binding.

---

[225] Wang and Ben-Naim (1997).

(ii) *Indirect, or solvent-induced, interaction between two globular proteins*

The quantity δG referred to as the solvent-induced interaction can be expressed as

$$\delta G(r) = \Delta G^*_{"D"}(r) - \Delta G^*_1 - \Delta G^*_2$$

$$= -k_B T \ln \frac{\langle \exp[-\beta B(r)]\rangle_0}{\langle \exp[-\beta B_1]\rangle_0 \langle \exp[-\beta B_2]\rangle_0}, \qquad (4.3.23)$$

where ΔG^*_α is the solvation Gibbs energy of the species α, and B_α is the binding energy, i.e. the total interaction energy between the species α and all the solvent molecules at some specified configuration $\mathbf{X}_1, \ldots, \mathbf{X}_N$. $B(r)$ is the binding energy of the "dimer" D at a distance r. The average is taken over all the configurations of the solvent molecules with the probability density of the solvent configurations *before* the insertion of the solute α, i.e.

$$P(\mathbf{X}_1, \ldots, \mathbf{X}_N) = \frac{\exp[-\beta U_N(\mathbf{X}_1, \ldots, \mathbf{X}_N)]}{\int \cdots \int d\mathbf{X}^N \exp[-\beta U_N(\mathbf{X}_1, \ldots, \mathbf{X}_N)]}.$$

$$(4.3.24)$$

Each of the solvation Gibbs energies ΔG^*_α can have a component corresponding to the three parts of the pair potential as described in Eq. (4.3.23). We can now estimate the various contributions to the solvent-induced part of the driving force. First, we calculate the hard (H) part. The hard part of the solvation Gibbs energy is simply the work required to form a cavity in the solvent, the radius of which is R_{CAV}, given by[226]

$$R_{CAV} = \frac{\sigma_P + \sigma_W}{2}. \qquad (4.3.25)$$

---

[226]Reiss (1966), Ben-Naim (1992). See also Appendix K.

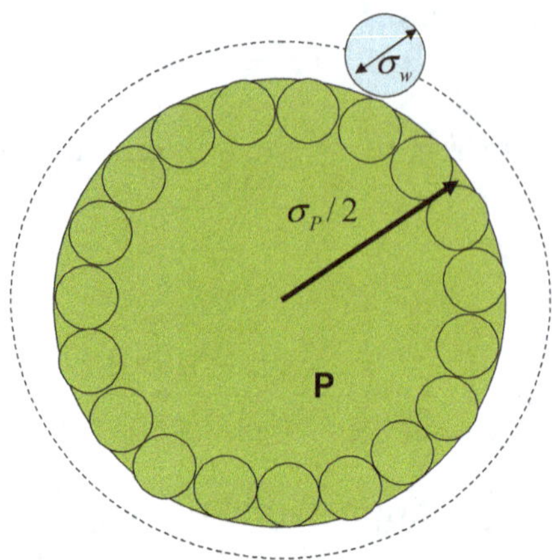

Fig. 4.8 The excluded volume formed by a protein of diameter σ_p and a solvent molecule of diameter σ_w (4.3.25).

The cavity is nothing but a region in the solvent that is excluded for a solvent molecule (Fig. 4.8). When the two proteins are at a large separation, their excluded volume (with respect to the solvent molecules) is simply the sum of the excluded volumes of the two separate proteins. However, when they form a dimer (whether real or virtual), the excluded volume of the pair is always smaller than that of the sum of the two excluded volumes of the two monomers. This effect always contributes negatively to δG. For a large protein, the solvation Gibbs energy is simply the $P\Delta V$ work required to create the excluded volume,[227] i.e.

$$\Delta G_\alpha^{*H} = PV_\alpha^{EX}. \tag{4.3.26}$$

Hence, the contribution to δG from this part is

$$\delta G^H = \Delta G_D^{*H} - \Delta G_{M_1}^{*H} - \Delta G_{M_2}^{*H}$$

$$= P\left[V_D^{EX} - V_{M_1}^{EX} - V_{M_2}^{EX}\right] < 0. \tag{4.3.27}$$

---

[227] Reiss (1966), Ben-Naim (1992).

Since the change in the excluded volume is always negative, δG^H will be negative too. The volume in the square brackets is simply the volume of the overlapping region between the excluded volumes of the two monomers (Fig. 4.2). For a protein of radius $R = 100\,\text{Å}$, we can estimate δG^H to be at most of the order of $-0.025\,\text{kJ}\,\text{mol}^{-1}$, which clearly cannot contribute significantly to δG.

The second contribution to δG is due to the soft (S) part of the solute–solvent interaction. The soft part of the solvation Gibbs energy of a solute α is the *conditional* solvation of the soft part of the potential U^S given that the hard part has been "turned on," or has been solvated.[228] This part will be proportional to the surface area of the protein, i.e. (see Sec. 4.2)

$$\Delta G_\alpha^{*S/H} = -\varepsilon A_\alpha, \qquad (4.3.28)$$

where $\varepsilon > 0$ is a typical interaction energy, say, between two methyl groups, and A_α is the surface area of the protein.

When two proteins form a dimer, the contribution of this part of the interaction to δG is

$$\delta G^{S/H} = \Delta G_D^{*S/H} - \Delta G_{M_1}^{*S/H} - \Delta G_{M_2}^{*S/H}$$
$$= -\varepsilon[A_D - A_{M_1} - A_{M_2}] > 0. \qquad (4.3.29)$$

Since the surface of the dimer is always smaller than the surface of the two separated monomers, the quantity in the brackets is negative, and the contribution to δG from (4.3.30) is positive. Therefore, this cannot contribute negatively to δG.

We note again that the effects of both the hard and the soft parts of the potential depend only on the *geometry* of the protein and not on the specific distribution of the FGs on its surface. Therefore, this term cannot contribute to the specificity of the

---

[228] See Ben-Naim (2006).

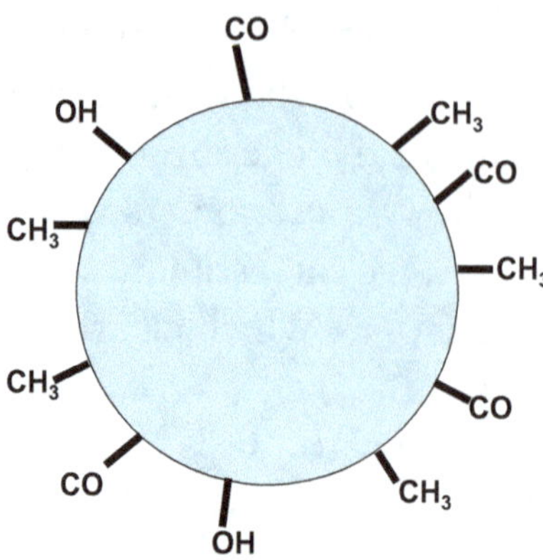

Fig. 4.9 Schematic description of a globular protein modeled as a sphere having functional groups protruding from its surface.

structure of the dimer (i.e. the orientation of one monomer relative to the other).

Finally, we discuss the contributions due to the specific functional groups on the surface of the protein. As we have done in Sec. 4.2, we view a globular protein as a large sphere having some functional groups (FGs) exposed to the solvent (Fig. 4.9). When two such proteins are brought to a close distance, the surroundings of some of these FGs may change. The exposed groups can be classified into three main classes[229]:

(a) FGs the solvation of which does not change upon association;

(b) FGs the solvation of which is completely eliminated in the process of association;

(c) FGs the solvation of which is only partially changed upon association.

Clearly, a FG in class (a) cannot contribute to δG. A FG in class (b) will contribute its conditional solvation Gibbs

---

[229]See Sec. 4.2 and Ben-Naim (1992).

energy.[230] It was estimated that the conditional solvation Gibbs energy of a *hydrophobic* group is about $-1.5\,\text{kJ}\,\text{mol}^{-1}$, whereas that of a hydrophilic group, such as a hydroxyl group, is of the order of about $-29\,\text{kJ}\,\text{mol}^{-1}$.

Thus, a hydrophobic group that loses it solvation will contribute negatively to δG. However, there cannot be too many of such groups in the interface between the two proteins at contact distance. Therefore, we can conclude that though these hydrophobic groups will contribute in the "right direction" to the driving force for association, we cannot expect that this contribution will be very large.

On the other hand, the elimination of the solvation of a hydrophilic group (such as OH, C=O, or NH) will contribute positively to δG (i.e. in the "wrong direction"). Therefore, this kind of hydrophilic effect could not contribute to the solvent-induced "driving force."

What remains is to consider FGs that belong to class (c), i.e. groups whose solvation spheres are initially uncorrelated but become correlated in the final configuration (either the *real* dimer or the virtual dimer). Two examples are shown as yellow disks in Fig. 4.3.

For simplicity, let us assume that there are only two kinds of FGs on the surface of the protein: A methyl group representing a hydrophobic FG, and a carbonyl (C=O) or a hydroxyl (OH) group representing a hydrophilic FG. In applying this methodology to a real specific protein, one needs to list all FGs and their specific location on the surface of the protein. Such a detailed listing has been done for some proteins.[231] Here, we shall assume that the two groups are randomly distributed on the surface of the protein, and that the mole fraction of the

---

[230]See Sec. 4.2.
[231]Wang and Ben-Naim (1997).

hydrophobic groups is $x(H\phi O)$, and the mole fraction of the hydrophilic groups is $x(H\phi I) = 1 - x(H\phi O)$. In Chapter 2, we found (and it is also clear intuitively) that the mole fraction of the hydrophilic groups that are exposed to the solvent must be larger than the mole fraction of hydrophobic groups (otherwise globular proteins would not be so soluble in water, see Chapter 2).

Clearly, FGs that belong to class (c) will contribute to δG only if they are close to each other at the configuration of the dimer. Whatever the relative concentration of the FGs on the surface of the proteins is, the number of such groups will grow as R, where R is the radius of the protein.

To estimate the contribution of the hydrophobic and the hydrophilic interactions, we assume that the maximum strength of the hydrophobic interaction is at contact distance between the two methyl groups (Fig. 4.10). Thus, if r_C is the distance between the centers of the carbon atoms, then the hydrophobic groups that are on the spherical strip of area $2\pi y d$, where

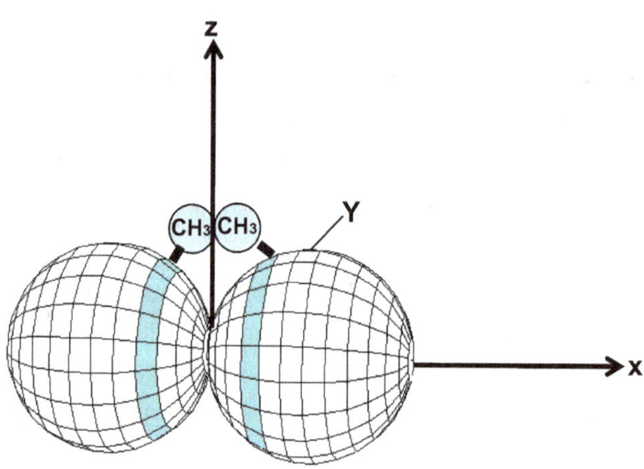

Fig. 4.10 Hydrophobic interaction between a pair of the methyl groups on the surface of the protein.

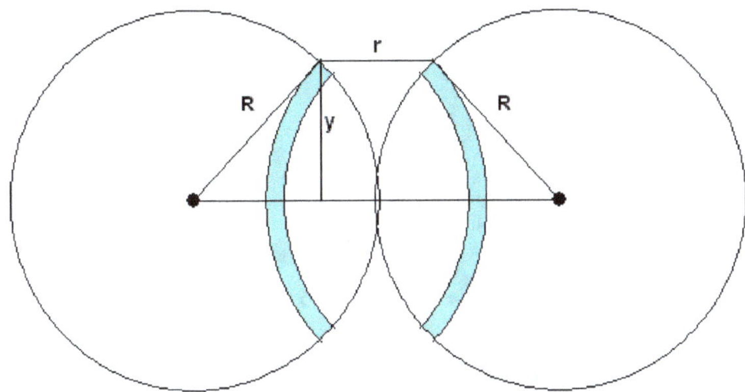

Fig. 4.11   The relevant geometry for the calculation of the contributions of the hydrophobic and hydrophilic interactions. r in the figure is either r_C in Eq. (4.3.30) or r_O in Eq. (4.3.31). The FGs that are in the two strips (blue) can be correlated.

$y = \frac{1}{2}\sqrt{r_C(4R - r_C)}$ and d is the width of the strip chosen as 0.5 Å, will contribute the maximum hydrophobic interaction (Fig. 4.11). Let ρ_C be the density of the hydrophobic groups (CH_3) per unit area of the surface. The number of such groups that contribute maximum hydrophobic interaction is

$$N(H\phi O) = \rho_C \pi d\sqrt{r_C(4R - r_C)}. \qquad (4.3.30)$$

Similarly, for the $H\phi I$ groups, we can estimate the number of groups in a strip of area $2\pi\, yd$ with $y = \frac{1}{2}\sqrt{r_O(4R - r_O)}$, where now r_O is the distance between two oxygens (of a hydroxyl or carbonyl group) that can form strong $H\phi I$ interactions. The number of such groups is

$$N(H\phi I) = \rho_O \pi d\sqrt{r_O(4R - r_O)}, \qquad (4.3.31)$$

where ρ_O is the average surface density of hydrophilic groups on the surface of the protein. r is either r_C for the hydrophobic case, or r_O for the hydrophilic case. It has been estimated that for each pair of such $H\phi I$ groups, one on each protein, that are at about a distance of 4.5 Å can be connected by HB bridge by a

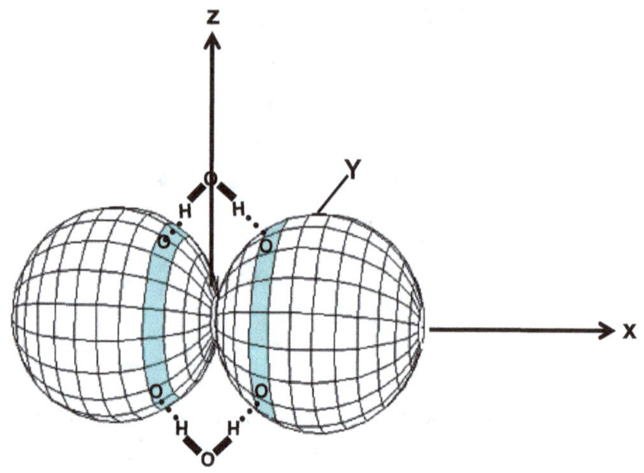

Fig. 4.12 Hydrophilic interaction between a pair of carbonyl groups. The main contribution to this effect is a hydrogen-bond bridge formed by a water molecule.

water molecule as indicated in Fig. 4.12. The Gibbs energy for one such $H\phi I$ interaction has been estimated to be about

$$\delta G(H\phi I) \approx -10.5\,\text{kJ}\,\text{mol}^{-1}, \tag{4.3.32}$$

and the corresponding contribution due to one pair of $H\phi O$ interaction is

$$\delta G(H\phi O) \approx -1.3\,\text{kJ}\,\text{mol}^{-1}. \tag{4.3.33}$$

Figure 4.13 summarizes the effect of both hydrophobic and hydrophilic interactions on the solvent-induced driving forces for protein–protein association. Here, we fixed the value of the temperature at $T = 298.15\,\text{K}$, and the average mass density of the protein at $\delta = 0.7\,\text{g/cm}^3$. The different curves correspond to different mole fractions of the hydrophilic groups on the surface of the protein.

For $x_O = 0.0$, all groups are hydrophobic. The curve is almost flat. This means that the hydrophobic groups contribute negatively to the association free energy; instead of a steeply

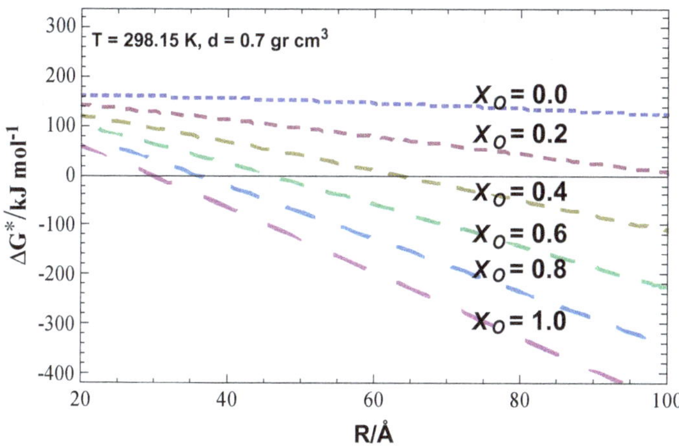

Fig. 4.13 Summary of the results of adding hydrophilic groups to the surface of the protein. x_O is the mole fraction of hydrophilic groups that are exposed to the solvent.

increasing function of R as in Figs. 4.4 and 4.5, the curve slightly decreases with R. However, the values of ΔG are still large and positive, indicating that association is very unfavorable.

For $x_O = 0.2$, we already see that ΔG becomes a slightly decreasing function of R. Even for $R = 100\,\text{Å}$, the values of ΔG are still positive. Once the mole fraction of the hydrophilic groups increases beyond 0.4, the values of the Gibbs energy become more and more negative, which means that for large R, there are more and more pairs of hydrophilic groups that are at the correct distance to form water bridges between them. It has been estimated by Kauzmann that on average, hydrophobic side chains make up about one-third of all side chains. Considering the fact that the backbone of the protein also contributes hydrophilic groups ($C{=}O$ and NH), we can estimate that an "average" protein might have a mole fraction of hydrophilic groups around 0.6 to 0.7. At these concentrations of hydrophilic groups, we find large negative Gibbs energies. The values of the negative Gibbs energies due to $H\phi I$ interactions are more than enough to compensate for the positive ΔG shown in Figs. 4.4 and 4.5. Thus, the net effect of adding the solvent-induced contribution is to turn a process which is highly unfavorable into a highly favorable one.

Finally, we note that $H\phi I$ interactions could be decisive in determining the specificity of the mode of association between two proteins. Consider, for example, two globular proteins, the surfaces of which are "decorated" with $H\phi O$ groups (represented as C in Fig. 4.14), and $H\phi I$ groups (represented as O in Fig. 4.14). Traditionally, the concept of molecular recognition was understood as requiring some kind of *geometrical* fitting between the two binding partners. However, viewed from the solvent-induced effect, a geometrical fit is not necessary. Two spheres with seemingly no specific structural features will bind even strongly and specifically when the relative configuration of the two proteins is such that maximum number of

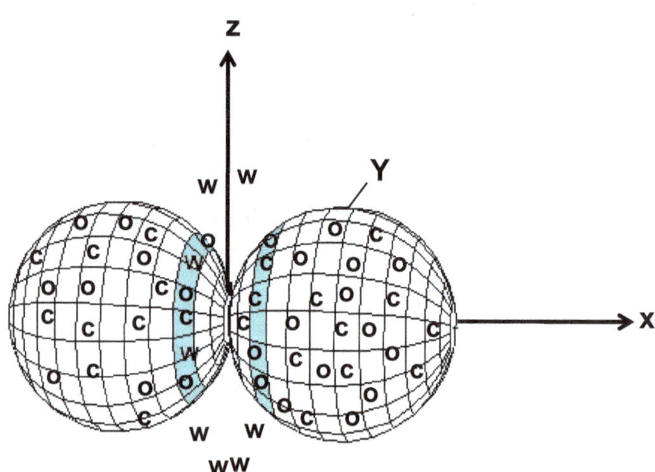

Fig. 4.14 Schematic illustration of two proteins with hydrophobic (C), and hydrophilic (O), groups on its surface. Water molecules (*w*) can "stitch" the two proteins by forming hydrogen-bond bridges between two hydrophilic groups.

hydrogen-bonded water bridges may be formed. Each water bridge can contribute a factor of about $\exp[-25/0.5] \approx 150$ to the concentration of the "dimer." This means that even at a very dilute solution of protein, these $H\phi I$ interactions can turn the rare event of dimer-formation into an almost certain event. The above argument is *a fortiori* true for binding proteins to DNA. The latter is abundantly "decorated" with $H\phi I$ groups on its surface. We shall discuss the question of molecular recognition in more detail in Sec. 4.6.

4.4. *$H\phi O$* or *$H\phi I$* Interaction: Which is More Important in the Association Process?

In this section, we examine the association between two model solutes. Each solute is a long molecule having one $H\phi I$ side and one $H\phi O$ side. The molecules are long enough such that when they bind in either the $H\phi O - H\phi O$ mode or the $H\phi I - H\phi I$ mode (see Fig. 4.15), all the FGs in the interface between the two solutes are effectively not exposed to the solvent, i.e. these groups belong to the **I** group. On the other hand, the FGs on the other side of the interface are presumed to belong to the **E** group.

Having a dilute solution of such solutes, we can, in principle, measure the association constant and the corresponding standard Gibbs energy. The experimental Gibbs energy takes into account all possible "modes" of association between **P** and **L**. In this section, we shall not be interested in the Gibbs energy of association, but only in comparing the two specific *modes* of *association*. Therefore, instead of the Gibbs energy of association between the two solutes, we shall examine the potential of mean force between the two solutes along two specific lines of approach. We are interested in comparing the Gibbs energy change for the process of bringing **P** and **L** from infinite

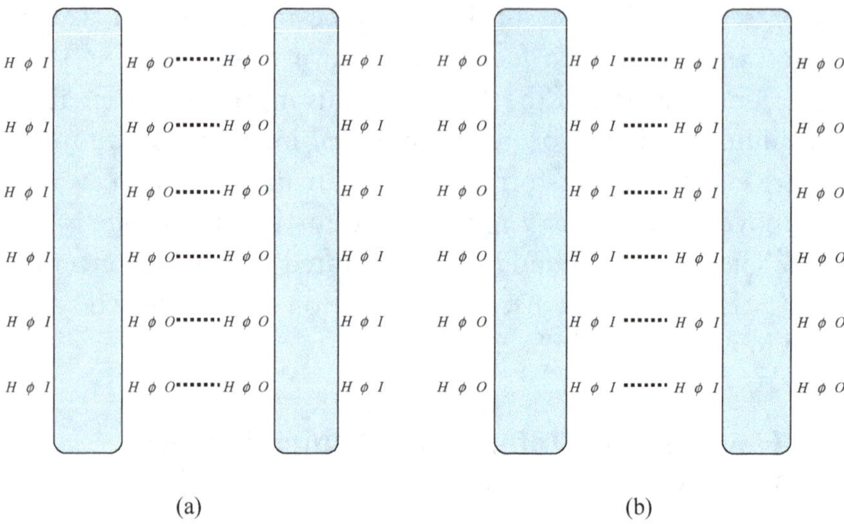

Fig. 4.15 Two rigid molecules having $H\phi O$ groups on one side, and $H\phi I$ groups on the other. We compare the probabilities of association in the (a) $H\phi O{-}H\phi O$ mode and (b) $H\phi I{-}H\phi I$ mode.

separation to the final configuration of **PL**, once when **PL** is in the $H\phi O{-}H\phi O$ mode (Fig. 4.15a), and once for the $H\phi I{-}H\phi I$ mode (Fig. 4.15b).

We denote the corresponding Gibbs energy changes by $\Delta G(H\phi O{-}H\phi O)$ and $\Delta G(H\phi I{-}H\phi I)$, respectively. The difference between the two is related to the ratio between the two probabilities, i.e.[232]

$$\frac{\Pr(H\phi O{-}H\phi O)}{\Pr(H\phi I{-}H\phi I)} = \frac{\exp[-\beta\Delta G(H\phi O{-}H\phi O)]}{\exp[-\beta\Delta G(H\phi I{-}H\phi I)]}. \tag{4.4.1}$$

Note carefully that each probability on the left-hand side of (4.4.1) might be extremely small, but we are interested only in the *ratio* of these two probabilities. When the ratio is much larger than 1, we can say that the $H\phi O$ mode of binding is

---

[232]See also Appendices C and P.

more favorable than the $H\phi I$ mode. On the other hand, when the ratio is very small, we say that the $H\phi I$ mode is more favorable.

The cases discussed below were designed in such a way that they correspond to the story of the rise and fall of the $H\phi O$ vis-à-vis $H\phi I$ effect, as discussed in the preface. Furthermore, there is also a pedagogical moral hidden in the design of these experiments: It is not always true that the observation of a $H\phi O$ (or a $H\phi I$) mode of binding implies that the $H\phi O$ (or the $H\phi I$) interactions are the more important part of the "driving force" for that process.

In all cases, the solutes **P** and **L** are linear molecules of roughly the same size. Each contains two sides: One $H\phi O$ and one $H\phi I$ side. When they form a complex **PL**, we can assume that both δG^H and $\delta G^{S/H}$ are the same for the two modes of association. Therefore, these terms cannot affect the ratio of probabilities in (4.4.1). The only factor that affects the probability ratio is the type of the FGs in the binding interface. We shall also assume that when **P** and **L** are separated, all the FGs on their surface are independently solvated.[233]

The following values of the Gibbs energies are used in the calculation of the probability ratio in (4.4.1):

(i) The direct van der Waals interaction energy between two methyl groups at contact distance is about $-1\,k_B T$.

(ii) The direct HB between one arm of a donor and one arm of an acceptor is about $-11\,k_B T$.

(iii) The conditional solvation Gibbs energy of a methyl group in water is about $+1\,k_B T$.

(iv) The conditional solvation Gibbs energy of one arm of a $H\phi I$ group in water is about $-3.7\,k_B T$.

---

[233]See Appendix F.

For simplicity and concreteness, we assume that there are only six FGs on each side of the solute molecules. It is clear that the number of such groups will not affect the ratio in (4.4.1).

4.4.1. *Association of* P *and* L *in an ideal gas phase*

In the first example, we compare the two "dimers" (a) and (b) shown in Fig. 4.15. For the $H\phi O$ mode, in the ideal gas phase, the interaction energy between the six methyl groups is of the order of $-6\,k_B T$. On the other hand, in the $H\phi I$ mode, we have six HBs with total interaction energy $-66\,k_B T$.

Thus, for this case we have

$$\Delta G(H\phi O - H\phi O) \approx -6\,k_B T,$$
$$\Delta G(H\phi I - H\phi I) \approx -66\,k_B T, \qquad (4.4.2)$$
$$\frac{\Pr(H\phi O - H\phi O)}{\Pr(H\phi I - H\phi I)} \approx \exp[-60] \approx 8.7 \times 10^{-27}. \qquad (4.4.3)$$

In this case, as expected, the $H\phi I$ mode is far more favorable than the $H\phi O$ mode. Note that the terms $H\phi O$ and $H\phi I$ do not strictly pertain to this case simply because there are no water molecules involved in the association. Nevertheless, we use the terms $H\phi O$ mode and $H\phi I$ mode since these involve the $H\phi O$ and $H\phi I$ groups, respectively, and it will also be convenient for comparison with other cases involving water, as discussed below. We conclude that in an ideal gas phase: *Hydrogen bonding dominates the driving force for the association process*

This conclusion is in accord with the commonly held opinion, led by Pauling, that HBs are the dominant driving forces in biochemical processes.

4.4.2. *Association in an organic liquid*

We use the same model compounds as in Sec. 4.4.2, and the same two possible modes of association. Now, however, the two solutes are in an organic liquid, say, hexane or benzene. In this case, we must also consider the loss of the conditional solvation Gibbs energies of the $H\phi O$ and the $H\phi I$ groups when forming the complex **PL**. However, since the solvent molecules do not form HBs with the $H\phi I$ groups, we can expect that the solvation Gibbs energy of a $H\phi O$ group and a $H\phi I$ group of roughly the same size would not differ very much. Hence, we expect that the total Gibbs energies of the association process would still be dominated by the *direct* interaction between the methyl groups in the $H\phi O$ mode, and the direct HBs between the hydroxyl and the carbonyl groups in the $H\phi I$ mode.[234]

Thus, although we expect the values in (4.4.2) to be modified for the association process in a solvent, we do not expect that such a modification will drastically change the ratio of the probabilities as in (4.4.3). Thus, in this case, we can still expect that *hydrogen bonding will dominate the driving force for the association process.*

4.4.3. *Association in aqueous solutions*

In calculating the Gibbs energy changes for the same processes in aqueous solutions, we must take into account both the direct interaction as well as the solvent-induced effect.

For the $H\phi O$ mode, we take about $-k_B T$ for each direct interaction. We also take about $-k_B T$ for the loss of the conditional solvation Gibbs energy of a pair of methyl groups. Therefore, we expect in this case to have

$$\Delta G(H\phi O - H\phi O) \approx -6 \times 2 = -12\, k_B T. \qquad (4.4.4)$$

On the other hand, for the $H\phi I$ mode, we have the *gain* of one direct HB, contributing about $-11\, k_B T$ per bond, and the

---

[234]Note again that since the solvent is not water, the terms $H\phi O$ and $H\phi I$ do not apply. We use these terms here for the same reason we use them in Sec. 4.4.1.

loss of conditional solvation *per arm* which is about $+3.7\,k_BT$. Thus, altogether we estimate for this case

$$\Delta G(H\phi I - H\phi I) \approx 6[-11 + 5 \times 3.7]\,k_BT \approx 45\,k_BT.$$

$$(4.4.5)$$

Thus, for the probability ratio we have

$$\frac{\Pr(H\phi O - H\phi O)}{\Pr(H\phi I - H\phi I)} = \frac{\exp[-\beta\Delta G(H\phi O - H\phi O)]}{\exp[-\beta\Delta G(H\phi I - H\phi I)]}$$

$$\approx [57] \approx 6 \times 10^{24}.$$

$$(4.4.6)$$

We see that in this case, the $H\phi O$ mode of association becomes far more favorable than the $H\phi I$ mode. This case seems to be in accordance with the hypothesis that the $H\phi O$ effect is the dominant driving force for the association process. In other words, we see that the $H\phi O$ groups are the ones which end up in the interface (the interior) of the complex **PL**, therefore we might conclude that it is the $H\phi O$ effect that is also responsible for the "driving force" leading to this mode of association.

Note however that we did not use the HB inventory argument to dismiss the contribution of direct hydrogen bonding. Instead, we have intentionally used the groups OH and C=O so that for each *direct* HB formed, we lose the solvation Gibbs energy of *five* arms (three of the OH and two of the C=O groups). It is this large loss of solvation Gibbs energy that discourages the $H\phi I$ mode of association, and favors the $H\phi O$ mode.

The moral we learn from this example is that one cannot reach any conclusion regarding the "driving force" simply from observation of what happens.

This example shows how important it is to *count* the number of *arms* the solvation of which was lost in the association process. In this particular example, five arms were lost, contributing about $55\,k_BT$ to ΔG, but only one HB was formed, which contributes only $-11\,k_BT$ to ΔG. Thus, the net effect is that the difference $\Delta G(H\phi O - H\phi O) - \Delta G(H\phi I - H\phi I)$ becomes positive.

Note carefully that the favoring of the $H\phi O$ mode of binding was due neither to the strength of the $H\phi O$ bonds formed by the two methyl groups, nor to the HB inventory argument. The reason that the probability ratio ended up favoring the $H\phi O$ mode is the *large number* of *arms* the solvation of which was lost.

Consider another extreme case, where for each pair of arms lost, we gain one direct HB (Fig. 4.16). Here, we assume that the amine group can form only one HB with water molecules. Thus, in the association of the molecule shown in Fig. 4.16, we have

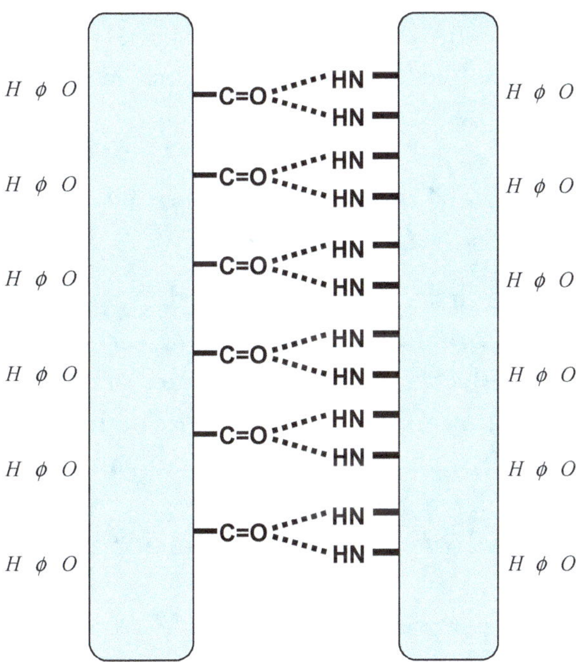

Fig. 4.16 Modified molecules as in Fig. 4.15. Here, however, we have six carbonyl groups on one molecule and 12 amine groups on the other molecule. The $H\phi O$ sides are unchanged as in Fig. 4.15.

12 direct HBs formed, but the number of arms the solvation of which is lost is 24. Therefore, in this example, the association will turn back to favoring the $H\phi I$ mode.

We can conclude from these two cases that the association in either the $H\phi O$ mode (in the first example), or in the $H\phi I$ mode (in the second example) is due to the balance between the energy gain due to the formation of a direct HB, and the solvation Gibbs energy change due to the loss of the solvated arms of the $H\phi I$ groups.

It should be stressed again that in the first example, one might be tempted to conclude that the $H\phi O$ mode of association is due to the strength of the $H\phi O$ effect and the "marginal contribution of hydrogen bonding" when they occur in aqueous solutions. Both of these arguments are incorrect. The $H\phi O$ effect in this case is not as large as it was believed, and the contribution of direct hydrogen bonding is not weakened by the HB inventory argument as was believed.

4.4.4. *Enhancement of the HϕO mode by strengthening the HϕI effects*

Look again at the first example discussed in Sec. 4.4.3. As we have seen, the favoring of the $H\phi O$ mode of binding was not due to the $H\phi O$ effect, but to the *reluctance* of the $H\phi I$ groups to lose the solvation Gibbs energy of so many arms. In the next example, we shall *strengthen* the conclusion reached in Sec. 4.4.3. We shall *enhance* the $H\phi O$ mode of association, not by enhancing the $H\phi O$ effect, but by manipulating the $H\phi I$ effect. We shall give two examples here. The reader will benefit from designing other possibilities.

(i) *Changing the relative distances between the HϕI groups*

In all the examples discussed in the previous sections, we assumed that the FGs were independently solvated. That assumption was necessary for the calculation of the total "driving force" as a sum of the contributions of the "diving force" of each pair of FGs, either $H\phi O$ or $H\phi I$.

Next, suppose that we start with the same two solutes as in Fig. 4.15. We do not change the *composition* of the surfaces of the two solutes, i.e. we keep the same number of $H\phi O$ groups and the same number of $H\phi I$ groups as in Fig. 4.15. All we do now is change the distances between a few pairs of $H\phi I$ groups. The modified molecule is shown in Fig. 4.17. For concreteness, suppose we bring a few pairs of $H\phi I$ groups on each solute to a distance of about 4.5 Å. This change will make the solvation of the two $H\phi I$ groups dependent and will lower the solvation Gibbs energy of each pair of $H\phi I$ groups by about $-5k_BT$.

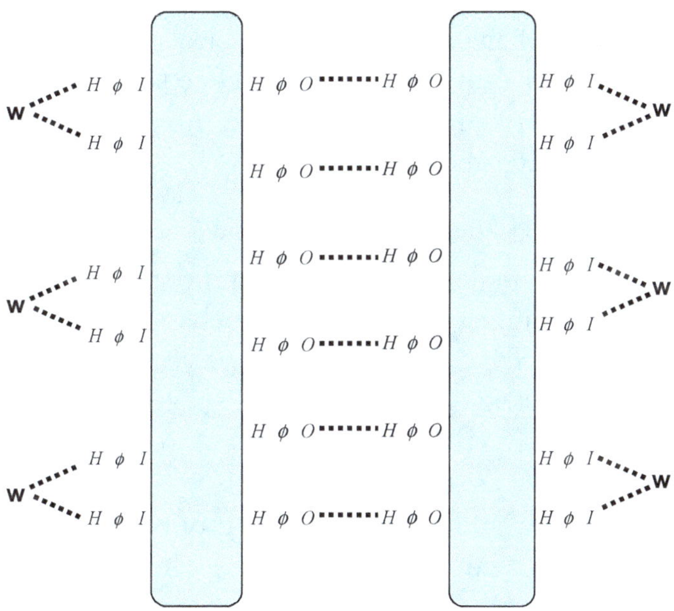

Fig. 4.17 Modified molecules as in Fig. 4.15. Here, however, the distances between the $H\phi I$ groups have been changed so that a water molecule can form a HB bridge between two $H\phi I$ groups. The $H\phi O$ side is not changed.

Thus, instead of the result in (4.4.6), we should have

$$\frac{\text{Pr}(H\phi O - H\phi O)}{\text{Pr}(H\phi I - H\phi I)} \approx 10^{24} \times \exp[10] \approx 22{,}000 \times 10^{24}.$$

$$(4.4.7)$$

This is for one pair of correlated $H\phi I$ groups. Clearly, the more pairs are correlated, the greater will be the tendency to bind in the $H\phi O$ mode.

Thus, by manipulating the $H\phi I$ *effects*, we have made the $H\phi O$ mode of association more favorable. Clearly, we cannot interpret this *enhancement* of the $H\phi O$ mode of association to a strengthening of the $H\phi O$ effect. Note that each pair of $H\phi I$ groups at the correct distance and orientation contributes about 150 to the probability ratio, and for the two pairs of $H\phi I$ groups, $150^2 \approx 22{,}000$.

We note also that if the two solutes are spheres or cubes, then one can obtain more dramatic $H\phi I$ effects by moving *three* groups to such configuration where the three groups can be bridged by one water molecule.[235]

(ii) *Adding a new $H\phi I$ effect*

In the previous example, we started with the same two solutes, with the same number of FGs, but we changed the distance between the two $H\phi I$ groups. Here, we leave the six $H\phi I$ groups as in the original example in Fig. 4.15, but we add to each solute two $H\phi I$ groups in the **J** region, as shown in Fig. 4.18.

Specifically, we add four carboxyl groups (two for each solute) in such a way that they can be bridged by water molecules. Thus, each pair of carboxyl groups (one on each solute) contributes an additional negative Gibbs energy change

---

[235]See also Sec. 4.8 and Appendix G.

Fig. 4.18 Modified molecules as in Fig. 4.15. Here, however, we add two pairs of $H\phi I$ groups in such a way that in the $H\phi O$ mode of association, these $H\phi I$ groups can form a one-water bridge.

to ΔG of the $H\phi O$ mode, leading to an enhancement of the $H\phi O$ mode by the same amount as in case (i).

Clearly, this enhancement was not caused by the strengthening of the $H\phi O$ effect, but by adding a new $H\phi I$ effect.

The reader is urged to consider other possible $H\phi I$ effects. For instance, changing the distances between a few pairs of $H\phi I$ groups, or changing the configuration of a triplet of $H\phi I$ groups so that they form an equilateral triangle with an edge of 4.5 Å. Starting with the $H\phi I$ mode example discussed in this section and Fig. 4.18, can you manipulate the $H\phi I$ interactions in such a way that the two solutes will switch the mode of association from the $H\phi I$ mode to the $H\phi O$ mode?

Instead of linear molecules, as discussed in the above examples, try to design examples of such planes or cubes with different distributions of $H\phi O$ and $H\phi I$ groups on their surfaces. What happens when one changes the distances between one or more pairs of $H\phi I$ groups?

It would be instructive for the reader to go through the examples in this section, and compare the arguments we used here with the three stages of my "conversion" to favoring $H\phi I$ over $H\phi O$ effects, as discussed in the preface.

4.5. Association by the Complete Absorption of a Small Solute into a Big Solute

In this section, we discuss an extreme example of an association process. A ligand **L** is completely absorbed by the bigger solute **P** (Fig. 4.19). We shall discuss two cases: The first, where the absorption of **L** does not affect the solvation of **P**, and the second, where **P** has a few conformations at equilibrium, and the absorption of **L** might affect the equilibrium concentrations of the two conformations.

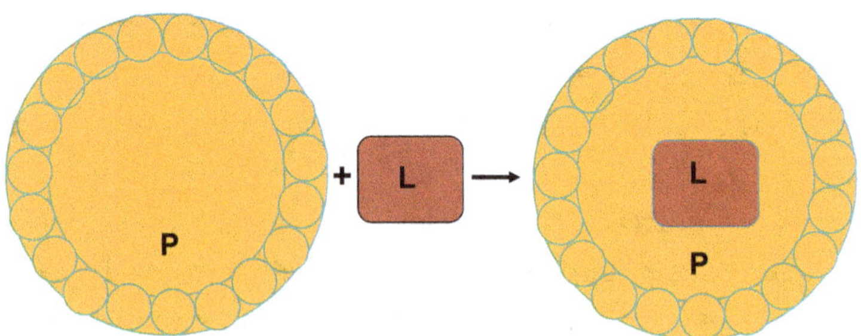

Fig. 4.19   Association by absorption of L into P.

In both cases, the solvent-induced part of the Gibbs energy of association involves only the loss of the solvation Gibbs energy of the ligand **L**.

4.5.1. *Absorption without conformational changes in* **P**

Consider the process of dimerization depicted in Fig. 4.19. As in previous sections, we shall be interested in the process of bringing the two solutes **L** and **P** from infinite separation to the final configuration of **PL**. As we have seen in Sec. 4.2, the Gibbs energy change for this process may be written as

$$\Delta G(\mathbf{P} + \mathbf{L} \to \mathbf{PL}) = \Delta U + \delta G, \qquad (4.5.1)$$

where ΔU is simply the interaction energy between **L** and **P** at the final configuration. The second term on the right-hand side of (4.5.1) may be written as

$$\delta G = \Delta G^*_{PL} - \Delta G^*_P - \Delta G^*_L. \qquad (4.5.2)$$

If we assume that the solute **L** is completely absorbed by **P** without any effect on the conformation of **P**, then (4.5.2) reduces to

$$\delta G = -\Delta G^*_L. \qquad (4.5.3)$$

Thus, the only contribution to the solvent-induced part of the "driving force" (4.5.1) is simply the loss of the solvation Gibbs energy of the solute **L**. If **L** is a $H\phi O$ solute, then ΔG^*_L is positive, and this corresponds to the solvation Gibbs energy of a $H\phi O$ solute in water, as used in Kauzmann's model. Note, however, that the solvation Gibbs energy of the solute **L** in an organic liquid does not feature in (4.5.2).[236] This is unlike Kauzmann's model of the transfer of a solute from water to an organic

---

[236]See also Appendix H.

liquid, the latter representing the interior of the protein. Instead of the solvation Gibbs energy in an "organic liquid," we have the total interaction ΔU between the solutes **L** and **P** at the final configuration **PL**.

It should also be noted that the assumption we have made in proceeding from (4.5.2) to (4.5.3) may be weakened. We do not need to assume that **L** does not have any effect on the conformation of **P**. The only requirement is that the solvation Gibbs energy of **P** does not change in this process. The result (4.5.3) also holds true when the solute **L** does have an effect on the conformation of **P**. This is discussed in the following section.

4.5.2. *Absorption with conformational changes in* P

The case of the absorption of a ligand in a protein with conformational changes is important in understanding the cooperativity of the binding process. The most studied example is the binding of oxygen to hemoglobin.[237]

For simplicity, we assume that **P** has two conformations, denoted by $\mathbf{P}_1$ and $\mathbf{P}_2$ at equilibrium

$$\mathbf{P}_1 \rightleftarrows \mathbf{P}_2. \tag{4.5.4}$$

Suppose also that the polymer **P** is very diluted in water, so that the equilibrium constant might be expressed in terms of the mole functions of $\mathbf{P}_1$ and $\mathbf{P}_2$, i.e.

$$K = \left(\frac{x_1}{x_2}\right)_{\text{eq}}. \tag{4.5.5}$$

When **L** is absorbed into **P**, its interaction energy ΔU is different for the different conformations. Denoting by ΔU_1 and ΔU_2 the total interaction energy between **L** and the polymers

---

[237]See, for example, Antonini *et al.* (1971), Imai (1982), DiCera (1996), Perutz (1970), Wyman and Gill (1990) and Ben-Naim (2001).

P_1 and P_2 respectively, we can write instead of the first term on the right-hand side of (4.5.1)

$$\Delta U = x_1 \Delta U_1 + x_2 \Delta U_2. \qquad (4.5.6)$$

This means that the interaction between L and the polymer is simply the average of the interaction energies ΔU_1 and ΔU_2 with the distribution (x_1, x_2). The reason is that when L approaches P, it sees two populations of the polymer, say N_1 in the P_1 conformer and N_2 in the P_2 conformer. ($N = N_1 + N_2$ being the total number of the P molecules.) Therefore, the interaction energy between L and P is the same as in the case where we have two *different* polymers P_1 and P_2, not necessarily two conformations of the same polymer P.

Thus, although the absorption of L in P might change the distribution (x_1, x_2), say, from (x_1, x_2) before the absorption to (x_1', x_2') after the absorption, the only distribution that matters in calculating ΔU is the initial distribution, i.e. (x_1, x_2) before the absorption occurs.

Regarding the solvent-induced part, we can use the following result[238] for any solute α having two conformations:

$$\exp[-\beta \Delta G_\alpha^*] = y_1^g \exp[-\beta \Delta G_1^*] + y_2^g \exp[-\beta \Delta G_2^*], \qquad (4.5.7)$$

where ΔG_1^* and ΔG_2^* are the solvation Gibbs energies of the two conformers, and y_1^g and y_2^g are the mole fractions of the two conformers in the gaseous phase.

Applying (4.5.7) to both P and PL, we have

$$\exp[-\beta \Delta G_{PL}^*] = y_1^g \exp[-\beta \Delta G_{P_1 L}^*] + y_2^g \exp[-\beta \Delta G_{P_2 L}^*], \qquad (4.5.8)$$

$$\exp[-\beta \Delta G_P^*] = y_1^g \exp[-\beta \Delta G_{P_1}^*] + y_2^g \exp[-\beta \Delta G_{P_2}^*]. \qquad (4.5.9)$$

---

[238]See Ben-Naim (2006) and Appendix B.

Note that the solvation Gibbs energies of P_1L and P_1 are the same, and also the solvation Gibbs energies of P_2L and P_2 are the same. The reason is that the solvation Gibbs energy of any solute depends only on its surface and not on its internal content (i.e. whether it has absorbed an L molecule or not).

The mole fractions and y_1^g and y_2^g depend on the rotational partition functions of the conformers P_1 and P_2. Since L is presumed to be a very small molecule compared to the polymer P, one can assume that y_1^g and y_2^g are nearly the same in (4.5.8) and (4.5.9). Therefore, we can conclude that the two solvation Gibbs energies ΔG_P^* and ΔG_{PL}^* are nearly the same. It follows that the solvent-induced contribution δG in (4.5.2) is again

$$\delta G \approx \Delta G_L^*. \qquad (4.5.10)$$

The argument presented above is very similar to the argument used in the entropy–enthalpy compensation for the solvation Gibbs energy of a solute in water.[239] The idea is that when a solute α is solvated in water, it might change the distribution of the *types* of water molecules. Within the mixture model approach to water, we know that whatever effect the solute has on the distribution of water species, this might affect the entropy and the enthalpy of solvation. However, the Gibbs energy of solvation will not be affected.

4.6. Specificity of the Binding Mode; Molecular Recognition

In Chapter 3, we have discussed two aspects of protein folding in which the water molecules can play a decisive role. One is the stability of the native, or the folded form of the protein. The second is the specific pathways along which the protein

---

[239]See Ben-Naim (2009) and Appendix O.

folds into the native structure. We have shown that water affects the *stability* of the protein through the solvent-induced *driving force*. On the other hand, the *specificity* of the folding pathways is determined by the solvent-induced *forces*, which also affect the kinetics of the folding process.

In the binding process discussed in this chapter, we can also distinguish between two aspects of the binding process in which the solvent-induced *driving force* plays a decisive role. One is the stability of the end product: The dimer or the oligomer. The second is the specificity of the binding mode. Both of these aspects involve the solvent-induced "driving force."

In the previous sections of this chapter, we discussed the relative importance of $H\phi O$ and $H\phi I$ effects on the stability of the complex **PL**. In this section, we shall focus on the factors that affect the preference of a specific binding site, say of a ligand **L** on the protein **P**.

Binding processes are ubiquitous in biological systems.[240] These range from the binding of small molecules like drugs to protein or to DNA, to the binding of proteins to DNA.[241] In all of these processes, the binding mode is highly specific. We start with the binding of a small ligand to a large macromolecule. In subsequent sections, we shall discuss the more general process of self-assembly of macromolecules.

4.6.1. *The lock-and-key model for molecular recognition*

How does a small ligand **L** select a specific binding site on the relatively large surface of the adsorbent molecule **P**?

---

[240]Ritter (1996), Tropp (1997), Stryer (1975), Ben-Naim (2001).

[241]Hard and Lunback (1996), Ptashne (1967), Helene and Lancelot (1982), von Hippel (1994), von Hippel and Goldberger (1979), von Hippel and Berg (1986).

Ever since Emil Fischer (1897) suggested the so-called lock-and-key model, the "recognition" of the binding site has been believed to be achieved through direct interaction between the ligand and the site. In this view, the selection of the binding site is according to the criterion

$$\underset{i}{\text{Min}} \quad [\Delta U(i)], \qquad (4.6.1)$$

where $\Delta U(i)$ is the direct interaction between the ligand and the ith site on the polymer **P**. The minimum is over all possible binding sites on **P**.

The simplest means of recognition is through the weak van der Waals interaction. The criterion (4.6.1) is fulfilled whenever there are more groups on **L** that interact with groups on **P**. This is equivalent to the largest surface of contact between **L** and **P**, hence the *geometrical* fit, which characterized the lock-and-key model (Fig. 4.20a).

Another means of achieving the criterion (4.6.1) is not by geometrical fit, but by a pattern of matching groups such as charged or hydrogen bonding groups on **L** and **P** (Fig. 4.20b). This kind of recognition does not depend on the geometry of the sites, but on the specific pattern of functional groups on the

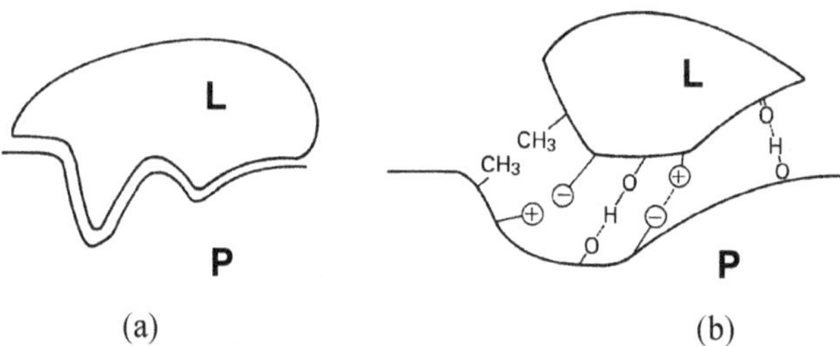

Fig. 4.20   Two variations of the lock-and-key model for recognition. (a) Geometrical fit; (b) complementary pattern of FGs preferential binding due to the lock-and-key model.

surfaces of **L** and **P**.[242] We shall refer to the lock-and-key model in both cases mentioned above, whether the fit is achieved by geometry or by a particular pattern of FGs.

Another variation of the lock-and-key model is the so-called induced-fit model,[243] which is essentially the same as the ones described in Fig. 4.20. Here, however, the fit is achieved after the binding has occurred. This case may be dealt with in the same way we dealt with the absorption of a ligand to a protein with conformational induced changes (Sec. 4.5).

The criterion (4.6.1) may be translated into probabilities, provided the process of binding is carried out in the gaseous phase.[244] For instance, when there are two sites (a) and (b), as in Fig. 4.21, we can say that the probability of binding to site (a) is larger than the probability of binding to site (b). The relationship between the probability ratio and the difference in the binding energies is

$$\frac{\Pr(a)}{\Pr(b)} = \frac{\exp[-\beta\Delta U(a)]}{\exp[-\beta\Delta U(b)]} = \exp[-\beta\{\Delta U(a) - \Delta U(b)\}].$$

$$(4.6.2)$$

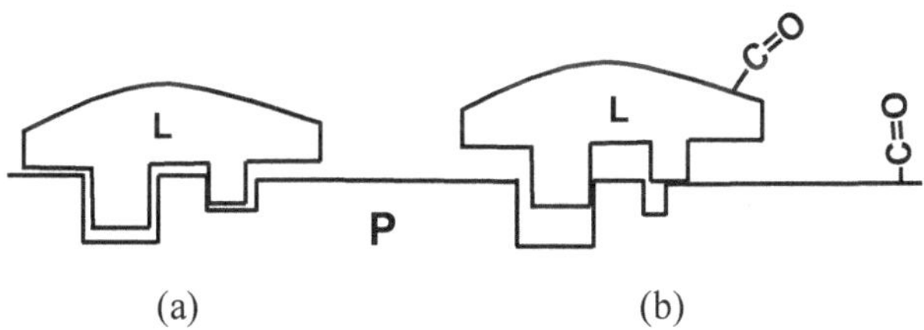

Fig. 4.21   The ligand **L** prefers to bind on site (a) of **P** rather than on site (b).

---

[242]Stillinger and Wasserman (1978), Tabushi and Mizutani (1987), Rebek *et al.* (1988), Lightner *et al.* (1987).
[243]Koshland (1958).
[244]Ben-Naim (2001).

This ratio is also equal to the ratio of the two measurable binding constants.[245] Note that the two carbonyl groups are far apart, and therefore the interaction between them is negligible when the system is in the gaseous phase.

4.6.2. *Molecular recognition through the solvent*

When the binding occurs in a solvent, the criterion for the selection of the preferable binding site is not (4.6.1), but

$$Min_i[\Delta G(i)], \qquad (4.6.3)$$

where $\Delta G(i)$ is the Gibbs energy change for binding to the ith site. The difference between $\Delta G(i)$ and $\Delta U(i)$ is the solvent-induced contribution to the binding Gibbs energy, and this is related to the solvation Gibbs energies of the three molecules **L**, **P** and **PL**, i.e. (here **PL** is understood to be the complex **PL** at the binding site i)

$$\delta G(i) = \Delta G^*_{PL} - \Delta G^*_P - \Delta G^*_L. \qquad (4.6.4)$$

In simple non-aqueous solvents, the main contributions to $\delta G(i)$ come from the hard (H) and soft (S) parts of the solute–solvent interactions. As we have seen in Sec. 4.3, these two contributions depend on the volumes and on the surface areas of the three solutes involved in the binding process. It is commonly believed that even when $\delta G(i)$ is large compared to $\Delta U(i)$, the difference in the values of $\delta G(i)$ for different sites i might not be very large. In other words, the modified criterion (4.6.3) might be very different from the criterion (4.6.1), but the *difference* between the two might not be sensitive to the specific site. As an extreme example, suppose that for each site i, we have

$$\Delta G(i) = \Delta U(i) + \delta G, \qquad (4.6.5)$$

---

[245]Ben-Naim (2001).

where δG is independent of i. In such a case, no matter how large δG might be, its effect on the probability ratio for binding to any two sites (a) and (b) would be negligible, i.e.

$$\frac{\Pr(a)}{\Pr(b)} = \frac{\exp[-\beta\Delta G(a)]}{\exp[-\beta\Delta G(b)]} \approx \frac{\exp[-\beta\Delta U(a)]}{\exp[-\beta\Delta U(b)]}. \qquad (4.6.6)$$

Thus, although the binding constant might be considerably modified in the presence of the solvent, the relative preference for the binding site might be the same as in the gaseous phase. This is equivalent to the statement that the lock-and-key model is still valid; the binding Gibbs energy is modified, but the preferential binding site is still determined by the total direct interaction energy $\Delta U(i)$.

The argument given above, whether made explicitly or implicitly, still dominates the thinking of scientists in the pharmaceutical sciences who are engaged in drug design. However, it was recently pointed out that in some binding processes occurring in aqueous solutions, the $H\phi I$ effect might be so profound that the lock-and-key model might be rendered completely irrelevant to the selection of preferential site.[246] In some cases, it can thoroughly modify the way one approaches the problem of drug design.[247]

We present here one example where a solvent-induced effect based on $H\phi I$ interaction can *reverse* the preference for the binding sites.

Consider again the two sites (a) and (b) in Fig. 4.22. P and L are exactly the same as in Fig. 4.21. Clearly, the ligand L fits better to site (a) than to site (b). Thus, in accordance with the lock-and-key model, site (a) will be preferred over site (b); in

---

[246]Ben-Naim (1987), Ben-Naim (2002).
[247]Wang and Ben-Naim (1996).

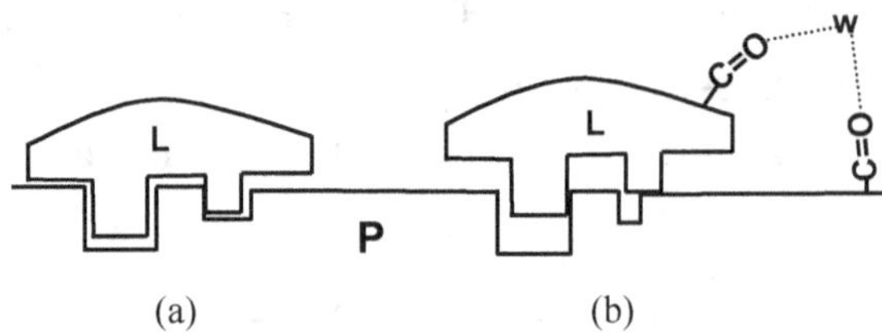

Fig. 4.22 Preferential binding due to $H\phi I$ effect. The same two sites as in Fig. 4.21, and the same ligand L, but now in water. Site (b) would be preferred in spite of the better fit of L to site (a).

probability terms,

$$r^{(g)} = \frac{\Pr(a)}{\Pr(b)} = \frac{\exp[-\beta\Delta U(a)]}{\exp[-\beta\Delta U(b)]} > 1. \qquad (4.6.7)$$

Clearly, in the gaseous phase, the ligand will prefer binding to site (a). Equivalently, because of the stronger binding energies, the binding constant to site (a) will be larger than the binding constant to site (b).

The preferential binding site might be reversed in aqueous solution. Suppose that the ligand L and the polymer P each have a FG that can form a HB with water. Note that these functional groups are not in the binding interface between L and P. Therefore, even in the presence of these FGs, the preferential binding site in the gaseous phase is still the site (a). This preferential site will probably be maintained if we add an organic solvent, say, hexane or benzene. However, in aqueous solutions, the whole story is quite different. Suppose that the FGs on L and P are such that when binding to (b), they can be bridged by a water molecule. In this case, the binding Gibbs energy to site (b) will be modified,

$$\Delta G(b) = \Delta U(b) + \delta G(b). \qquad (4.6.8)$$

In binding to site (a), the two FGs are far apart so that they do not interact directly nor indirectly. Therefore, the binding

Gibbs energy to (a) is

$$\Delta G(a) = \Delta U(a) + \delta G(a), \qquad (4.6.9)$$

where we have

$$|\delta G(a)| \ll |\delta G(b)|. \qquad (4.6.10)$$

The probability ratio in this case $r^{(w)}$ is

$$r^{(w)} = \frac{\Pr(a)}{\Pr(b)} = \frac{\exp[-\beta\Delta G(a)]}{\exp[-\beta\Delta G(b)]} = r^{(g)}\frac{\exp[-\beta\delta G(a)]}{\exp[-\beta\delta G(b)]}.$$
$$(4.6.11)$$

For simplicity, we assume that $\delta G(a)$ is negligible, and that $\delta G(b)$ is due to one $H\phi I$ correlation, which amounts to about $-5k_B T$. In this case, we shall have

$$r^{(w)} \approx \frac{r^{(g)}}{150}. \qquad (4.6.12)$$

Thus, although we have started with $r^{(g)} > 1$, i.e. the preferential site is (a), the addition of one $H\phi I$ interaction could change the probability ratio by a factor of 150 in favor of site (b). Clearly, in such a case, the whole lock-and-key argument becomes irrelevant to the problem of preferential binding site.

The finding that $H\phi I$ interactions can change the preferential binding site has far reaching consequences for the problem of drug design,[248] either for designing new drugs or for modifying existing drugs to improve their efficacy. Some specific examples were discussed recently.[249] We shall not present these highly technical examples here. The interested reader should consult the article by Wang and Ben-Naim (1996).

---

[248] See, for example, Kuntz (1992), Perun and Propst (1989), Propst and Perun (1992), Greer *et al.* (1994) and Roerding and Kroon (1989).
[249] Wang and Ben-Naim (1996).

In connection with the solvent-induced effect on preferential binding, it should be noted that a protein's surface is very inhomogenous, both in its structure and its chemical constituency. Therefore, it is likely that in most binding processes to a protein, both the direct (i.e. lock-and-key model), and the indirect interactions are of comparable magnitudes. There might be crevices on the surfaces of the proteins that provide tightly fitted binding sites, as well as $H\phi I$ interaction through $H\phi I$ groups that do not belong to the binding interface. The situation is quite different in DNA, where the surface is much more homogenous, both in terms of the structure as well as in terms of the distribution of FGs. Therefore, in the binding of ligands, drugs or even proteins to DNA, it is more likely that the $H\phi I$ interaction will dominate the binding Gibbs energy, and hence also the selection of the binding site.[250,251]

We next turn to a hypothetical example, where we have a seemingly featureless surface, for instance a flat surface with $H\phi I$ groups randomly distributed on it. Looking superficially at this surface, we might not detect any preferred binding site for the ligand **L** (Fig. 4.23a). Therefore, from the point of view of the lock-and-key model, there exists no preferential binding site.

However, a ligand approaching this surface in water might "feel" very different affinities to different regions on this apparently homogenous surface. A ligand hovering above the surface might find one region far more favorable for binding than any other region. An illustration of such an example is shown in Fig. 4.23b.

Let Pr(0) be the probability of binding of the ligand to an arbitrary region on the surface of the protein which does not contain $H\phi I$ groups [examples (1), (2) and (3) in Fig. 4.23a]. In

---

[250]Schwabe (1997).
[251]Privalov *et al.* (2007).

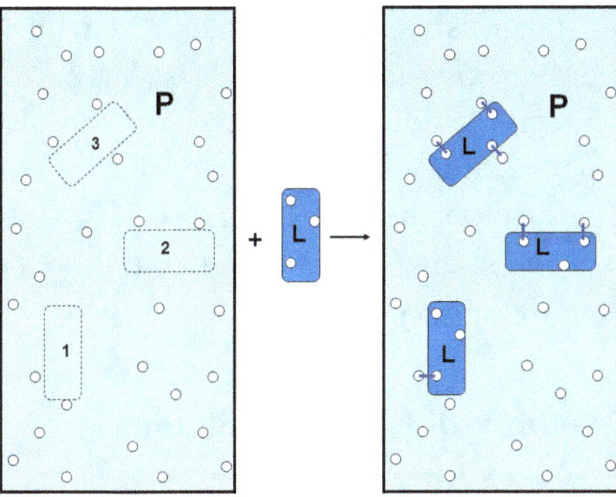

Fig. 4.23   A macromolecule **P** with many $H\phi I$ groups (shown as white circles) distributed on its surface. A ligand **L** has one side $H\phi O$, and the other side having three $H\phi I$ groups (only this side is shown). In the gaseous phase, the ligand will bind preferentially to any part of the surface which does not contain $H\phi I$ groups with equal probability. Three possible sites are indicated by (1), (2) and (3). In water, the same three sites become preferential binding sites. The ratios of the probabilities for binding on these three sites are given in (4.6.13). The possible HB bridges are shown by blue lines connecting two $H\phi I$ groups.

the gaseous phase, the Gibbs energy of binding to any point on these regions on the surface of the protein is nearly independent of the location of the binding site. However, in water, the ligand might find sites at which the binding Gibbs energy will be much larger due to the formation of one, two or three $H\phi I$ interactions by means of a water bridge. The relative probabilities for binding to sites (1), (2) and (3) (Fig. 4.23b) are

$$\frac{\mathrm{Pr}(1)}{\mathrm{Pr}(0)} \approx \exp[5] \approx 150,$$

$$\frac{\mathrm{Pr}(2)}{\mathrm{Pr}(0)} \approx \exp[10] \approx 2.2 \times 10^4, \qquad (4.6.13)$$

$$\frac{\mathrm{Pr}(3)}{\mathrm{Pr}(0)} \approx \exp[15] \approx 3.3 \times 10^6.$$

It is clear from this example that a seemingly featureless surface might provide binding sites, such as (1), (2) and (3), with significantly different binding constants. We can also imagine that a ligand approaching a DNA molecule might "see" a nearly homogenous surface on the DNA. However, looking through the water, the ligand might "see" some sites as more preferential for binding than the others.

4.7. Self-Assembly of Macromolecules

The term self-assembly is used in biochemistry for many processes, including protein folding, replication of DNA, micelle and membrane formation, and the synthesis of proteins on the ribosomes. In this chapter, we restrict the usage of this term to the spontaneous aggregation of proteins to form multi-subunit molecules. Basically, these processes are generalizations of the binding or the association of two solutes to form dimers, as discussed in Secs. 4.1–4.5.

As in the case of dimerization processes, there are essentially two aspects of these processes that we are concerned with: The factors that contribute to the thermodynamic "driving forces" and the specificity of the binding mode. Focusing on these two aspects from the solvent point of view, we expect that the same factors that contribute to the driving forces in the association between two proteins will be operative in the case of aggregation of a larger number of subunits. Based on the discussion in the preceding sections, we can expect that $H\phi I$ interactions should play a dominant role in determining the stability of the aggregates.

We next turn to the specificity of the binding mode of many subunits. As in the case of two solutes, the binding Gibbs energy has two components: The direct interactions which are responsible for the recognition by the lock-and-key mechanism,

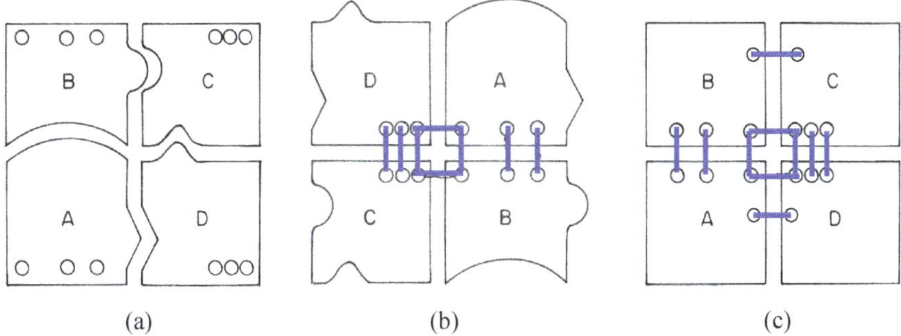

Fig. 4.24   An illustration of preferential binding mode in two dimensions. In (a), the four subunits recognize each other by the lock-and-key mechanism (*A* identifies *B*, *B* identifies *C*, *C* identifies *D* and *D* identifies *A*). This mode of binding will be preferable in the gaseous phase. The same four subunits in water could bind differently (b). In (c), we show four subunits which from the point of view of the lock-and-key model have no preferential mode of binding. However, in water there is a preferential mode of binding.

and the indirect, solvent-induced part, which is likely to be dominated by $H\phi I$ interaction.

There is no difficulty in extending the lock-and-key model for the case of multi-subunit aggregation. For instance, in Fig. 4.24a, we have four subunits (here in two dimensions), where *A* recognizes one surface of *B*, *B* recognizes one surface of *C*, *C* recognizes one surface of *D* and *D* recognizes one surface of *A*. Thus, in the absence of solvent, the most likely mode of binding is the one with the maximum geometrical fit, which is the same as maximizing the direct interactions between each pair of subunits.

In water, the criterion for recognition might change quite drastically. In Fig. 4.24b, we show the same four subunits as in Fig. 4.24a. But in aqueous solutions, the preferential mode of binding might be changed. If there are $H\phi I$ groups that can be bridged by water molecules, then the preferred mode of binding will be the one which minimizes the Gibbs energy. If there are sufficiently large numbers of such pairs of $H\phi I$ groups, then the $H\phi I$ effect will cause a switch of preference of the binding mode — compare Figs. 4.24a and 4.24b.

Another example is shown in Fig. 4.24c. The four subunits (again in two dimensions, but the argument can easily be extended to three-dimensional macromolecules) are featureless, and from the point of view of the lock-and-key model, there exists no single preferred binding mode. Each possible configuration of the aggregate has the same likelihood of occurrence.

However, the same four subunits in water will behave quite differently. As in the example shown in Fig. 4.23, the solvent might create a recognition code based on pairs of $H\phi I$ groups. For instance, in Fig. 4.24c, A recognizes B, B recognizes C, C recognizes D and D recognizes A. We see that the subunits, having no means of recognition through direct interactions, can recognize each other through solvent-induced effects.[252]

The extension to aggregation of many subunits of the same kind or of different kinds is quite straightforward. Figure 4.25 shows a possible self-assembly of identical subunits (a) and of two different subunits (b). With a little imagination, one can construct a recognition code for any type of self-assembly for any number of identical or different subunits.[253]

We conclude this section by noting that the mechanism of molecular recognition through solvent-induced effects has some evolutionary advantage. Clearly, the recognition by means of the lock-and-key model is a built-in feature of the molecules involved and cannot be regulated by changes in the solvent composition. On the other hand, changes in the solvent composition can either weaken or strengthen the $H\phi I$ effects. It is therefore

---

[252] See Appendix G.

[253] This figure is only for illustration. In Fig. 4.25b, we assume that all modes of binding are weak when they do not involve $H\phi I$ interactions. The mode of binding shown in the figure is preferred because of the $H\phi I$ interactions.

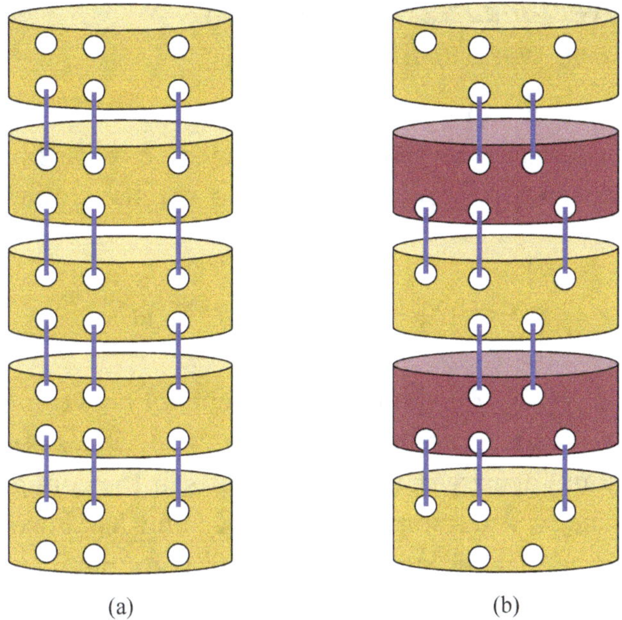

Fig. 4.25 Self-assembly of (a) identical subunits; (b) different subunits.

likely, that evolution has "exploited" this means of controlling vital biological processes to its advantage.

4.8. Strong Solvent-induced Forces between Macromolecules

In Sec. 3.8, we have discussed one type of $H\phi I$ effect, i.e. the solvent-induced force between two $H\phi I$ groups bridged by one water molecule. In this section, we present a few other possible $H\phi I$ effects between surfaces having many $H\phi I$ groups distributed on their surface. These effects are expected to be important in determining the forces between two macromolecules with many $H\phi I$ groups on their surface, such as DNA, RNA, large micelles and membranes.

4.8.1. *Solvent-induced force by means of one-water bridges*

Consider two macromolecules denoted **L** and **P**. These can be the same or different molecules. The general expression for the solvent-induced force on, say, **L** given a specific configuration of **L** and **P** is[254]

$$\mathbf{F}_L^{(SI)} = \int -\nabla_L U(\mathbf{X}_L, \mathbf{X}_w)\rho(\mathbf{X}_w/\mathbf{X}_L, \mathbf{X}_P)d\mathbf{X}_w. \qquad (4.8.1)$$

Here, $U(\mathbf{X}_L, \mathbf{X}_w)$ is the interaction energy between **L** and a water molecule at the configuration $(\mathbf{X}_L, \mathbf{X}_w)$. $\rho(\mathbf{X}_w/\mathbf{X}_L, \mathbf{X}_P)$ is the conditional density of water molecules at the configuration $\mathbf{X}_w$, given the two macromolecules at $\mathbf{X}_L$ and $\mathbf{X}_P$. For simplicity, we assume that the surfaces of **L** and **P** are parallel to each other, and we are interested in the component of the force perpendicular to these surfaces. This simplification is, of course, not essential — with a little effort, especially notational, one can discuss any type of surface.

In this section, we are interested only in the various types of $H\phi I$ forces due to a single water bridge. In all the examples below we shall estimate the solvent-induced force on **L**, given **P** at some specific configuration.

(i) *A water molecule bridging two $H\phi I$ groups*

Figure 4.26a shows two $H\phi I$ groups bridged by a water molecule. If the distance between the two $H\phi I$ groups is about 4.5 Å, we may expect strong solvent-induced forces. This case was discussed in Sec. 3.8, where we have distinguished between different types of FGs on the two solutes.

We shall always assume that the main contribution to the integral (4.8.1) comes from a small "volume" of the

---

[254]See Sec. 3 and Appendix C.

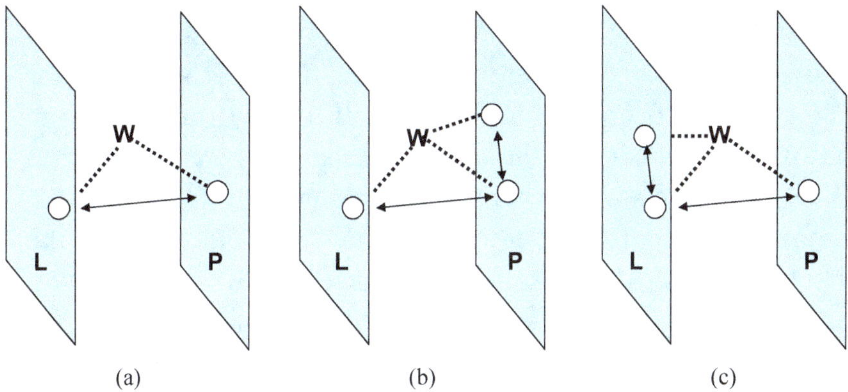

Fig. 4.26   (a) A one-water bridge connecting one *HϕI* group on **P** and one *HϕI* group on **L**; (b) a one-water bridge connecting one *HϕI* group on **L** and two *HϕI* groups on **P**; (c) a one-water bridge connecting two *HϕI* groups on **L** and one *HϕI* group on **P**. *HϕI* groups are shown as white circles; a double arrow indicates a distance of about 4.5 Å.

configurations of the water molecules — locations and orientations at which both factors of the integrand are the largest. We denote this volume by $\Delta\mathbf{X}_w$ and rewrite (4.8.1) as

$$\mathbf{F}_L^{(SI)} \approx -\nabla_L U(\mathbf{X}_L, \mathbf{X}_w)\rho(\mathbf{X}_w/\mathbf{X}_L, \mathbf{X}_P)\Delta\mathbf{X}_w. \qquad (4.8.2)$$

In Sec. 3.8, we found that the solvent-induced force between *two HϕI* groups is the largest among the four possible pairs of FGs. Here, we shall discuss other possible forces, still mediated by a single water molecule, but bridging more than two *HϕI* groups. As we have noted in Sec. 3.8, all these forces are *statistical forces*, resulting from averaging over all possible configurations of a water molecule. The solvent-induced force (4.8.2) does not imply that a water molecule is permanently hydrogen-bonded to the two *HϕI* groups. Rather, it is the value of the force exerted by a water molecule being at configurations from which it can exert maximal force, and at which the conditional density is largest. With this comment in mind we turn to the next case.

(ii) *A water molecule bridging three HφI groups*

Figure 4.26 shows two more examples where a single water molecule can form a bridge between three *HφI* groups. Note that we are considering the force on **L**, given the configuration $\mathbf{X}_L, \mathbf{X}_P$. We also assume that the configuration of $\mathbf{X}_w$ is such that it maximizes the force we are concerned with. In case (b), the direct force $-\nabla_L U(\mathbf{X}_L, \mathbf{X}_w)$ is the same as in (4.8.2). However, the conditional density of water molecules at $\mathbf{X}_w$, $\rho(\mathbf{X}_w/\mathbf{X}_L, \mathbf{X}_P)$ is enhanced by *three HφI* groups, not only by *two* as in the case discussed in (i) above. Therefore, we expect the force in this case to be stronger than in case (i). In case (c) of Fig. 4.26, we have a stronger direct force by a water molecule on **L**. We can roughly estimate that the component of the force perpendicular to the plane of **L** will be about twice the force in (4.8.2). The conditional density in this case is roughly the same as in case (b), i.e. a density of water molecules from which it can form three HBs with the *HφI* groups is enhanced by the *three HφI* groups. Therefore, we expect that the force exerted on **L** in case (c) will be larger than in case (b), and this is in turn larger than in case (a) discussed in (i). This order of forces is schematically shown in Fig. 4.27. Of course, we must have *HφI* groups on **L** and **P** such that we have one donor and two acceptors, or one acceptor and two donors for the hydrogen bonding with a water molecule at $\mathbf{X}_w$.

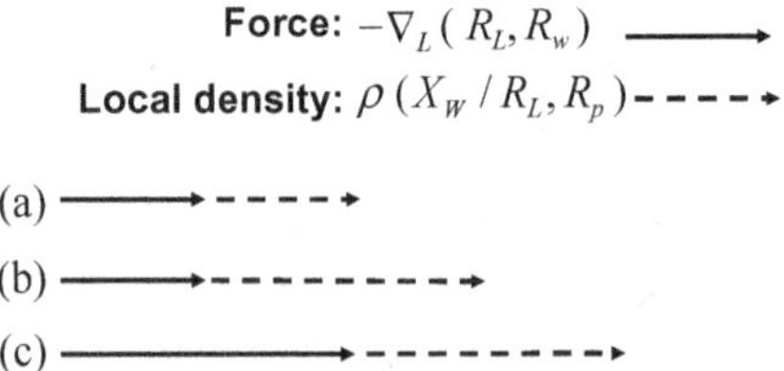

Fig. 4.27 The order of the strength of the *HφI* forces in the three cases shown in Fig. 4.26.

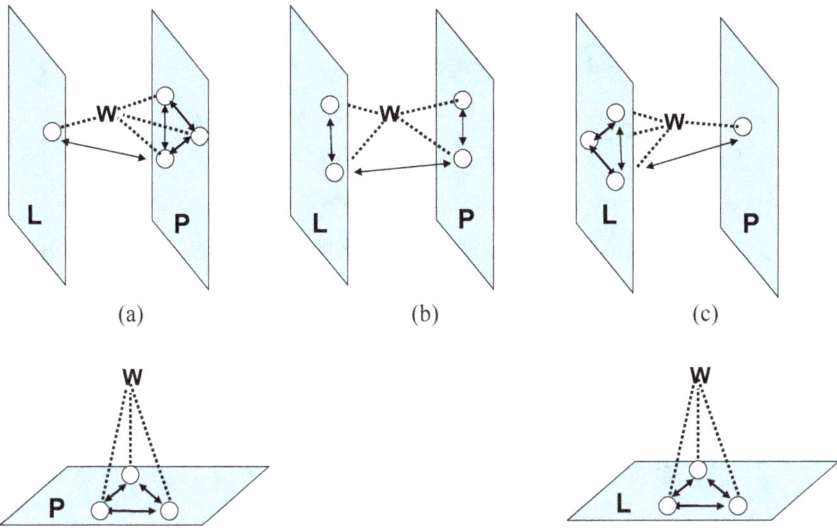

Fig. 4.28   (a) A one-water bridge connecting one *HϕI* group on L and three *HϕI* groups on P. The surface of P is also shown under (a); (b) a one-water bridge connecting two *HϕI* groups on L and two *HϕI* groups on P; (c) a one-water bridge connecting three *HϕI* groups on L and one *HϕI* group on P. The surface of L is shown under (c).

(iii) *A water molecule bridging four HϕI groups*

The strongest force that we can expect from a *single* water molecule is when it can bridge four *HϕI* groups. Figure 4.28 shows three possible cases: (a) when one *HϕI* group is on L and three on P; (b) when two *HϕI* groups are on L and two on P; (c) when three *HϕI* groups are on L and one on P. Note that since a water molecule can form at most four HBs, in these cases, we have exhausted all the "arms" of a single water molecule.

We also assume that the locations and orientations of all the *HϕI* groups are such that they can form a strong HB (roughly, the four oxygen atoms of the *HϕI* groups must be in the corner of a regular tetrahedron as in the case of ice). Also, the *HϕI* groups must be such that two of them are acceptors and two are donors for hydrogen bonding.

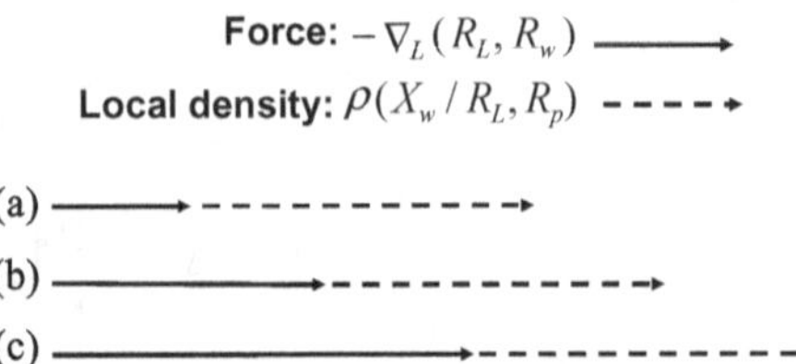

Fig. 4.29 The order of the strength of the $H\phi I$ forces for cases (a), (b) and (c) in Fig. 4.28, compared with Fig. 4.27.

Following similar reasoning as in cases (i) and (ii), we expect that the conditional density $\rho(\mathbf{X}_w/\mathbf{X}_L, \mathbf{X}_P)$ is the highest in these cases, since four $H\phi I$ groups are contributing to enhance the local density at the required configurations $\mathbf{X}_w$.[255] The component of the force on L perpendicular to the plane of L will be different in the three cases. Case (a) here will be similar to case (a) of Fig. 4.26, case (b) will be stronger and case (c) even stronger. Hence, the qualitative diagram of the solvent-induced force is as shown in Fig. 4.29.

4.8.2. *Solvent-induced force by means of two-water bridges*

We discuss briefly a few possibilities of a two-water bridge between two $H\phi I$ groups on **L** and **P**.

(i) *A two-water bridge connecting two $H\phi I$ groups*

The simplest case is shown in Fig. 4.30a. If we use either the expression (4.8.1) or its approximate version (4.8.2), it is clear that the direct force $-\nabla_L U(\mathbf{X}_L, \mathbf{X}_w)$ can be made as large as in (4.8.1). However, the local density at $\mathbf{X}_w$ is now much smaller simply because its distance from the $H\phi I$ group on **P** is much greater than in the case of Sec. (4.8.1).

---

[255]See Appendix S.

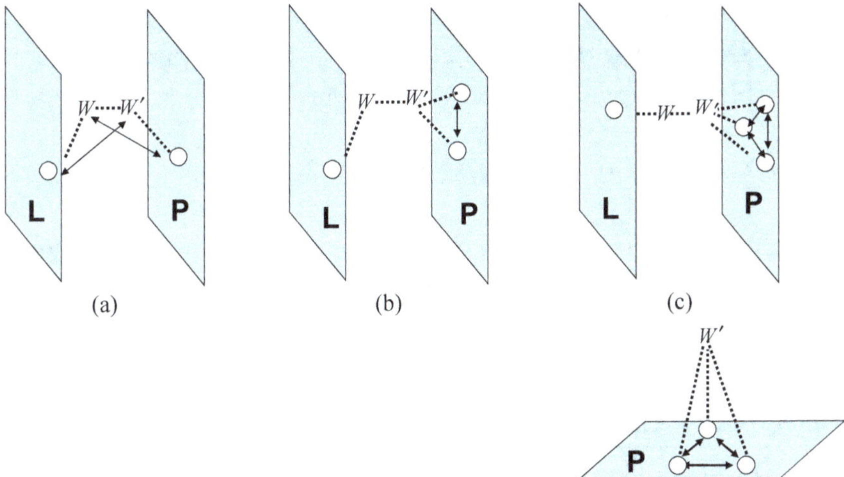

Fig. 4.30   (a) A two-water bridge connecting one *HϕI* group on **L** and one *HϕI* group on **P**; (b) a two-water bridge connecting one *HϕI* group on **L** and two *HϕI* groups on **P**; (c) a two-water bridge connecting one *HϕI* group on **L** and three *HϕI* groups on **P**. The surface of **P** is shown under (c).

We therefore use an intermediate water molecule at $\mathbf{X}_{w'}$ which is at a favorable configuration to form a HB with a *HϕI* group on **P**, and hence can affect the local density of water molecules at $\mathbf{X}_w$.

To do this, we use the following identity

$$\rho(\mathbf{X}_w/\mathbf{X}_L,\mathbf{X}_P) = \frac{\rho(\mathbf{X}_w,\mathbf{X}_L,\mathbf{X}_P)}{\rho(\mathbf{X}_L,\mathbf{X}_P)} = \int \frac{\rho(\mathbf{X}_w,\mathbf{X}_{w'},\mathbf{X}_L,\mathbf{X}_P)}{\rho(\mathbf{X}_L,\mathbf{X}_P)} d\mathbf{X}_{w'}$$

$$= \int \rho(\mathbf{X}_w/\mathbf{X}_{w'},\mathbf{X}_L,\mathbf{X}_P)\rho(\mathbf{X}_{w'}/\mathbf{X}_L,\mathbf{X}_P)d\mathbf{X}_{w'}.$$

$$(4.8.3)$$

In the first equality, we use the definition of the conditional probability. This is based on the interpretation of $\rho(\mathbf{X}_w/\mathbf{X}_L,\mathbf{X}_P)d\mathbf{X}_w$ as the probability of finding a water molecule in a small "volume" $d\mathbf{X}_w$ about $\mathbf{X}_w$, given the "condition" $\mathbf{X}_L,\mathbf{X}_P$. In the second equality, we rewrite the triplet density [or the corresponding probability of the "event" $(\mathbf{X}_w, \mathbf{X}_L, \mathbf{X}_P)$] by adding the "event" $\mathbf{X}_{w'}$ and integrating over all possible outcomes of the added "event," i.e.

$$\rho(\mathbf{X}_w, \mathbf{X}_L, \mathbf{X}_P) = \int \rho(\mathbf{X}_w, \mathbf{X}_{w'}, \mathbf{X}_L, \mathbf{X}_P) dX_{w'}. \qquad (4.8.4)$$

In probabilistic terms, (4.8.4) simply means that when we sum (or integrate) over all possible values of $\mathbf{X}_{w'}$, we get the probability density of the event $(\mathbf{X}_w, \mathbf{X}_L, \mathbf{X}_P)$.[256] Finally, we use again the definition of the conditional probability to write

$$\frac{\rho(\mathbf{X}_w, \mathbf{X}_{w'}, \mathbf{X}_L, \mathbf{X}_P)}{\rho(\mathbf{X}_L, \mathbf{X}_P)} = \frac{\rho(\mathbf{X}_w/\mathbf{X}_{w'}, \mathbf{X}_L, \mathbf{X}_P)\rho(\mathbf{X}_{w'}, \mathbf{X}_L, \mathbf{X}_P)}{\rho(\mathbf{X}_L, \mathbf{X}_P)}$$

$$= \rho(\mathbf{X}_w/\mathbf{X}_{w'}, \mathbf{X}_L, \mathbf{X}_P)\rho(\mathbf{X}_{w'}/\mathbf{X}_L, \mathbf{X}_P).$$

$$(4.8.5)$$

Hence, we get the final expression on the right-hand side of (4.8.3). The meaning of the equality (4.8.3) is quite simple. The conditional density (or probability) of finding a water molecules at $\mathbf{X}_w$, given the condition $\mathbf{X}_L, \mathbf{X}_P$, is the same as the product of the conditional density at $\mathbf{X}_{w'}$, given $\mathbf{X}_L, \mathbf{X}_P$, times the conditional density at $\mathbf{X}_w$, given the condition $\mathbf{X}_{w'}, \mathbf{X}_L, \mathbf{X}_P$, integrated over all possible values of the configurations of the intermediary molecule w'.[257]

As we have done in Eq. (4.8.2), we can now write the approximate version of the solvent-induced force on L, given the condition $\mathbf{X}_L, \mathbf{X}_P$. Thus, instead of (4.8.2), we have in this

---

[256]This is essentially the definition of the marginal probability. See also Papoulis and Pillai (2002).

[257]The result (4.8.3) is easier to understand for the discrete case

$$P(i/j) = \frac{P(i,j)}{P(j)} = \sum_k \frac{P(i,k,j)}{P(j)} = \sum_k \frac{P(i/k,j)P(k,j)}{P(j)}$$

$$= \sum_k P(i/k,j)P(k/j).$$

case the approximate expression

$$\mathbf{F}_L^{(SI)} \approx - \left[\nabla_L U(\mathbf{X}_L, \mathbf{X}_w)\right] \rho(\mathbf{X}_w/\mathbf{X}_L, \mathbf{X}_p)\Delta\mathbf{X}_w$$

$$\approx - \left[\nabla_L U(\mathbf{X}_L, \mathbf{X}_w)\right] \rho(\mathbf{X}_w/\mathbf{X}_{w'}, \mathbf{X}_L, \mathbf{X}_p)\Delta\mathbf{X}_w$$

$$\times \rho(\mathbf{X}_{w'}/\mathbf{X}_L, \mathbf{X}_p)\Delta\mathbf{X}_{w'}. \tag{4.8.6}$$

In (4.8.6), $\Delta\mathbf{X}_w$ and $\Delta\mathbf{X}_{w'}$ are the sizes of the "volume" of the configurations from which w can form a HB with the $H\phi I$ group on **L**, and from which w' can form a HB with the $H\phi I$ group on **P**, respectively. The qualitative meaning of (4.8.6) is that the $H\phi I$ group on **P** affects the density of water at the intermediary point $\mathbf{X}_{w'}$, this density of water molecules affects the density at $\mathbf{X}_w$, which exerts the force on **L**. However, unlike the case of Eq. (4.8.2), where $\Delta\mathbf{X}_w$ was small but the local density $\rho(\mathbf{X}_w/\mathbf{X}_L, \mathbf{X}_P)$ was relatively large, here we have the product of $\Delta\mathbf{X}_w$ and $\Delta\mathbf{X}_{w'}$. In addition, the local density $\rho(\mathbf{X}_{w'}/\mathbf{X}_L, \mathbf{X}_P)$ is not expected to be as large as in (4.8.2) because of the greater distance **L** and **P**.

We can therefore conclude that a force produced by a chain of two water molecules connecting the $H\phi I$ groups on **L** and **P** is not expected to be large.

Before we move on to the next case of a two-water bridge, it is important to compare the force in (4.8.2) and (4.8.6) and to understand why the latter is much weaker than the former.

In both cases, the direct force $-\nabla_L U(\mathbf{X}_L, \mathbf{X}_w)$ is the same. In (4.8.2), we have a local density at $\mathbf{X}_w$ which is enhanced by both **L** and **P**. This is so because we have chosen the configurations of the $H\phi I$ groups on **L** and **P** to be such that they can form a HB with a water molecule at $\mathbf{X}_w$. On the other hand, in (4.8.6), the distance between **L** and **P** is greater. Therefore, we cannot expect the same enhancement of the local density at $\mathbf{X}_w$ from the $H\phi I$ groups on **L** and **P**. In the next examples, we shall examine situations for which an enhancement of the local density at $\mathbf{X}_w$ is achieved as before by water molecules at $\mathbf{X}_{w'}$. However, the density at $\mathbf{X}_{w'}$ is now enhanced by *two HϕI* groups on $\mathbf{P}$.

(ii) *A two-water bridge connecting one HϕI group on* $\mathbf{L}$ *and two HϕI groups on* $\mathbf{P}$

The situation is depicted in Fig. 4.30b. The solvent-induced force on $\mathbf{L}$ has the same form as in (4.8.6), with one important difference. The local density of water molecules at $\mathbf{X}_{w'}$ is again conditioned on $\mathbf{X}_L, \mathbf{X}_P$, but unlike case (i) in Fig. 4.30a, the enhancement of the local density at $\mathbf{X}_{w'}$ is due to *two HϕI* groups on $\mathbf{P}$. These groups are supposed to be at the right distance and orientation so that they can form *two* HBs with a water molecule at $\mathbf{X}_{w'}$. This means that there is a larger probability of finding a water molecule at $\mathbf{X}_{w'}$, hence its effect on the local density of water molecules at $\mathbf{X}_w$ will be larger than in the previous case. Therefore, we can expect a larger force on $\mathbf{L}$. Thus, although the force on $\mathbf{L}$ has the same expression here as in (4.8.6), the *magnitude* of the local density at $\mathbf{X}_w$ is enhanced because of the enhancement of the local density at $\mathbf{X}_{w'}$ due to the *two HϕI* groups on $\mathbf{P}$.

(iii) *A two-water bridge connecting one HϕI group on* $\mathbf{L}$ *and three HϕI groups on* $\mathbf{P}$

This case is an "upgrade" of the previous case. Again, we have one *HϕI* group on $\mathbf{L}$. Therefore, the maximal direct force that can be exerted by a water molecule at $\mathbf{X}_w$ is the same as before. However, in the present example, there are *three HϕI* groups on $\mathbf{P}$, at a configuration such that they can form three HBs with water molecules (see Fig. 4.30c).

Again, the expression for the solvent-induced force is the same as in (4.8.6). The only difference is that now the density of water at $\mathbf{X}_{w'}$ is "supported" by *three*, rather than two *HϕI* groups on $\mathbf{P}$. This means that there is a very large probability of finding a water molecule at $\mathbf{X}_{w'}$. This in turn means that the density of water molecules at $\mathbf{X}_{w'}$ will have a more pronounced enhancement on the density of water molecules at $\mathbf{X}_{w'}$.

Therefore, the solvent-induced force in this case is expected to be larger than in case (ii) discussed above.

Another way of interpreting the origin of this stronger force is as follows: When there are three *HφI* groups on **P**, properly configured so that they can form *three* HBs with a water molecule, the probability of finding a water molecule at the configuration $\mathbf{X}_{w'}$ is almost unity. This is equivalent to saying that *there is* a water molecule at $\mathbf{X}_{w'}$, which is also tantamount to saying that we have effectively moved the wall **P** to the location of $\mathbf{X}_{w'}$ (see Fig. 4.31). This is equivalent to the case where there is *one HφI* group on **P** which can form a one-water bridge with a *HφO* group on **L**, and therefore the force is expected to be as strong as the one-water bridge discussed in Sec. 4.8.1.

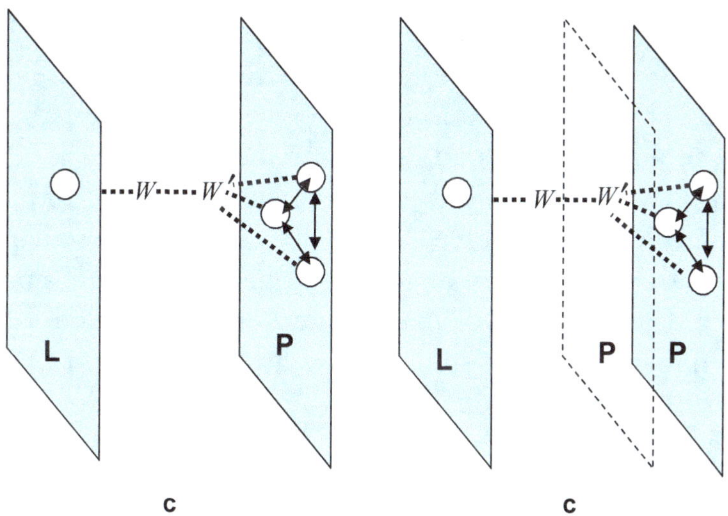

Fig. 4.31 The enhanced density of water molecules at the point $X_{w'}$ due to the pattern of *HφI* groups on the surface of **P**. This is the same as in Fig. 4.30c.

4.8.3. *Stronger forces between HøI surfaces*

In the previous sections, we have given various examples of *HøI* forces. All of them are a result of *HøI* groups on the surface of the macromolecules. Of course, we did not exhaust all possible solvent-induced forces. A host of new forces may be obtained by three-water bridges, four-water bridges, etc. Within each kind of bridge, we can have different *strengths* of the force, as well as different *ranges*. For instance, if **P** contains a pattern of *HøI* groups such that they can induce a high local density of water molecules at some distance from the surface of **P**, this would be equivalent to moving the surface of **P** towards L. An example was shown in Fig. 4.31. This argument can be extended: If there exists a pattern of water molecules on a plane **P'**, then these water molecules will produce a high density of water molecules on a plane **P''**, and so on. An example is shown in Fig. 4.32, and more examples are discussed in the original article.[258]

These kinds of forces may be operative in processes such as protein folding, protein–protein association and self-assembly of multi-subunit macromolecules.

It should be noted that whenever the *HøI* force is long-ranged, it will contribute to the dynamical aspect of the process in which it is involved. These forces can be oscillatory with alternating regions of repulsion and attraction, similar to the PMF between two small solutes, or the observed forces between mica surfaces. Similar forces are expected to be operating between biological membranes, micelles or between cells and clay minerals.[259]

---

[258]Ben-Naim (1990).

[259]Israelachvili (1985, 1987, 1991), Israelachvili and McGuiggan (1988), Parsegian and Gingell (1973), Forslind and Jacobsson (1975), Henderson and Lozada-Cassou (1986).

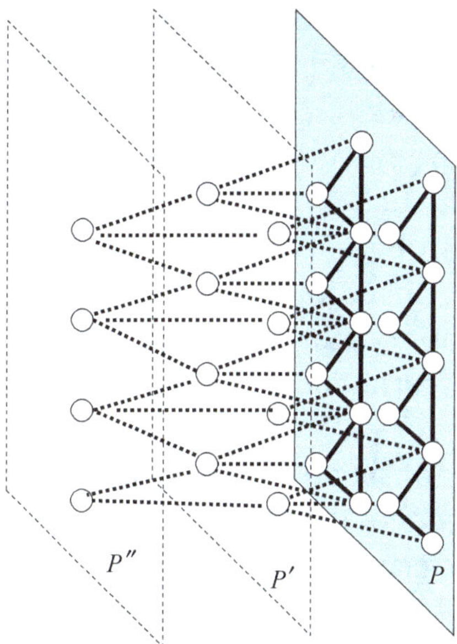

Fig. 4.32   An extension of the phenomenon shown in Fig. 4.31. A high density of water molecules on a plane P' will produce a high density of water molecules on a plane P''.

As we have noted earlier, solvent-induced forces are advantageous from an evolutionary point of view. A small change in the composition of the solvent could either accelerate or decelerate a process, and the corresponding $H\phi I$ effects could stabilize or destabilize a specific structure or an aggregate. Thus, by manipulation of the concentration of solutes, Nature can gain control over biochemical processes.

4.9. Concluding Remarks and Suggestions for Future Research

The elucidation of the factors involved in the processes of self-assembly of macromolecules is no less important than the examination of the corresponding process of protein folding. There are many aspects of this problem which deserve to be studied. Regarding the association between two proteins, I would suggest to extend the work discussed in Sec. 4.4 in two directions. First, to examine the thermodynamics of the association of simple model proteins such as cubes, long cylinders, etc. Second, to examine the distribution of FGs on the surfaces of both the monomers and the dimer (or oligomer) of real proteins.[260] This could lead to a more realistic estimate of the various contributions to the solvent-induced part of the driving forces in the association process.[261]

Also, the study of single factors such as the conditional solvation Gibbs energies of the various side chains of amino acids, as well as pair correlation between those side chains would be helpful for the problem of self-assembly, as much as it is helpful for the problem of protein folding.

---

[260]See Ben-Naim *et al.* (1989). Some experimental evidence for water bridges between subunits of real protein are discussed by Finney (1979). See also Finney *et al.* (1982) and Thanki *et al.* (1988).

[261]Xu *et al.* (1997).

The General Statistical Mechanical Expression for the Chemical Potential and the Pseudo-Chemical Potential

In this appendix, we derive the general expression of the chemical potential. We shall assume that the system is classical in the sense that classical statistical mechanics applies to the system. The derivation is somewhat lengthy, but it is essentially quite simple: Take the ratio of the partition functions of the system *before* and *after* the addition of one particle. We shall do the derivation for a pure one-component system and leave the extension to a mixture as an exercise for the reader. It is instructive to follow the derivations of the expression for the chemical potential for the three different ensembles, reaching essentially the same final expression.

(i) The T, V, N ensemble

For simplicity, we start with the system of spherical particles. This assumption is not necessary, but it simplifies the notations. We consider a system characterized by the variables T, V, N, i.e. N particles contained in a volume V and maintained at a temperature T. The chemical potential in such a system is defined as[262]

$$\mu = \left(\frac{\partial A}{\partial N}\right)_{T,V}$$

$$= \lim_{\Delta N \to 0} \frac{A(T, V, N + \Delta N) - A(T, V, N)}{\Delta N}$$

$$= \lim_{\Delta N \to 0} \left[A\left(T, \frac{V}{\Delta N}, \frac{N}{\Delta N} + 1\right) - A\left(T, \frac{V}{\Delta N}, \frac{N}{\Delta N}\right)\right]$$

$$= \lim_{\substack{V' \to \infty \\ N' \to \infty}} [A(T, V', N' + 1) - A(T, V', N')], \qquad (A.1)$$

where A is the Helmholtz Gibbs energy.

The first two equalities on the right-hand side of relation (A.1) follow from the *definition* of the derivative with respect to the variable N, viewed as a continuous variable.

The third equality follows from the extensive character of the Helmholtz energy. The last equality states that instead of fixing V and N and adding the "infinitesimal" quantity ΔN, we may let V' and N' increase to infinity, keeping their ratio N'/V' fixed and adding *one* to N'. The reason for doing this step is that N is a discrete variable, while the derivative in (A.1) is defined for the addition of an *infinitesimal* amount of molecules.

The fundamental link between the Helmholtz energy A and statistical mechanics is through the relation

$$A = -k_B T \ln Q, \qquad (A.2)$$

where $Q(T, V, N)$ is the canonical partition function. For our simple system, this partition function has the form

$$Q(T, V, N) = \frac{q^N}{N! \Lambda^{3N}} \int \cdots \int_V dR_1 \cdots dR_N$$

$$\times \exp[-\beta U_N(R_1, \ldots, R_N)]. \qquad (A.3)$$

---

[262]Note that here we define the chemical potential per *molecule*. In thermodynamics, it is defined per mole of the species.

Here, q is the *internal* partition function of a single particle. It may include contributions from rotational, vibrational, electronic, and nuclear degrees of freedom. In Eq. (A.3), we have assumed that all these internal degrees of freedom are separable, i.e. they are independent of the environment surrounding each molecule.

The quantity Λ^3 is the momentum partition function. It arises from the integration over all possible translational momenta of the molecule. It depends on the molecular mass m and the temperature T as follows:

$$\Lambda^3 = \frac{h^3}{(2\pi m k_B T)^{3/2}},\tag{A.4}$$

where h is the Planck constant and k_B is the Boltzmann constant. Λ is sometimes referred to as the de Broglie thermal wavelength, $\mathbf{R}_i$ is the locational vector describing the location of the center of the ith molecule, while $d\mathbf{R}_i$ stands for an element of volume $dx_i dy_i dz_i$ located at $\mathbf{R}_i$. The integration in Eq. (A.3) is extended for each molecule over the entire volume V of the system. The symbol $U_N(\mathbf{R}_1,\ldots,\mathbf{R}_N)$ denotes the total potential energy of the N molecules at the specific configuration $\mathbf{R}_1,\ldots,\mathbf{R}_N$. The factor $N!$ arises from the indistinguishability of the N molecules, and $\beta = (k_B T)^{-1}$.

The expression for the chemical potential is obtained by taking the ratio of the canonical partition functions for $N+1$ and N particles. Thus, using Eqs. (A.1)–(A.3), we obtain

$$\mu = A(N+1) - A(N)$$
$$= -k_B T \ln[Q(N+1)/Q(N)]$$
$$= -k_B T \ln\left\{\left[\frac{q^{N+1}}{(N+1)!\Lambda^{3(N+1)}}\int\cdots\int d\mathbf{R}_0 d\mathbf{R}_1 \cdots d\mathbf{R}_N\right.\right.$$
$$\left.\left. \times \{\exp[-\beta U_{N+1}(R_0,\ldots,R_N)]\}\right]\right.$$

$$\times \left\{ \frac{q^N}{N!\Lambda^{3N}} \int \cdots \int d\mathbf{R}_1 \cdots d\mathbf{R}_N \right.$$

$$\left. \times \exp[-\beta U_N(\mathbf{R}_1,\ldots,\mathbf{R}_N)] \right\}^{-1} \Bigg\}$$

$$= -k_B T \ln \left\{ \frac{q}{(N+1)\Lambda^3} \int \cdots \int d\mathbf{R}_0 \cdots d\mathbf{R}_N \, \Pr(\mathbf{R}_1,\ldots,\mathbf{R}_N) \right.$$

$$\left. \times \exp[-\beta B_0(\mathbf{R}_0,\ldots,\mathbf{R}_N)] \right\}. \tag{A.5}$$

In the last form on the right-hand side of Eq. (A.5), we used the notation $\Pr(\mathbf{R}_1,\ldots,\mathbf{R}_N)$ for the fundamental probability density of finding the configuration $\mathbf{R}_1,\ldots,\mathbf{R}_N$ of the N particles:

$$\Pr(\mathbf{R}_1,\ldots,\mathbf{R}_N)$$

$$= \frac{\exp[-\beta U_N(\mathbf{R}_1,\ldots,\mathbf{R}_N)]}{\int \cdots \int d\mathbf{R}_1 \cdots d\mathbf{R}_N \exp[-\beta U_N(\mathbf{R}_1,\ldots,\mathbf{R}_N)]}. \tag{A.6}$$

Also, the total interaction energy among the $N+1$ molecules is written as

$$U_{N+1}(\mathbf{R}_0,\mathbf{R}_1,\ldots,\mathbf{R}_N) = U_N(\mathbf{R}_1,\ldots,\mathbf{R}_N) + B_0(\mathbf{R}_0,\ldots,\mathbf{R}_N). \tag{A.7}$$

Equation (A.7) may be viewed as a definition of the *binding energy* B_0 of the newly added particle at $\mathbf{R}_0$ to all the other N particles of the system being at a specific configuration $\mathbf{R}_1,\ldots,\mathbf{R}_N$.

Because of the invariance of the integral in Eq. (A.5) to the choice of the origin of our coordinate system, we can measure all the vectors $\mathbf{R}_i$ with respect to an origin chosen at $\mathbf{R}_0$. By effecting this change of variables, we may transform Eq. (A.5) into a simpler form, i.e.

$$\mu = -k_B T \ln \left\{ \frac{qV}{(N+1)\Lambda^3} \int \cdots \int d\mathbf{R}'_1 \cdots d\mathbf{R}'_N \Pr(\mathbf{R}'_1, \ldots, \mathbf{R}'_N) \right.$$

$$\left. \times \exp[-\beta B_0(\mathbf{R}'_1, \ldots, \mathbf{R}'_N)] \right\}, \tag{A.8}$$

where the two functions in the integrand depend on the (relative) coordinates $\mathbf{R}'_1, \ldots, \mathbf{R}'_N$. The integration over $\mathbf{R}_0$ produced the volume V. We may now put $\rho = (N+1)/V \cong N/V$ (macroscopic system) and rewrite Eq. (A.8) in the more compact form

$$\mu = -k_B T \ln \left[\frac{q}{\rho \Lambda^3} \langle \exp(-\beta B_0) \rangle \right], \tag{A.9}$$

where the symbol $\langle \rangle$ signifies an average over all the configurations of the N particles (excluding the newly added one) using the probability density $\Pr(\mathbf{R}_1, \ldots, \mathbf{R}_N)$ defined in Eq. (A.6).

We now repeat the same derivation as above, but with one additional restriction — that the newly added particle be placed at a fixed position $\mathbf{R}_0$. The pseudo-chemical potential (PCP) is defined by

$$\mu^* = A(N+1, \mathbf{R}_0) - A(N)$$

$$= -k_B T \ln \left\{ \frac{\frac{q^{N+1}}{N!\Lambda^{3N}} \int d\mathbf{R}_1 \cdots d\mathbf{R}_N \exp[-\beta U_{N+1}(\mathbf{R}_0, \ldots, \mathbf{R}_N)]}{\frac{q^N}{N!\Lambda^{3N}} \int d\mathbf{R}_1 \cdots d\mathbf{R}_N \exp[-\beta U_N(\mathbf{R}_1, \ldots, \mathbf{R}_N)]} \right\}$$

$$= -k_B T \ln \left\{ q \int \cdots \int d\mathbf{R}_1 \cdots d\mathbf{R}_N \Pr(\mathbf{R}_1, \ldots, \mathbf{R}_N) \right.$$

$$\left. \times \exp[-\beta B_0(\mathbf{R}_1, \ldots, \mathbf{R}_N)] \right\}$$

$$= -k_B T \ln[q \langle \exp(-\beta B_0) \rangle]. \tag{A.10}$$

It is instructive to examine carefully the differences between Eqs. (A.5) and (A.10). First, in Eq. (A.5), we have integration over $N+1$ locations versus N locations in Eq. (A.10). This produces the factor V in Eq. (A.8), which is absent from Eq. (A.10). In relation (A.5), we have a momentum partition function Λ^3

which is absent in Eq. (A.10). Finally, in Eq. (A.5), we have $(N + 1)$, which is absent in Eq. (A.10). All these differences arise from the fact that in Eq. (A.10), we have constrained the $(N + 1)$th particle to a fixed position. Such a particle cannot wander in the entire volume V, does not have momentum partition function Λ^3 and, being at a fixed position, is *distinguishable* from the rest of the N members of the same species.[263]

Combining Eq. (A.9) with Eq. (A.10), we arrive at the fundamental expression

$$\mu = \mu^* + k_B T \ln \rho \Lambda^3, \qquad (A.11)$$

which corresponds to a splitting of the process of adding one particle to the system (at constant T and V) in two steps. First, we place the particle at a fixed position, say $\mathbf{R}_0$. The corresponding change in the Helmholtz energy is μ^*. Next, we release the constraint imposed on the particle — this step always lowers the Helmholtz energy (since our system is classical $\rho\Lambda \ll 1$)[264] by the amount $k_B T \ln \rho \Lambda^3$, to which we refer as the *liberation Helmholtz energy.*

We can now also appreciate the various ingredients that contribute to the last term on the right-hand side of (A.11). First, the particle that is released from its fixed position may now wander in the entire volume V. Second, the released particle gains momentum and hence acquires a momentum partition function Λ^3. Finally, as long as the new particle is at a fixed position, it is distinguishable from all the other particles. Once it has been released, it becomes indistinguishable from the rest of the N particles. We say that this particle is being *assimilated* by the

---

[263]For more details, see Ben-Naim (2006, 2008).
[264]See Hill (1956, 1960).

other members of the same species.[265] Altogether, these three factors form the liberation Helmholtz energy.

Two important generalizations of (A.11) are: First, for non-spherical molecules, one can repeat the same steps as above and reach the same expression as in (A.11), with the same significance of each of the terms in (A.11). The important assumption is made that the internal partition function of each molecule is not affected by the interactions among the molecules. This assumption is made in most cases of a *rigid* molecule. If there are internal rotational degrees of freedom, one must re-derive the expression of the pseudo-chemical potential for each conformer. We shall discuss this case in Appendix B.

The second generalization is for mixtures of any number of components. Again, the derivation of the expression for the chemical potential is essentially the same except for more cumbersome notations. For a two-component system of A and B, one defines the chemical potential of, say, A as

$$\mu_A = A(T, V, N_A + 1, N_B) - A(T, V, N_A, N_B), \qquad \text{(A.12)}$$

where we added one A molecule to a system characterized by the variables T, V, N_A, N_B.[266] Repeating the same steps as in the derivation of (A.11), we obtain the final expression

$$\mu_A = \mu_A^* + k_B T \ln \rho_A \Lambda_A^3. \qquad \text{(A.13)}$$

Note again that (A.13) is valid for molecules which are not necessarily spherical.

---

[265] The word *assimilation* is used here to distinguish it from the process of mixing. The latter is used whenever two distinguishable species are mixed. For more details, see Ben-Naim (2008).

[266] Note that the letter A is used both for the species A, and for the Helmholtz energy.

(ii) The T, P, N ensemble

We now derive the expression for the chemical potential in the T, P, N ensemble. The set of variables T, P, N is the most important one in actual experiments. The derivation here will be quite condensed, since most of the steps are similar to those carried out in the T, V, N ensemble.

For a one-component system, the chemical potential in the T, P, N ensemble is defined by

$$\mu = (\partial G/\partial N)_{T,P} = G(T, P, N + 1) - G(T, P, N), \qquad (A.14)$$

where G is the Gibbs energy of the system.

As in the previous derivation, we first express the chemical potential in terms of the ratio of the corresponding partition functions. This leads to the result

$$\mu = -k_B T \ln \left\{ \left[q \int dV \int d\mathbf{R}_0 \cdots d\mathbf{R}_N \right.\right.$$

$$\times \exp[-\beta U_{N+1}(\mathbf{R}_0, \ldots, \mathbf{R}_N) - \beta PV] \Big]$$

$$\times \left[\Lambda^3 (N + 1) \int dV \int d\mathbf{R}_1 \cdots d\mathbf{R}_N \right.$$

$$\left.\left. \times \exp[-\beta U_N(\mathbf{R}_1, \ldots, \mathbf{R}_N) - \beta PV] \right] \right\}. \qquad (A.15)$$

We now use the conditional distribution function of finding a configuration $\mathbf{R}_1, \ldots, \mathbf{R}_N$, given that the system has a volume V, i.e.

$$\Pr(\mathbf{R}_1, \ldots, \mathbf{R}_N/V)$$

$$= \frac{\exp[-\beta U_N(\mathbf{R}_1, \ldots, \mathbf{R}_N)]}{\int_V \cdots \int d\mathbf{R}_1 \cdots d\mathbf{R}_N \exp[-\beta U_N(\mathbf{R}_1, \ldots, \mathbf{R}_N)]}. \qquad (A.16)$$

Using this distribution function, Eq. (A.15) may be rewritten as

$$\mu = -k_B T \ln\left[\frac{q}{\Lambda^3(N+1)} \int_0^\infty dV \, \mathrm{Pr}(V)\right.$$

$$\left. \times \int_V d\mathbf{R}_0 \cdots d\mathbf{R}_N \mathrm{Pr}(\mathbf{R}_1,\ldots,\mathbf{R}_N/V) \exp\left(-\beta B_0\right)\right]. \quad (A.17)$$

The essential difference between Eqs. (A.17) and (A.5) is the additional integration over all possible volumes, with the probability distribution $\mathrm{Pr}(V)$. Assuming that the system is macroscopically large, the probability density $\mathrm{Pr}(V)$ should have a single sharp peak at $\langle V \rangle$. For simplicity, we may assume that $\mathrm{Pr}(V)$ behaves as a Dirac delta function

$$\mathrm{Pr}(V) = \delta(V - \langle V \rangle), \quad (A.18)$$

where $\langle V \rangle$ is the average volume of the system characterized by the variables T, P, N.

This assumption enables us to simplify expression (A.17) as follows:

$$\mu = -k_B T \ln\left[\frac{q\langle V \rangle}{\Lambda^3(N+1)}\right.$$

$$\left. \times \int_{\langle V \rangle} \cdots \int d\mathbf{R}_1 \cdots d\mathbf{R}_N \, \mathrm{Pr}(\mathbf{R}_1,\ldots,\mathbf{R}_N/\langle V \rangle) \exp\left(-\beta B_0\right)\right]$$

$$= -k_B T \ln\left[\frac{q}{\rho\Lambda^3}\langle\exp\left(-\beta B_0\right)\rangle\right], \quad (A.19)$$

which is identical in form to expression (A.9). The difference is in the definition of the density $\rho = (N+1)/\langle V \rangle$ and the meaning of the average sign $\langle \; \rangle$, which in relation (A.19) is over all configurations of the N particles with the conditional distribution given in expression (A.16).

We can repeat exactly the same procedure, but instead of adding a free particle to the system, we place it at a fixed position (within the boundaries of the system). In this way we derive the corresponding expression for the PCP:

$$\mu^* = -k_B T \ln[q\langle\exp(-\beta B_0)\rangle]. \tag{A.20}$$

The last two equations are now combined to yield

$$\mu = \mu^* + k_B T \ln \rho \Lambda^3, \tag{A.21}$$

which is identical in form to Eq. (A.11). Note that from the two definitions

$$\mu(T, P, N) = \left(\frac{\partial G}{\partial N}\right)_{P,T} = G(T, P, N+1) - G(T, P, N) \tag{A.22}$$

and

$$\mu(T, V, N) = \left(\frac{\partial A}{\partial N}\right)_{V,T} = A(T, V, N+1) - A(T, V, N), \tag{A.23}$$

it follows that if we choose two systems characterized by the variables T, P, N and T, V, N, respectively, such that the *average volume* $\langle V\rangle$ in the former is equal to the *volume* of the latter, the two chemical potentials in Eqs. (A.22) and (A.23) will have equal values. (The same is true if we require that the pressure in the T, V, N ensemble be the same as the pressure in the T, P, N ensemble).

If we require that $V = \langle V\rangle$, then the densities $\rho = N/V$ in Eq. (A.11) and $\rho = N/\langle V\rangle$ in Eq. (A.21) will also be the same. Hence, the liberation Helmholtz energy in Eq. (A.11) and the liberation Gibbs energy in Eq. (A.21) are equal, and therefore the PCPs are also equal, namely

$$\mu^*(T, V, N) = \mu^*(T, P, N), \tag{A.24}$$

with the requirement that $V = \langle V\rangle$.

(iii) The T, V, μ ensemble

For completeness, we add here one more expression for the chemical potential in an open system characterized by the variables T, V, μ. Here, the chemical potential is one of the independent variables used to specify our system. The corresponding partition function, the so-called grand partition function, is given for a one-component classical system by[267]

$$\Xi(T, V, \mu) = \sum_{N \geq 1} \frac{z^N}{N!} \int \cdots \int d\mathbf{R}^N \exp[-\beta U_N(\mathbf{R}^N)], \quad (A.25)$$

where $z = \exp(\beta\mu)/\Lambda^3$, and $\mathbf{R}^N$ is a shorthand notation for $\mathbf{R}_1, \ldots, \mathbf{R}_N$.

Likewise, we write a grand partition function characterized by the same variables T, V, μ but with the additional particle at a fixed position, say, $\mathbf{R}_0$ within the volume V, i.e.

$$\Xi(T, V, \mu; \mathbf{R}_0) = \sum_{N \geq 1} \frac{z^N}{N!} \int \cdots \int d\mathbf{R}^N \exp[-\beta U_{N+1}(\mathbf{R}^N, \mathbf{R}_0)].$$

$$(A.26)$$

We note that Eq. (A.26) may be viewed as a partition function for a system subjected to an "external" field of force created by a particle placed at $\mathbf{R}_0$.

The average number of particles in the open system is provided by the well-known relation[268]

$$\langle N \rangle = k_B T \left(\frac{\partial \ln \Xi}{\partial \mu} \right)_{T,V}$$

$$= \frac{z}{\Xi} \sum_{N \geq 1} \frac{N z^{N-1}}{N!} \int \cdots \int d\mathbf{R}^N \exp[-\beta U_N(\mathbf{R}^N)]. \quad (A.27)$$

---

[267]For details, see Ben-Naim (1992).
[268]For details, see Ben-Naim (1992).

Each integral on the right-hand side of Eq. (A.27) may be rewritten as

$$\int \cdots \int d\mathbf{R}^N \exp[-\beta U_N(\mathbf{R}^N)]$$

$$= V \int d\mathbf{R}^{N-1} \exp[-\beta U_N(\mathbf{R}^{N-1}, \mathbf{R}_1)]. \qquad (A.28)$$

where we have transformed to coordinates relative to, say, $\mathbf{R}_1$ and integrated over $\mathbf{R}_1$ to obtain the volume. From Eqs. (A.27) and (A.28), we obtain

$$\langle N \rangle = \frac{zV}{\Xi} \sum_{N \geq 1} \frac{z^{N-1}}{(N-1)!} \int \cdots \int d\mathbf{R}^{N-1} \exp[-\beta U_N(\mathbf{R}^{N-1}, \mathbf{R}_1)]$$

$$= \frac{zV}{\Xi} \sum_{N \geq 0} \frac{z^N}{N!} \int \cdots \int d\mathbf{R}^N \exp[-\beta U_{N+1}(\mathbf{R}^N, \mathbf{R}_1)]$$

$$= \frac{zV \, \Xi(T, V, \mu; \mathbf{R}_1)}{\Xi(T, V, \mu)}, \qquad (A.29)$$

where we have changed the summation index and identified the resulting sum with Eq. (A.26). [Note that the choice of the fixed position $\mathbf{R}_0$ in Eq. (A.26) or $\mathbf{R}_1$ in Eq. (A.29) produces the same quantities $\Xi(T, V, \mu; \mathbf{R}_0) = \Xi(T, V, \mu; \mathbf{R}_1)$.]

Equation (A.29) may now be rewritten as

$$\mu = k_B T \ln \rho \Lambda^3 - [k_B T \ln \Xi(T, V, \mu; \mathbf{R}_0) - k_B T \ln \Xi(T, V, \mu)]$$

$$= k_B T \ln \rho \Lambda^3 - [P(T, V, \mu; \mathbf{R}_0)V - P(T, V, \mu)V]$$

$$= k_B T \ln \rho \Lambda^3 + \mu^*, \qquad (A.30)$$

where the second term on the right-hand side is the (PV) work required to place a particle at a fixed position $\mathbf{R}_0$ in the system. The density ρ is defined here as $\rho = \langle N \rangle / V$. Thus, relation (A.30) has the same form as our basic expression (A.11) except for a reinterpretation of the density and the work term according to the specific variables used to characterize our system.

The second term on the right-hand side of relation (A.30) may be rewritten, noting (A.25) and (A.26), in the form

$$\mu = k_B T \ln \rho \Lambda^3 - k_B T \ln \langle \exp\left(-\beta B_0\right)\rangle, \qquad (A.31)$$

which has the same form as Eq. (A.9) with the appropriate reinterpretation of the average quantities in the T, V, μ.

Finally, we note that by a straightforward generalization of the arguments given above, one may obtain essentially the same expressions for the chemical potential of a non-spherical particle and in multi-component systems.

The Pseudo-Chemical Potential and the Solvation Helmholtz Energy of a Molecule Having Internal Rotational Degrees of Freedom

Let s be a molecule with internal rotational degrees of freedom. A simple example is butane. In this case, the conformation of the molecule is defined by the dihedral angle ϕ (see, for example, Fig. 3.17). We ignore the rotations about the 1–2 bond and the 3–4 bond. In more general cases such as a protein, the conformation $\mathbf{P}_S$ comprises all the rotational angles $\phi_i \psi_i$ in the molecule. We assume that the vibrational, electronic, and nuclear partition functions are separable and independent of the configuration of the molecules in the system. We define the pseudo-chemical potential (PCP) of a molecule with a fixed conformation $\mathbf{P}_S$ as the change in the Helmholtz energy for the process of introducing s into the system l (at fixed T, V) in such a way that its center of mass is at a fixed position $\mathbf{R}_S$ and its conformation $\mathbf{P}_S$ is "frozen-in." If we now release the constraint on the fixed position of the center of the mass, we may define the chemical potential of the conformer $\mathbf{P}_S$ in the gas and liquid phases as follows:

$$\mu_S^g(\mathbf{P}_S) = \mu_S^{*g}(\mathbf{P}_S) + k_B T \ln \rho_S^g \Lambda_S^3, \qquad (\text{B.1})$$

$$\mu_S^l(\mathbf{P}_S) = \mu_S^{*l}(\mathbf{P}_S) + k_B T \ln \rho_S^l \Lambda_S^3. \qquad (\text{B.2})$$

We note that the rotational partition function of the entire molecule and the internal partition functions of s are included in the PCP. In classical systems, the momentum partition function Λ_S^3 is independent of the environment, whether it is a gas or a liquid phase.

The solvation Helmholtz energy of the conformer $\mathbf{P}_S$ is defined as

$$\Delta\mu_S^*(\mathbf{P}_S) = \mu_S^{*l}(\mathbf{P}_S) - \mu_S^{*g}(\mathbf{P}_S) = -k_B T \ln\langle\exp[-\beta B_S(\mathbf{P}_S)]\rangle,$$

$$(B.3)$$

i.e. this is the Helmholtz energy of transferring an s molecule, being frozen in its $\mathbf{P}_S$ conformation from a fixed position in an ideal gas g, into a fixed position in l. Clearly, since $\mathbf{P}_S$ is fixed, the orientation of the entire molecule $\mathbf{\Omega}_S$ does not affect the solvation Helmholtz energy. If we also assume that all vibrational, electronic, and nuclear degrees of freedom are not affected by this transfer from g to l, we can write the second equality in Eq. (B.3), where we have the same average quantity as in Eq. (A.9), with the additional constraint that the conformation of the molecule $\mathbf{P}_S$ is "frozen in."

Next, we wish to find the relation between $\Delta\mu_S^*(\mathbf{P}_S)$ and the experimental Helmholtz energy of solvation of the molecule s.

We carry out the derivations in two steps. For convenience, we use the T, V, N ensemble. Suppose first that s can attain only two conformations A and B, say the cis and trans conformations of a given molecule at equilibrium (Fig. 3.17). The PCP of A is the change in the Helmholtz energy for placing an A molecule at a fixed position in l. The corresponding statistical mechanical expression is

$$\exp\left(-\beta\mu_A^{*l}\right)$$
$$= \frac{q_A \int d\mathbf{X}^N d\mathbf{\Omega}_A \exp[-\beta U_N(\mathbf{X}^N) - \beta B_A(\mathbf{X}^N) - \beta U^*(A)]}{(8\pi^2)\int d\mathbf{X}^N \exp[-\beta U_N(\mathbf{X}^N)]},$$

$$(B.4)$$

where $B_A(\mathbf{X}^N)$ is the binding energy of A to the rest of the system of N molecules at configuration $\mathbf{X}^N$ (note that N is the sum of all molecules in the system including any s molecules but excluding only the added A molecule). $U^*(A)$ denotes the intramolecular potential or the internal rotation potential function of S at the state A.

Integration over $\boldsymbol{\Omega}_A$ produces $8\pi^2$, and hence Eq. (B.4) may be rewritten as

$$\exp\left(-\beta\mu_A^{*l}\right) = q_A \exp[-\beta U^*(A)]\langle\exp\left(-\beta B_A\right)\rangle, \qquad \text{(B.5)}$$

where the average is over all configurations of the N molecules in the system. Likewise, for the gaseous phase we have

$$\exp\left(-\beta\mu_A^{*g}\right) = q_A \exp[-\beta U^*(A)]. \qquad \text{(B.6)}$$

Hence, the solvation Helmholtz energy of A is obtained from Eqs. (B.5) and (B.6) in the form

$$\exp\left(-\beta\Delta\mu_A^{*l}\right) = \langle\exp\left(-\beta B_A\right)\rangle, \qquad \text{(B.7)}$$

and a similar expression holds true for B.

To obtain the connection between $\Delta\mu_A^{*l}$, $\Delta\mu_B^{*l}$ and $\Delta\mu_S^{*l}$, we start with the equilibrium condition

$$\mu_S^l = \mu_A^l = \mu_B^l, \qquad \text{(B.8)}$$

or, equivalently

$$\mu_S^{*l} + k_B T \ln \rho_S^l \Lambda_S^3 = \mu_A^{*l} + k_B T \ln \rho_A^l \Lambda_S^3$$
$$= \mu_B^{*l} + k_B T \ln \rho_B^l \Lambda_S^3, \qquad \text{(B.9)}$$

where $\rho_S^l = \rho_A^l + \rho_B^l$; ρ_A^l and ρ_B^l are the densities of A and B at equilibrium. Equation (B.9) may be rearranged to yield

$$\exp\left(-\beta\mu_S^{*l}\right) = \frac{\rho_S^l}{\rho_A^l} \exp\left(-\beta\mu_A^{*l}\right), \qquad \text{(B.10)}$$

and similarly

$$\exp\left(-\beta\mu_S^{*l}\right) = \frac{\rho_S^l}{\rho_B^l}\exp\left(-\beta\mu_B^{*l}\right). \tag{B.11}$$

Multiplying Eq. (B.10) by x_A^l and Eq. (B.11) by x_B^l and adding the resulting two equations (where $x_A^l = \rho_A^l/\rho_S^l$ and $x_B^l = 1 - x_A^l$), we get

$$\exp\left(-\beta\mu_S^{*l}\right) = \exp\left(-\beta\mu_A^{*l}\right) + \exp\left(-\beta\mu_B^{*l}\right). \tag{B.12}$$

This equation is equivalent to the statement that the partition function of a system with one additional S particle at a fixed position is the sum of the partition function of the same system with one A particle at a fixed position and the partition function of the same system with one B particle at a fixed position.

We now write the corresponding expression for the ideal gas phase, namely

$$\exp\left(-\beta\mu_S^{*g}\right) = \exp\left(-\beta\mu_A^{*g}\right) + \exp\left(-\beta\mu_B^{*g}\right). \tag{B.13}$$

Taking the ratio between expressions (B.12) and (B.13), we obtain

$$\exp\left(-\beta\Delta\mu_S^{*l}\right)$$
$$= \frac{q_A\exp[-\beta U^*(A)]\exp\left(-\beta\Delta\mu_A^{*l}\right) + q_B\exp[-\beta U^*(B)]\exp\left(-\beta\Delta\mu_B^{*l}\right)}{q_A\exp[-\beta U^*(A)] + q_B\exp[-\beta U^*(B)]},$$
$$\tag{B.14}$$

or equivalently

$$\exp\left(-\beta\Delta\mu_S^{*l}\right) = y_A^g\exp\left(-\beta\Delta\mu_A^{*l}\right) + y_B^g\exp\left(-\beta\Delta\mu_B^{*l}\right), \tag{B.15}$$

where y_A^g and y_B^g are the equilibrium mole fractions of A and B in the gaseous phase. These are defined by:

$$y_A^g = \frac{q_A\exp[-\beta U^*(A)]}{q_A\exp[-\beta U^*(A)] + q_B\exp[-\beta U^*(B)]}$$

and

$$y_B^g = 1 - y_A^g. \tag{B.16}$$

Generalization to the case with n discrete conformations is straightforward:

$$\exp\left(-\beta\Delta\mu_S^{*l}\right) = \frac{\sum_{i=1}^{n} q_i \exp[-\beta U^*(i)]\langle\exp\left(-\beta B_i\right)\rangle}{\sum_i q_i \exp[-\beta U^*(i)]}$$

$$= \sum_{i=1}^{n} y_l^g \exp\left(-\beta\Delta\mu_i^{*l}\right). \tag{B.17}$$

And the generalization to the continuous case is

$$\exp\left(-\beta\Delta\mu_S^{*l}\right) = \frac{\int d\mathbf{P}_S q(\mathbf{P}_S) \exp[-\beta U^*(\mathbf{P}_S)]\langle\exp[-\beta B(\mathbf{P}_S)]\rangle}{\int d\mathbf{P}_S q(\mathbf{P}_S) \exp[-\beta U^*(\mathbf{P}_S)]}$$

$$= \int d\mathbf{P}_S y^g(\mathbf{P}_S) \exp[-\beta\Delta\mu^{*l}(\mathbf{P}_S)]$$

$$= \langle\langle\exp[-\beta B(\mathbf{P}_S)]\rangle\rangle, \tag{B.18}$$

where $q(\mathbf{P}_S)$ denotes the rotational, vibrational, etc. partition function of a single s molecule at a specific conformation, $\mathbf{P}_S$, $y^g(\mathbf{P}_S)d\mathbf{P}_S$ is the mole fraction of s molecules at conformations between $\mathbf{P}_S$ and $\mathbf{P}_S + d\mathbf{P}_S$, and in the final expression on the right-hand side of relation (B.18), we have rewritten the integral as a double average quantity: One over all configurations of the N molecules, and the other over all conformations of the s molecule with distribution function $y^g(\mathbf{P}_S)$.

In this section we have treated the T, V, N system. A similar treatment can be applied to the T, P, N system (see Appendix A).

In most of the book, we shall discuss processes involving one conformation only. It is important to realize that at the end of each calculation, when we discuss real proteins, we must take the appropriate average over all possible conformations of the molecules involved in the process.

APPENDIX C

The Potential of Mean Force (PMF) and the Solvent-Induced Force

---

Traditionally, the PMF is defined in terms of the molecular correlation functions.[269] Here, we shall take a short cut. We shall define the PMF in terms of the Helmholtz (or Gibbs) energy of a system having two particles at some fixed positions in the liquid.

For simplicity, we deal with a one-component system of N simple spherical particles having no internal degrees of freedom contained in a volume V and at temperature T. The Helmholtz energy of the system is

$$\exp[-\beta A(T, V, N)] = \frac{1}{N! \Lambda^{3N}} \int \cdots \int d\mathbf{R}^N \exp[-\beta U_N(\mathbf{R}^N)].$$

$$(C.1)$$

Next, suppose that we fix the locations of two specific particles at, say, $\mathbf{R}_1$ and $\mathbf{R}_2$. The Helmholtz energy in this case is

$$\exp[-\beta A(T, V, N; \mathbf{R}_1, \mathbf{R}_2)]$$

$$= \frac{1}{(N-2)! \Lambda^{3(N-2)}} \int d\mathbf{R}^{(N-2)} \exp[-\beta U_N(\mathbf{R}^N)]. \quad (C.2)$$

Note that in (C.2), there is $(N-2)!$ instead of $N!$ as in (C.1), and we have $\Lambda^{3(N-2)}$ instead of Λ^{3N}. Also, the integrations in (C.2) are over all the locations of $N-2$ particles, rather than over N particles as in (C.1). The total potential energy of the

---

[269]See, for example, Hill (1956), Ben-Naim (2006).

system $U_N(\mathbf{R}^N)$ is the same in (C.1) and (C.2), namely

$$U_N(\mathbf{R}^N) = U_N(\mathbf{R}_1, \ldots, \mathbf{R}_2) = U(\mathbf{R}_1, \mathbf{R}_2) + \sum_{i=3}^{n} U(\mathbf{R}_1, \mathbf{R}_i)$$

$$+ \sum_{i=3}^{n} U(\mathbf{R}_2, \mathbf{R}_1) + U_{N-2}(\mathbf{R}_3, \ldots, \mathbf{R}_N). \qquad \text{(C.3)}$$

Thus, $U_N(\mathbf{R}^N)$ is the total interaction energy among *all* the N particles in the system being at some configuration $\mathbf{R}^N = \mathbf{R}_1, \ldots, \mathbf{R}_N$. This is written in four terms in (C.3). The first term is the interaction energy between the two particles 1 and 2 at $\mathbf{R}_1, \mathbf{R}_2$. The second and third terms on the right-hand side of (C.3) are the total binding energy of the two particles 1 and 2 to the rest of the $N - 2$ particles. The last term on the right-hand side of (C.3) is simply the total interaction energy among the $N - 2$ particles at the configuration $\mathbf{R}_3, \ldots, \mathbf{R}_N$.

The change in Helmholtz energy for the process of bringing the two particles 1 and 2 from infinite separation to the final configuration $\mathbf{R}_1, \mathbf{R}_2$ is denoted by ΔA, for which we have:

$$\exp[-\beta \Delta A]$$

$$= \exp[-\beta A(T, V, N; \mathbf{R}_1, \mathbf{R}_2) - \beta A(T, V, N; \mathbf{R}_{12} = \infty)]$$

$$= \exp[-\beta U(\mathbf{R}_1, \mathbf{R}_2)]$$

$$\times \frac{\int \cdots \int d\mathbf{R}^{(N-2)} \exp[-\beta B_{12}(\mathbf{R}_1, \mathbf{R}_2) - \beta U_{N-2}]}{\int \cdots \int d\mathbf{R}^{(N-2)} \exp[-\beta B_{12}(R = \infty) - \beta U_{N-2}]}.$$

$$\text{(C.4)}$$

In (C.4), we have extracted the factor $\exp[-\beta U(\mathbf{R}_1 \mathbf{R}_2)]$ from the integral sign (note that $U(R_{12} = \infty) = 0$), and we denote by B_{12} the total binding energy of the two particles 1 and 2 to the rest of the system.

Dividing both the numerator and denominator in (C.4) by the configurational integral

$$Z_N = \int \cdots \int d\mathbf{R}^N \exp[-\beta U_N(\mathbf{R}^N)], \qquad (C.5)$$

we can rewrite C.4 as

$$
\begin{aligned}
\Delta A &= A(\mathbf{R}_1, \mathbf{R}_2) - A(\mathbf{R}_{12} = \infty) \\
&= U(\mathbf{R}_1, \mathbf{R}_2) - k_B T \ln\langle \exp[-\beta B_{12}(\mathbf{R}_1, \mathbf{R}_2)]\rangle_0 \\
&\quad + k_B T \ln\langle \exp[-\beta B_{12}(R_{12} = \infty)]\rangle_0 \\
&= U(\mathbf{R}_1, \mathbf{R}_2) + \Delta A^*(\mathbf{R}_1, \mathbf{R}_2) - \Delta A^*(R_{12} = \infty) \\
&= U(\mathbf{R}_1, \mathbf{R}_2) + \Delta A^*(\mathbf{R}_1, \mathbf{R}_2) - 2\Delta A^*(\mathbf{R}_0) \\
&= U(\mathbf{R}_1, \mathbf{R}_2) + \delta A(\mathbf{R}_1, \mathbf{R}_2) = W(\mathbf{R}_1, \mathbf{R}_2). \qquad (C.6)
\end{aligned}
$$

Note carefully the steps we have taken in (C.6). First, we have rewritten the two integrals in (C.4) as two average quantities, here over all the configurations of the $N-2$ particles in the T, V, N ensemble. Second, we identified the two averages as the solvation Helmholtz energy of the two particles at $\mathbf{R}_1, \mathbf{R}_2$ and of the two particles at infinite separation. The latter was written as twice the solvation Helmholtz energy of the one particle in the system, which is denoted by $\Delta A^*(\mathbf{R}_0)$, where $\mathbf{R}_0$ is any arbitrarily chosen location in the liquid. This quantity is independent of $\mathbf{R}_0$, but the notation $\Delta A^*(\mathbf{R}_0)$ is used to remind us of the process of solvation (see Appendices A and B). Finally, we identified the difference of the solvation Helmholtz energies as the solvent-induced part of the potential of mean force.[270]

Thus, we have in fact defined in (C.6) the PMF as the work of bringing the two particles from infinite separation to some

---

[270]Note again that by "solvent," we mean all particles in the system except the two particles placed at fixed locations.

final configuration $\mathbf{R}_1, \mathbf{R}_2$ within the liquid, keeping the volume and the temperature of the system fixed.

We note here a few generalizations of (C.6) that may be obtained straightforwardly. First, the system can have any number of components and not necessarily simple spherical particles, in which case we select the two specific particles at fixed configuration $\mathbf{X}_1, \mathbf{X}_2$ rather than at fixed locations $\mathbf{R}_1, \mathbf{R}_2$. Second, we could have done the same process keeping the pressure and temperature fixed — the corresponding work would be the Gibbs energy change for the same process as in (C.6).

The force acting on particle 1 given the two particles 1 and 2 at $\mathbf{R}_1, \mathbf{R}_2$, averaged over all the configurations of the other particles in the system, is defined by

$$\mathbf{F}_1 = -\nabla_1 W(\mathbf{R}_1, \mathbf{R}_2) - \nabla_1 \delta A(\mathbf{R}_1, \mathbf{R}_2). \qquad \text{(C.7)}$$

The first term on the right-hand side of (C.7) is the direct force exerted on particle 1 by particle 2. The second is the *indirect*, solvent-induced, part of the force acting on particle 1, given that the two particles are at the configuration $\mathbf{R}_1, \mathbf{R}_2$.

Next, we express the solvent-induced force as an average quantity. The derivation is standard but lengthy.[271] The result is

$$\mathbf{F}_1 = -\nabla_1 U(\mathbf{R}_1, \mathbf{R}_2) - \int d\mathbf{R}_3 [\nabla_1 U(\mathbf{R}_1, \mathbf{R}_3)] \rho(\mathbf{R}_3/\mathbf{R}_1, \mathbf{R}_2).$$

$$\text{(C.8)}$$

Here, the solvent-induced force is expressed as an average of the quantity $-\nabla_1 U(\mathbf{R}_1, \mathbf{R}_3)$, which is the *direct* force acting on particle 1 by any "solvent" particle at $\mathbf{R}_3$. The average is taken with the conditional density $\rho(\mathbf{R}_3/\mathbf{R}_1, \mathbf{R}_2)$, i.e. the local density at $\mathbf{R}_3$, given two particles at $\mathbf{R}_1, \mathbf{R}_2$.

Note carefully that if the two particles 1 and 2 are spherical, then the density $\rho(\mathbf{R}_3/\mathbf{R}_1, \mathbf{R}_2)$ is symmetrical about the line

---

[271] For details, see Ben-Naim (2006).

connecting particles 1 and 2. This ensures that the average force is along the line connecting particles 1 and 2, but the direction of the force could either be positive or negative, i.e. towards or away from particle 2. Note also that this is only an *average* force. This means that the instantaneous direction of motion of particle 1 is not necessarily in the direction of the average force, not an instantaneous force as described in Sec. 1.2.

An important generalization of (C.8) is for any mixture of components. Take any two particles (suppose they are still spherical, but this assumption is not essential), call them A and B, at $(\mathbf{R}_A, \mathbf{R}_B)$. The average force acting on particle A at $\mathbf{R}_A$, given another particle B at $\mathbf{R}_B$ is

$$\mathbf{F}_A(\mathbf{R}_A, \mathbf{R}_B) = -\nabla_A U(\mathbf{R}_A, \mathbf{R}_B)$$
$$- \sum_{k=1}^{c} \int d\mathbf{R}_k [\nabla_A U_{Ak}(\mathbf{R}_A, \mathbf{R}_k)] \rho_k(\mathbf{R}_k / \mathbf{R}_A, \mathbf{R}_B).$$

$$(C.9)$$

Note carefully that the direct force in (C.9) is as in (C.8). The indirect force is a sum over all species $k = 1, \ldots, c$, where $\rho_k(\mathbf{R}_k / \mathbf{R}_A, \mathbf{R}_B)$ is the local density of the species k at $\mathbf{R}_k$, given one particle of type A at $\mathbf{R}_A$ and one particle of type B at $\mathbf{R}_B$ (Fig. C.1).

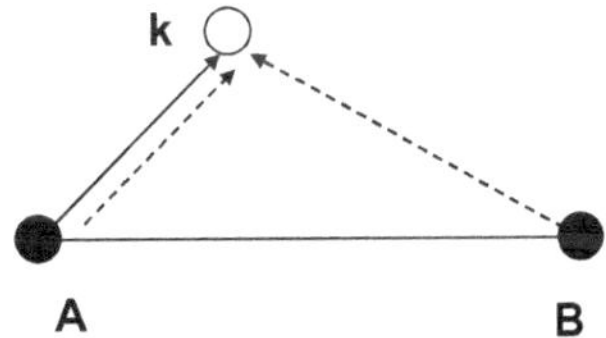

Fig. C.1   Schematic description of the two components of the indirect force on A. The full arrow indicates the direct force of a molecule at $\mathbf{R}_k$, and the dashed arrows indicate the conditional density at $\mathbf{R}_k$ due to the two particles at $\mathbf{R}_A$ and $\mathbf{R}_B$.

It should also be noted that all the results in this appendix are valid in the T, P, N ensemble, where the Gibbs energy replaces the Helmholtz energy whenever it appears.

Finally, we note an important relationship between ΔA or W and probabilities.[272]

$$\text{Pr}(\mathbf{R}_1, \mathbf{R}_2)d\mathbf{R}_1 d\mathbf{R}_2 = \exp[-\beta \Delta A(\mathbf{R}_1, \mathbf{R}_2)]d\mathbf{R}_1 d\mathbf{R}_2$$

$$= \exp[-\beta W(R_1, R_2)]d\mathbf{R}_1 d\mathbf{R}_2, \quad \text{(C.10)}$$

where $\text{Pr}(\mathbf{R}_1, \mathbf{R}_2)d\mathbf{R}_1 d\mathbf{R}_2$ is the probability of finding one particle in a small volume $d\mathbf{R}_1$ at $\mathbf{R}_1$, and a second particle in a small volume $d\mathbf{R}_2$ at $\mathbf{R}_2$.

---

[272]For details, see Ben-Naim (2006).

APPENDIX D

Conditional Solvation and Conditional Correlation

We start with the general expression for the solvation Helmholtz energy of a solute α in a solvent:

$$\exp[-\beta \Delta A_\alpha^*] = \langle \exp[-\beta B_\alpha] \rangle_0, \qquad (D.1)$$

where B_α is the binding energy of α to all solvent molecules (Appendix A). The average on the right-hand side of (D.1) is over all the configurations of the solvent molecules in the T, V, N ensemble.

We now want to explain the difference in the two conditional solvation Helmholtz energies of the hydroxyl groups in Eqs. (1.5.10) and (1.5.11).[273]

We start with the solvation Helmholtz energy of an ethanol molecule, which we write as

$$\exp[-\beta \Delta A_{CH_3CH_2OH}^*] = \langle \exp[-\beta B_{CH_3CH_2OH}] \rangle_0. \qquad (D.2)$$

Here, CH_3CH_2OH represents the *real* ethanol molecules, before we perform any splitting or modification.

The two conditional solvation quantities in Eqs. (1.5.10) and (1.5.11) are obtained by different ways of splitting the binding energy of the ethanol molecule. In one, we split the interaction

---

[273]Here, we discuss the Helmholtz energy of solvation. The final results may also be applied for the Gibbs energy provided we interpret all the average quantities in the T, P, N ensemble rather than in the T, V, N ensemble. See Appendix A.

energy between an ethanol molecule and a solvent molecule, say, water, into two parts: The interaction between the ethyl group and a water molecule, and that between the hydroxyl group and a water molecule. We write the corresponding split in the binding energy of ethanol as

$$B(\text{ethanol}) = B(\text{ethyl}) + B(\text{hydroxyl}). \qquad (D.3)$$

In the second, we split the interaction energy between ethanol and water in two parts: The interaction of the Lennard-Jones part, which is approximated by the interaction between *propane* and water, and the interaction between the hydrogen-bonding "arms" of the hydroxyl group and a water molecule. The corresponding split of the binding energy of ethanol is[274]

$$B(\text{ethanol}) = B(\text{propane})$$
$$+ B(\text{HBing of the arms of the hydroxyl group}).$$
$$(D.4)$$

Thus, in the first method (D.3), we view the interaction energy between ethanol and water as consisting of two interactions between the two *parts* of the ethanol with water. In the second method (D.4), we split the total interaction into two types, say, van der Waals and hydrogen bonding.

The two ways of splitting the binding energy of ethanol when used in (D.2) provides two different splits of the solvation Helmholtz energy

$$\exp-\beta\Delta A^*_{\text{CH}_3\text{CH}_2\text{OH}}$$
$$= \langle \exp[-\beta B(\text{ethyl})]\rangle_0 \langle \exp[-\beta B(\text{hydroxyl})]\rangle_{\text{ethyl}}, \qquad (D.5)$$

---

[274]A more detailed discussion of these two ways of splitting of the interaction energy may be found in Ben-Naim (2006), Sec. 7.7.

$$\exp[-\beta \Delta A^*_{\mathrm{CH_3CH_2OH}}]$$

$$= \langle \exp[-\beta B(\text{propane})]\rangle_0 \langle \exp[-\beta B(\text{arms of OH})]\rangle_{\text{propane}},$$

$$(\mathrm{D}.6)$$

or equivalently

$$\Delta A^*_{\mathrm{CH_3CH_2OH}} \approx \Delta A^*_{\mathrm{CH_3CH_3}} + \Delta A^*_{\mathrm{OH/CH_3CH_2}}, \qquad (\mathrm{D}.7)$$

$$\Delta A^*_{\mathrm{CH_3CH_2OH}} \approx \Delta A^*_{\mathrm{CH_3CH_2CH_3}} + \Delta A^*_{\mathrm{HB\ of\ OH/propane}}. \qquad (\mathrm{D}.8)$$

This is essentially the same as the two Eqs. (1.5.10) and (1.5.11), except that here, we have done the solvation at constant volume instead of constant pressure as in Sec. 1.5.

The meaning of the two conditional Helmholtz energies of solvation is quite simple. In the first case, we first transfer the ethyl group of ethanol into the liquid. The corresponding work (at T, V constant) is approximately equal to the solvation Helmholtz energy of ethane. In the second step, we solvate the hydroxyl group by bringing it to the location next to the ethyl group. This is the *conditional* Helmholtz energy of the hydroxyl group, given the ethyl backbone (BB). In the second case, we first "switch off" the hydrogen-bonding capability of the ethanol molecule. We transfer the resulting "molecule" into the liquid. The corresponding work is approximately equal to the solvation Helmholtz energy of a propane molecule. In the next step, we "turn on" the hydrogen-bonding arms of the hydroxyl group. The corresponding work is the *conditional* solvation Helmholtz energy of the hydrogen bonding of the hydroxyl group, given the propane molecule.

Clearly, these two splits of the process of solvation are different. However, for a hydroxyl or a carbonyl group, the *values* of the conditional solvation Helmholtz (or Gibbs) energies are nearly the same.

In Chapter 1, we have used model compounds to estimate the Gibbs or Helmholtz energy of solvation. Strictly, the conditional solvation Helmholtz energy of a group k is defined by

$$\Delta A^*_{k/\mathrm{BB}} = \Delta A^*_{\mathrm{BB}-k} - \Delta A^*_{\mathrm{BB}} \approx \Delta A^*_{\mathrm{BB}-k} - \Delta A^*_{\mathrm{BB}-\mathrm{H}}, \qquad (\mathrm{D}.9)$$

where BB means the backbone, and BB–k means the backbone with the group k, and BB–H means the backbone with a hydrogen replacing the group k.

When the BB is a protein, none of the quantities on the right-hand side of (D.9) is available. Therefore, we need to find a small model compound to replace the BB.

Recall the definition of the solvation Helmholtz (or Gibbs) energy in (D.1). It is the average of the quantity $\exp[-\beta B_\alpha]$ over all the configurations of the solvent molecules. Similarly, the conditional solvation Helmholtz energy is defined as an average of the same function, but with the *conditional* distribution of solvent configurations, i.e.

$$\exp[-\beta \Delta A^*_{k/\mathrm{BB}}] = \int d\mathbf{X}^N \Pr(\mathbf{X}^N/\mathrm{BB}) \exp[-\beta B_k], \qquad (\mathrm{D}.10)$$

where the conditional distribution is

$$\Pr(\mathbf{X}^N/\mathrm{BB}) = \frac{\int d\mathbf{X}^N \exp[-\beta U_N - \beta B_{\mathrm{BB}} - \beta B_k]}{\int d\mathbf{X}^N \exp[-\beta U_N - \beta B_k]}, \qquad (\mathrm{D}.11)$$

where B_k and B_{BB} are the binding energies of the functional group (FG), k and the backbone, respectively.

Clearly, the integrand in (D.10) depends on the binding energy of the FG k, and on the conditional distribution of solvent molecules in the vicinity of the group k.

We can now clarify the condition we impose on the model compound M to replace the entire backbone (BB) of the protein. Suppose that we cut out a small segment of the protein (Fig. D.1) in such a way that the distribution of the solvent molecules in

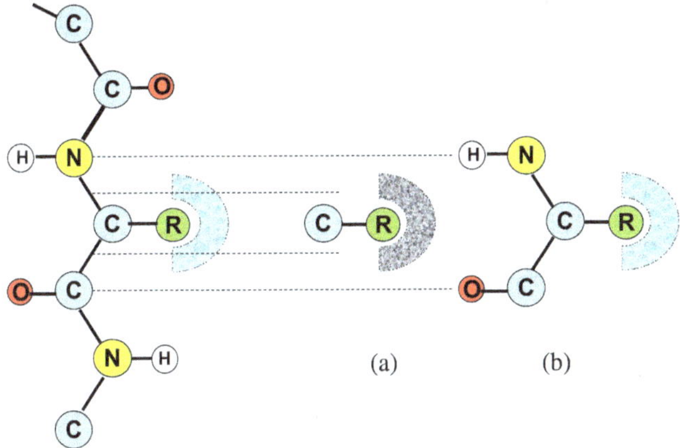

Fig. D.1   (a) A small segment of the protein next to the FG R. The environment around this FG is very different from the one in the original protein. (b) A larger segment of the protein carrying the FG R. Here, the environment around the FG is almost the same as in the original protein.

the region next to k is not affected. In this case we can replace the conditional distribution in (D.11) with a conditional distribution given the segment, i.e. $\Pr(X^N/segment)$.

The next step is to replace the segment of the protein with a model compound M such that the value of the conditional probability distribution of solvent molecules in the vicinity of the group k is the same as in the vicinity of the segment. Once we do that, we can write instead of (D.10)

$$\exp[-\beta \Delta A^*_{k/BB}] \approx \exp[-\beta \Delta A^*_{k/M}]$$

$$= \int d\mathbf{X}^N \Pr(\mathbf{X}^N/M) \exp[-\beta B_k], \quad (D.12)$$

and therefore we can also replace (D.9) with

$$\Delta A^*_{k/BB} \approx \Delta A^*_{k/M} \approx \Delta A^*_{M-k} - \Delta A^*_{M-H}, \quad (D.13)$$

where on the right-hand side of (D.13), we have the difference in the solvation Gibbs energies of the model compound M with and without the FG k. Note that by "without k," we mean the model compound in which the group k has been replaced by a hydrogen atom.

To summarize, we start with the protein with a FG, k. We wish to estimate the conditional solvation Helmholtz (or Gibbs) energy of the k group, given the backbone BB. We cannot use Eq. (D.9) since we do not know the solvation Helmholtz energies of the entire protein with or without the group k. Instead, we choose a model compound M having the same group k such that the distribution of the solvent molecules around the group k is approximately the same as in the original protein. This choice ensures that the integral (D.10) will not be affected by the replacement of the BB by M. On the other hand, we must also choose M to be as small as possible so that the solvation Helmholtz energies of the molecules M–k and M–H can be measured.

In Chapter 1, we have shown that this methodology works very well for small backbones (see Tables 1.4 to 1.6). We therefore expect that it will also work for macromolecules such as proteins.

All the discussion so far has been concerned with methods of estimating the conditional solvation Helmholtz (or Gibbs) energy of a FG by experimental means. It is straightforward to extend the application of the same method to estimate the solvent-induced contribution to the driving force in the process of bringing two FGs from infinite separation to a close distance. The reason is quite simple. Consider the process of bringing two FGs k and l to a close distance. The indirect part of the driving force for this process is

$$\delta A = \Delta A^*(k, l/\text{BB}) - \Delta A^*(k/\text{BB}) - \Delta A^*(l/\text{BB})$$

$$\approx \Delta A^*(k, l/M) - \Delta A^*(k/M) - \Delta A^*(l/M), \quad \text{(D.14)}$$

where $\Delta A^*(k, l/M)$ is the conditional solvation Helmholtz energy of the two groups k and l at some close distance, and

$\Delta A^*(k/M)$ is the conditional solvation Helmholtz energy of the group k given the model compound.

Once we have a model compound, we can use a relationship similar to (D.9) to rewrite (D.14) in an even simpler form. For instance, consider the two isomers as shown in Fig. 1.23a with the same two groups k and l once in the 1, 2 positions, and once in the 1, 4 positions (in the latter case, we assume that the distance between the two groups is large enough so that the two groups may be considered to be independently solvated). We rewrite (D.14) as

$$\delta A(1,4 \to 1,2) \approx \Delta A^*(1,2/M) - \Delta A^*(1,4/M)$$

$$\approx \Delta A^*_{M-1,2} - \Delta A^*_{M-H} - [\Delta A^*_{M-1,4} - \Delta A^*_{M-H}]. \quad (D.15)$$

Since ΔA^*_{M-H} appears twice on the right-hand side of (D.15) we can simplify this expression

$$\delta A(1,4 \to 1,2) \approx \Delta A^*_{M-1,2} - \Delta A^*_{M-1,4}, \quad (D.16)$$

where on the right-hand side of (D.16), we have the solvation Helmholtz energies of the two isomers, not the *conditional* solvation Helmholtz energies. This expression, or the analogous expression with the Gibbs energies, was used in Sec. 1.7.

In the rest of this appendix, we present some examples examining the accuracy of the approximation in (D.13). We examine the conditional solvation Gibbs energy of a $H\phi I$ group and a $H\phi O$ group.

(i) The conditional solvation Gibbs energy of a $H\phi I$ group.

Table 1.4 shows some numerical values of the conditional Gibbs energy of solvation of a hydrophilic group ($H\phi I$).

Suppose we have a protein with one hydroxyl group (OH) on its surface. We want to know the conditional solvation Gibbs energy of such a group. As noted above, we do not have the solvation Gibbs energies of the protein with and without such a FG. In Table 1.4, we show a cyclohexane molecule with and without a hydroxyl group.

The cyclohexane molecule is used here as a model (M) for the protein (α). Fortunately, we have the values of the Gibbs energies of both cyclohexanol and cyclohexane. Therefore, we can calculate the difference

$$\Delta G^{*OH/M} = \Delta G^*_{M-OH} - \Delta G^*_{M-H}. \qquad (D.17)$$

We find the value of the conditional solvation Gibbs energy of a hydroxyl group given the model compound M is about $-28.0\,\text{kJ mol}^{-1}$. Now, suppose we could not measure the solvation Gibbs energies of M–OH and M–H, and we want to estimate the *difference* on the right-hand side of (D.17). We use a small model compound such as methane, ethane or propane as *model* compounds for M. Recall that M itself is a model for the protein α. We see from Table 1.4 that if we take a short segment of the cyclohexane — three methylene groups, such as propane — and add on OH, we get a value of the conditional solvation Gibbs energy of about

$$\Delta G^{*OH/C-C-C} \cong \Delta G^*_{C-C-C-OH} - \Delta G^*_{C-C-C}$$

$$\approx -28.4\,\text{KJ mol}^{-1}. \qquad (D.18)$$

This value is very close to the value obtained in (D.17) for the backbone of cyclohexane, or cycloheptane. Thus, it seems that the methodology of estimating the conditional solvation Gibbs energy works quite well.

(ii) The conditional solvation Gibbs energy of a $H\phi O$ group.

Table 1.5 shows some values of the solvation Gibbs energies of cyclohexane and methyl cyclohexane. Again, we see that by choosing a small segment of the cyclohexane consisting of three methylene groups, we get a value of the conditional solvation Gibbs energy which is of the same order of magnitude as the conditional solvation Gibbs energy of a methyl group next to cyclohexane.

Note that in the case of a methyl group, the *conditional solvation* Gibbs energy is positive; it is relatively small compared to the *solvation* Gibbs energy of methane (about $7.9\,\text{kJ mol}^{-1}$), and compared to the solvation Gibbs energy of a hydroxyl group as shown in Table 1.8.

Table 1.6 shows similar values as in Table 1.4, but we now replace the cyclohexane with an aromatic molecule, such as benzene or naphthalene. Here, we see that the conditional solvation Gibbs energy of a methyl group next to an aromatic molecule is an order of magnitude smaller compared to the conditional solvation Gibbs energy of the same group next to a saturated hydrocarbon.

Note carefully that the addition of a methyl group next to a double bond of propene has a very different value of the conditional solvation Gibbs energy compared to methyl group next to a single bond in a molecule such as butene or butane.

The few values given in Tables 1.4, 1.5 and 1.6 are sufficient to demonstrate the validity of the methodology of estimating the conditional solvation Gibbs energy of a FG using small model compounds. Of course, one needs to have similar data on backbones that produce environments similar to that of a protein such as diketopiperazine (Fig. 1.36). Furthermore, to deal with real proteins, we shall also need to have the relevant data on model compounds with all the side chains of amino acids.

The conditional correlation between two FGs is simply the difference between two conditional solvation Gibbs energies [e.g. Eq. (D.16)]. Therefore, the same methodology and the same approximation apply for this quantity.

Non-Additivity of the Potential of Mean Force and of the Solvation Gibbs Energy

In this appendix, we shall discuss an important aspect of the potential of mean force (PMF) and of the solvation Gibbs (or Helmholtz) energy. The question posed on the PMF is the following: Suppose we have three simple particles at configuration $\mathbf{R}_1, \mathbf{R}_2, \mathbf{R}_3$. The PMF is defined as the change in the Helmholtz energy (in the T, V, N ensemble), or the Gibbs energy (in the T, P, N ensemble) associated with the process of bringing the three particles from infinite separation to the final configuration. Thus, in the T, P, N ensemble,

$$W(\mathbf{R}_1, \mathbf{R}_2, \mathbf{R}_3)$$
$$= G(T, P, N; \mathbf{R}_1, \mathbf{R}_2, \mathbf{R}_3) - G(T, P, N; R_{ij} = \infty). \qquad (E.1)$$

The question we pose is: Under what conditions can one approximate $W(\mathbf{R}_1, \mathbf{R}_2, \mathbf{R}_3)$ as a sum over pairs of PMFs?

$$W(\mathbf{R}_1, \mathbf{R}_2, \mathbf{R}_3) = W(\mathbf{R}_1, \mathbf{R}_2) + W(\mathbf{R}_1, \mathbf{R}_3) + W(\mathbf{R}_2, \mathbf{R}_3). \qquad (E.2)$$

This is known as the Kirkwood supervision approximation.[275] It was originally introduced by Kirkwood in the theory of liquids to achieve a relatively simple integral equation for the

---

[275]Kirkwood (1935).

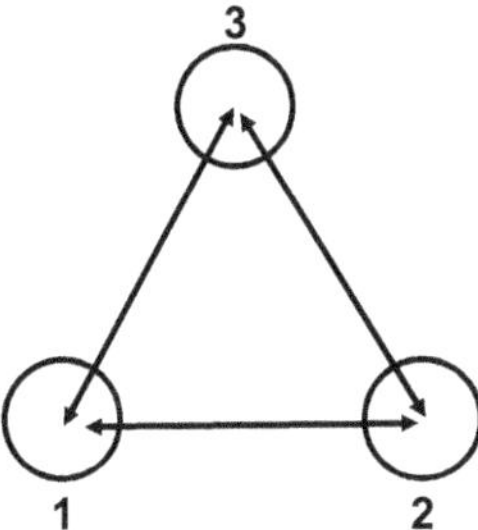

Fig. E.1   The energy change for bringing the three particles to a configuration $\mathbf{R}_1, \mathbf{R}_2, \mathbf{R}_3$ is the sum of the three energy changes for the process of bringing two particles to a distance as indicated by the double arrows.

pair correlation function.[276] This approximation is intuitively very appealing. The reason is that such an additive assumption is a good approximation for the potential energy, i.e. (Fig. E.1)

$$U(\mathbf{R}_1, \mathbf{R}_2, \mathbf{R}_3) \approx U(\mathbf{R}_1, \mathbf{R}_2) + U(\mathbf{R}_1, \mathbf{R}_3) + U(\mathbf{R}_2, \mathbf{R}_3). \quad \text{(E.3)}$$

In fact, the pairwise additivity of the potential function is exact for some systems, e.g. hard spheres, point charges, point dipoles, etc. It is also a good approximation for the total interaction energy among three (or more) simple, non-polar molecules. Note, however, that this approximation is not always true for direct interactions. For example, the interaction energy among three polarizable molecules such as water is not additive. However, much of the progress in the theory of liquids as well as water was achieved by using this pairwise additivity [for more details, see Ben-Naim (2009).]

Perhaps Kirkwood was inspired by the additivity of the potential energy (E.3) to suggest the additivity of the PMF (E.1). Today, it is not uncommon to encounter, especially in the biochemical literature, reference to the PMF as a *potential energy function*. Thinking of it as a "potential energy," one is inclined

---

[276]Hill (1956, 1960).

to also apply the additivity assumption (E.3) to the PMF without giving a second thought to its validity. This nomenclature, though common, is unfortunate. The PMF is different from a potential energy in some fundamental aspects. One is that the PMF is temperature dependent, whereas the potential energy is approximately independent of temperature. Another is the non-additivity of the PMF, even when there is an exact additivity of the potential energy. Note that whenever a term corresponding to "hydrophobic interactions" is included in the "energy function" of a protein, it effectively converts the "energy function" into a "free energy function" or a PMF of the protein. This is a very common practice in the study of protein folding (see for instance Bryngelson and Wolynes (1987).)

In the rest of the appendix, we shall present some arguments showing the source of the non-additivity of the PMF. The general argument is the following:

For a classical system we can write

$$W(\mathbf{R}_1, \mathbf{R}_2, \mathbf{R}_3)$$

$$= U(\mathbf{R}_1, \mathbf{R}_2, \mathbf{R}_3) + \delta G(\mathbf{R}_1, \mathbf{R}_2, \mathbf{R}_3)$$

$$= U(\mathbf{R}_1, \mathbf{R}_2, \mathbf{R}_3) + \Delta G^*(\mathbf{R}_1, \mathbf{R}_2, \mathbf{R}_3) - 3\Delta G_S^*. \qquad (E.4)$$

Equation (E.4) is a simple extension of Eq. (C.6). It states that the work (here, at T, P, N constants) of bringing three particles 1, 2, 3 from infinite separation to the final configuration $\mathbf{R}_1$, $\mathbf{R}_1$, $\mathbf{R}_3$ can be written in two terms: A *direct interaction energy* $U(\mathbf{R}_1, \mathbf{R}_2, \mathbf{R}_3)$, and a solvent-induced part $\delta G(\mathbf{R}_1, \mathbf{R}_2, \mathbf{R}_3)$. The latter can be rewritten as the difference in the solvation Gibbs energy of the triplet of particles in the configuration $\mathbf{R}_1, \mathbf{R}_2, \mathbf{R}_3$, and three times the solvation Gibbs energy of one particle in the same solvent.[277]

---

[277]Here, the "solvent" could be a real solvent and the particles are solute molecules, or the solvent could be all the molecules in the system except the ones which are placed at fixed positions.

To highlight the source of the non-additivity of the PMF, we shall assume that the potential energy in (E.4) is pairwise additive (approximately or exactly). We shall show below that the source of the non-additivity of the PMF is the *solvation* Gibbs energy $\Delta G^*(\mathbf{R}_1, \mathbf{R}_2, \mathbf{R}_3)$.

Note that when the solvent density tends to zero, all the solvation Gibbs energies on the right-hand side of (E.4) will tend to zero. At this limit, the PMF becomes identical with the potential energy. It is perhaps this limiting example that inspired Kirkwood as well as many others to adopt the supposition approximation for the PMF. Unfortunately, whenever a solvent is present, even a solvent consisting of a single molecule, the additivity assumption of the PMF is invalid.

Before turning to specific examples, it is instructive to first present the general argument: Why we cannot expect the PMF to be pairwise additive. Realizing that it is the *solvation* Gibbs energy in (E.4) which is the culprit for the non-additivity of the PMF, we write the statistical mechanical expression for the quantity $\Delta G^*(\mathbf{R}_1, \mathbf{R}_2, \mathbf{R}_3)$ as

$$\exp[-\beta \Delta G^*(\mathbf{R}_1, \mathbf{R}_2, \mathbf{R}_3)]$$
$$= \exp[-\beta B(\mathbf{R}_1, \mathbf{R}_2, \mathbf{R}_3)]$$
$$= \langle \exp[-\beta B(\mathbf{R}_1)] \exp[-\beta B(\mathbf{R}_2)] \exp[-\beta B(\mathbf{R}_3)] \rangle. \quad \text{(E.5)}$$

Note that we have written (E.5) for three simple solute particles, and we have also assumed that any internal degrees of freedom of these three particles are not affected by the presence of the solvent. We have also assumed that the binding energy of the three particles is additive, in the sense that

$$B(\mathbf{R}_1, \mathbf{R}_2, \mathbf{R}_3) = \sum_i U(\mathbf{R}_1, \mathbf{X}_i) + U(\mathbf{R}_2, \mathbf{X}_i) + U(\mathbf{R}_3, \mathbf{X}_i)$$
$$= B(\mathbf{R}_1) + B(\mathbf{R}_2) + B(\mathbf{R}_3), \quad \text{(E.6)}$$

where the sum over i is over all "solvent" molecules (i.e. all molecules in the system except the three particles at $\mathbf{R}_1$, $\mathbf{R}_2$, $\mathbf{R}_3$).

As can be seen in (E.5), the solvation Gibbs energy is an average of the *product* of three functions $\prod_{i=1}^{3} \exp[-\beta B(\mathbf{R}_i)]$. When the distances between all the pairs of particles are very large, the average of the two products may be factored into a product of three averages, i.e. when each $R_{ij} \to \infty$ $(i,j = 1,2,3)$, we can write

$$\langle \exp[-\beta B(\mathbf{R}_1, \mathbf{R}_2, \mathbf{R}_3)] \rangle \to \prod_{i=1}^{3} \langle \exp[-\beta B(\mathbf{R}_i)] \rangle. \qquad (E.7)$$

However, at this distance, the whole solvent-induced contribution δG in (E.4) is zero. It is also very likely that at these distances, $U(\mathbf{R}_1, \mathbf{R}_2, \mathbf{R}_3)$ is nearly zero. Therefore, this limit is of no interest.

At any configuration for which (E.7) is not valid, there exists no hint, nor any known condition under which the quantity on the right-hand side of (E.5) may be factored into pairwise terms, namely

$$\begin{aligned}
\exp[&-\beta B(\mathbf{R}_1, \mathbf{R}_2, \mathbf{R}_3)] \\
&\neq \langle \exp[-\beta B(\mathbf{R}_1, \mathbf{R}_2)] \rangle \langle \exp[-\beta B(\mathbf{R}_1, \mathbf{R}_3)] \rangle \\
&\times \langle \exp[-\beta B(\mathbf{R}_2, \mathbf{R}_3)] \rangle.
\end{aligned} \qquad (E.8)$$

The reason why such a factorization is not valid is a result of the fundamental difference between the *potential energy* function and the *solvation Gibbs energy* function. Figure E.2 illustrates this difference. The full bold lines represent the *direct interaction energy* between pairs of particles. The dashed lines represent interactions between the "solute" molecules and the "solvent" molecules. The solvation Gibbs energy depends on the averaging over the "solute–solvent" interaction (through the binding energy). This depends only on the "dashed lines" and

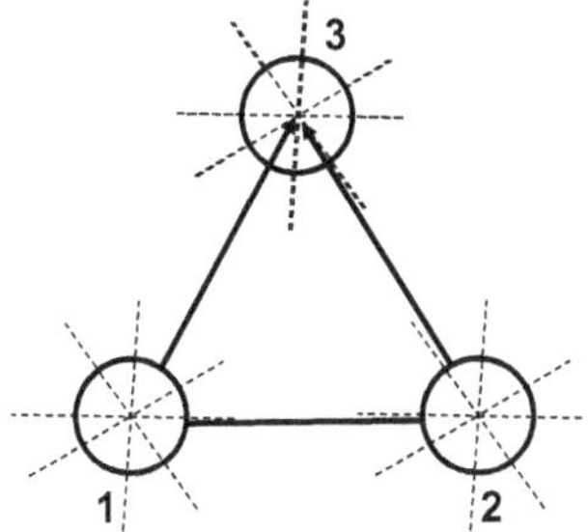

Fig. E.2 The solvation Gibbs energy does not include interactions along the line connecting the centers of each pair of particles, as in Fig. E.1. Instead, the interactions are between the solute molecules and the solvent molecules (dashed lines).

cannot produce "bold lines" or "lines of interaction" between the solute particles.

We next turn to some specific examples demonstrating the non-additivity of the Helmholtz (or Gibbs) energy of solvation.

(i) Three hard spheres (HS) at $\mathbf{R}_1, \mathbf{R}_2, \mathbf{R}_3$ and a solvent consisting of a single hard sphere

This is the simplest case of solvation of the triplet of HS particles in a one-hard-sphere "solvent" denoted w. The solvation Helmholtz energy in this case is

$$\exp[-\beta \Delta A^*(\mathbf{R}_1, \mathbf{R}_2, \mathbf{R}_3)]$$

$$= \langle \exp[-\beta B(\mathbf{R}_1, \mathbf{R}_2, \mathbf{R}_3)] \rangle$$

$$= \int \frac{d\mathbf{R}_W}{V} \exp[-\beta B(\mathbf{R}_1) - \beta B(\mathbf{R}_2) - \beta B(\mathbf{R}_3)]$$

$$= \frac{1}{V} \int d\mathbf{R}_W \exp[-\beta U(R_{1W}) - \beta U(R_{2W}) - \beta U(R_{3W})]$$

$$= \frac{V - V^{EX}(\mathbf{R}_1, \mathbf{R}_2, \mathbf{R}_3)}{V}. \tag{E.9}$$

In the second step on the right-hand side of (E.9), we wrote the average quantity explicitly. Here, the probability density of the solvent molecule is simply

$$\Pr(\mathbf{R}_W) = \frac{1}{V}. \tag{E.10}$$

Since there is only one solvent molecule, denoted w, the binding energy reduces to the solute–solvent interaction energy. We next use the property of the HS interaction potential, which makes the integrand zero whenever the solvent molecule penetrates into the excluded volume (V^{EX}) of the triplet of particles, and unity otherwise. The resulting expression is the right-hand side of (E.9).

Figure E.3 shows three configurations of the three particles and the corresponding excluded volumes. For simplicity, we assume that the configuration $\mathbf{R}_1$, $\mathbf{R}_2$, $\mathbf{R}_3$ is an equilateral triangle. Since V is much larger than V^{EX}, we can expand ΔA^* to first order in V^{EX}/V to obtain

$$\Delta A^*(\mathbf{R}_1, \mathbf{R}_2, \mathbf{R}_3) = -k_B T \ln\left(1 - \frac{V^{EX}(\mathbf{R}_1, \mathbf{R}_2, \mathbf{R}_3)}{V}\right)$$

$$= \frac{k_B T}{V} V^{EX}(\mathbf{R}_1, \mathbf{R}_2, \mathbf{R}_3), \tag{E.11}$$

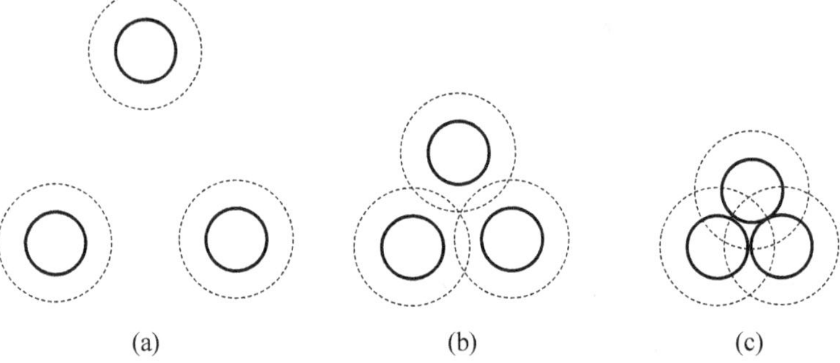

(a)   (b)   (c)

Fig. E.3   Three configurations of three spherical particles (here depicted as three disks). At large distances, there is no overlap in the excluded volume about each particle. At closed distances, there is some overlap (b), and the maximum overlap is shown in the configuration (c).

and similarly for a pair of particles at $\mathbf{R}_1, \mathbf{R}_2$.

$$\Delta A^*(\mathbf{R}_1, \mathbf{R}_2) \approx \frac{k_B T}{V} V^{EX}(\mathbf{R}_1, \mathbf{R}_2). \qquad \text{(E.12)}$$

The assumption of pairwise additivity is equivalent to the equality of

$$V^{EX}(\mathbf{R}_1, \mathbf{R}_2, \mathbf{R}_3)$$
$$= V^{EX}(\mathbf{R}_1, \mathbf{R}_2) + V^{EX}(\mathbf{R}_1, \mathbf{R}_3) + V^{EX}(\mathbf{R}_2, \mathbf{R}_3). \qquad \text{(E.13)}$$

Clearly, such an equality does not exist, as can be seen from Fig. E.3. Note that when the particles are sufficiently far apart, such that the excluded volume around the particles is the sum of the excluded volume of each particle, then we have

$$V^{EX}(\mathbf{R}_1, \mathbf{R}_2, \mathbf{R}_3) = 3 V^{EX}(\mathbf{R}_1), \qquad \text{(E.14)}$$

$$V^{EX}(\mathbf{R}_1, \mathbf{R}_2) = 2 V^{EX}(\mathbf{R}_1). \qquad \text{(E.15)}$$

In this case, the equality (E.13) certainly does not hold. However, the solvent-induced contribution δG and the interaction energy are zero, and this case is of no interest.

(ii) "Solvation" on an adsorbent molecule with conformational changes

This example is instructive because the "solvation" of the ligand on the adsorbent molecule can be solved exactly. We describe here the model and the results. The details of the calculations are quite lengthy and are available elsewhere.[278]

The solvent in our case is a system of adsorbent molecules. Each can be in two conformational states, and each has three adsorption sites (Fig. E.4a). The system is simple enough such that one can write down the partition function of the system and all the relevant thermodynamical properties. Specifically, we shall be interested in the analogues of the PMF, or

---

[278]See Ben-Naim (2001).

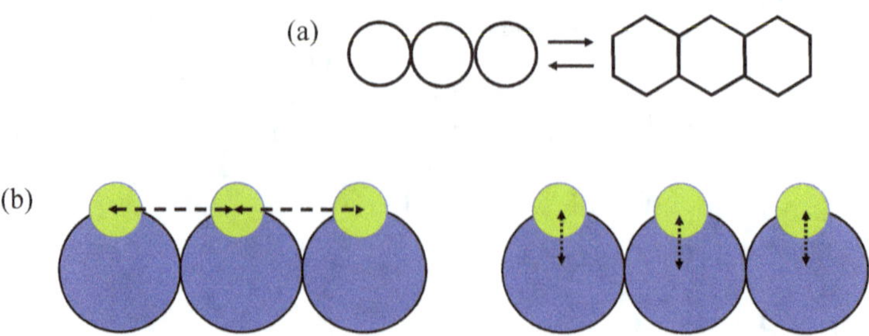

Fig. E.4 (a) An equilibrium between two configurations of an adsorbent molecule having three binding sites; (b) The analog of the PMF between the three ligands (black disks) has two components: The direct interactions between the ligands, and an indirect interaction. The latter depends only on the ligand–adsorbent interactions, not on the ligand–ligand interactions.

equivalently the Helmholtz energy change for the following two processes[279]

$$2[1,0,0] \rightarrow [1,1,0] + [0,0,0], \qquad (E.16)$$

$$3[1,0,0] \rightarrow [1,1,1] + 2[0,0,0]. \qquad (E.17)$$

In process (E.16), we start with two singly occupied adsorbent molecules, and transfer one ligand from one adsorbent molecule to the other to form a doubly occupied adsorbent molecule. In process (E.17), we start with three singly occupied adsorbent molecules, and transfer two ligands to the other to form a triply occupied adsorbent molecule.

The Helmholtz energy change for the process (E.16) is

$$W[1,1,0] = U[1,1] + \delta A \text{ (process E.16)}$$

$$= U[1,1] - k_B T \ln \frac{(1 + Kh^2)(1 + K)}{(1 + Kh^2)}. \qquad (E.18)$$

In (E.18), we wrote the PMF as a sum of two terms: The *direct interaction* energy between the two ligands occupying two sites on the adsorbent molecule, and the *indirect* part, or the

---

[279]For details, see Ben-Naim (2001), Chapters 4 and 5.

"solvent-induced" part of the Helmholtz energy change. The latter is expressed in terms of the two molecular quantities

$$K = \exp[-\beta(E_H - E_L)], \quad h = \exp[-\beta(U_H - U_L)], \quad (\text{E.19})$$

where E_H and E_L are the energy levels corresponding to the two conformations H and L, respectively, and U_H and U_L are the binding energies to the sites of H and L, respectively.

Similarly, the Helmholtz energy change for the process (E.17) is

$$W[1,1,1] = 3U[1,1] + \delta A \ (\text{process E.17})$$
$$= 3U[1,1] - k_B T \ln \frac{(1 + Kh^3)(1 + K)^2}{(1 + Kh)^3}, \quad (\text{E.20})$$

where we have assumed pairwise additivity for the direct interaction energy, i.e. $U[1,1,1] = 3U(1,1)$, and K and h have the same significance as in (E.19).

The essential difference between the direct and the indirect parts of the PMF is schematically shown in Fig. E.4b. This figure is a "reduction" of Fig. E.2, in the sense that the full lines connecting the ligands are the same as in Fig. E.2. On the other hand, the "solute–solvent" lines of interaction (the dashed lines in Fig. E.2b) are replaced by a single line connecting the ligand to the adsorbent molecule.

Since we have assumed the direct interaction energy between ligands is pairwise additive, we can focus on the indirect parts of $W(1,1,1)$ and $W(1,1)$ and examine the condition under which we might have additivity of the form

$$\delta A(\text{process E.17}) \approx 3\delta A(\text{process E.15}), \quad (\text{E.21})$$

or equivalently

$$\frac{(1 + Kh^3)(1 + K)^2}{(1 + Kh)^3} \stackrel{?}{=} \left[\frac{(1 + Kh^2)(1 + K)}{(1 + Kh)^2} \right]^3. \quad (\text{E.22})$$

Note that for either $K = 0$ or $h = 1$, the quantities on the two sides of (E.22) become equal to unity, hence the corresponding δA are zero. This is the case when there exists no indirect part of the PMF and is therefore of no interest to us. For any other values of $K \neq 0$ and $h \neq 1$, there exists no condition under which the equality sign applies to (E.22). The reader is referred to Ben-Naim (2001) for more details. Here, it is sufficient to conclude that because of the fundamental difference between the direct and the indirect parts of the PMF, as schematically depicted in Figs. E.2 and E.4b, one cannot expect to obtain pairwise additivity of the PMF.

(iii) Ising model in one-dimensional system

We discuss here the simplest 1-D Ising model. Particles occupying lattice points on a 1-D system can be in one of the two states, say, "up" or "down", or A and B. It is well known that the triplet correlation function in this system has the following property:

For any consecutive triplet of particles, $i, i+1$ and $i+2$, the triplet correlation function may be written as

$$g(S_i, S_{i+1}, S_{i+2}) = g(S_i, S_{i+1})g(S_{i+1}, S_{i+2}), \qquad \text{(E.23)}$$

where S_i is the state ("up" or "down", or A and B) of the site i. This property follows from the Markovian character of the 1-D Ising model.[280] Equation (E.23) is equivalent to the following equality of the analogue of the PMF:

$$W(S_i, S_{i+1}, S_{i+2}) = W(S_i, S_{i+1}) + W(S_{i+1}, S_{i+2}). \qquad \text{(E.24)}$$

This is sometimes referred to as the "Kirkwood superposition approximation."[281] However, the additivity expressed in

---

[280] For details, see Ben-Naim (1992) Chapter 4 and Ben-Naim (2001).
[281] Birshtein and Ptitsyn (1966).

(E.24) should be clearly distinguished from the Kirkwood super-position approximation, which, for the Ising model, should be written as

$$W(S_i, S_{i+1}, S_{i+2}) = W(S_i, S_{i+1}) + W(S_{i+1}, S_{i+2}) + W(S_i, S_{i+2}).$$
$$(E.25)$$

One can show that the third term on the right-hand side of (E.25) is not zero.[282] It follows that the Kirkwood superposition approximation does not hold for the 1-D Ising model. Thus, although an additivity of the form (E.24) exists, it is different from the Kirkwood superposition approximation.

(iv) Experimental evidence for non-additivity of the PMF

We present here an example of experimental evidence for the non-additivity of the PMF, as well as an approximate estimate of the extent of its non-additivity.

Consider the process of bringing three methane molecules from infinite separation from each other to some final con-figuration $(\mathbf{R}_1, \mathbf{R}_2, \mathbf{R}_3)$, as in cyclopropane (Fig. E.5). The potential of mean force, or the Gibbs energy change for this process is

$$W(\mathbf{R}_1, \mathbf{R}_2, \mathbf{R}_3) = U(\mathbf{R}_1, \mathbf{R}_2, \mathbf{R}_3) + \delta G(\mathbf{R}_1, \mathbf{R}_2, \mathbf{R}_3). \qquad (E.26)$$

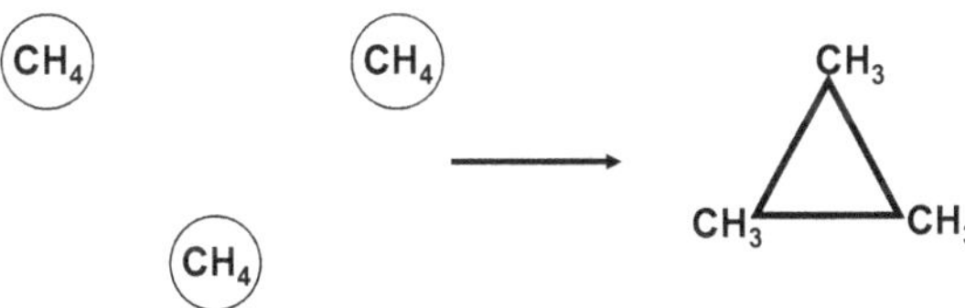

Fig. E.5   The process of bringing three methane molecules to the configuration of cyclopropane.

---

[282]Ben-Naim (1992), Chapter 4.

In the course of studying the $H\phi O$ interaction among three methane molecules, one focuses on δG, which is the relevant quantity. The direct interaction energy is presumed to be the same in any solvent. Focusing on δG has an advantage because δG does not depend on the direct interaction energy among the three methane molecules. Therefore, one can study the value of δG even for configurations which are inaccessible experimentally.

Specifically, suppose we choose the configuration $\mathbf{R}_1, \mathbf{R}_2, \mathbf{R}_3$ as that of a cyclopropane molecule. For this particular configuration, we can estimate δG (cyclopropane) by[283]

$$\delta G(\text{cyclopropane}) \approx \Delta G^*(\text{cyclopropane}) - 3\Delta G^*(\text{methane}).$$
$$(E.27)$$

The two quantities on the right-hand side of (E.27) are available. Therefore, we can estimate

$$\delta G(\text{cyclopropane}) \approx -14.6 \, \text{kJ} \, \text{mol}^{-1}. \qquad (E.28)$$

Similarly, we estimate δG for the process of bringing two methane molecules to the configuration of ethane. For this process, we have

$$\delta G(\text{ethane}) \approx \Delta G^*(\text{ethane}) - 2\Delta G^*(\text{methane})$$
$$\approx -9.04 \, \text{kJ} \, \text{mol}^{-1} \qquad (E.29)$$

Clearly, the indirect part of the work of bringing the three methane molecules to the final configuration of cyclopropane is not equal to three times the work associated with the process of bringing two methane molecules to the configuration of ethane.

---

[283]The details of this procedure for estimating the strength of the $H\phi O$ interaction may be found in Ben-Naim (2009), Chapter 4.

We can define the extent of non-additivity by the quantity

$$\delta G(\text{cyclopropane}) - 3\delta G(\text{ethane})$$

$$\approx -14.6 + 3 \times 9.04 = 12.52 \, \text{kJ mol}^{-1}. \quad \text{(E.30)}$$

We see the extent of deviations from pairwise additivity is of almost the same order of magnitude as $\delta G(\text{methane})$ or $\delta G(\text{cyclopropane})$.

The Statistical Mechanical Definition of Independence of Solvation and of Conditional Solvation

Consider the solvation process of two solutes S at some distance R. The statistical mechanical expression for the solvation Gibbs or Helmholtz energy is[284]

$$\exp[-\beta \Delta A_{SS}^*(\mathbf{R}_1, \mathbf{R}_2)] = \langle \exp[-\beta B_{SS}(\mathbf{R}_1, \mathbf{R}_2)] \rangle_0, \qquad \text{(F.1)}$$

where $B_{SS}(\mathbf{R}_1, \mathbf{R}_2)$ is the binding energy of the two solutes to all solvent molecules.

$$B_{SS}(\mathbf{R}_1, \mathbf{R}_2) = \sum_{i=1}^{N} U(\mathbf{R}_1, \mathbf{X}_i) + U(\mathbf{R}_2, \mathbf{X}_i) = B_1 + B_2, \qquad \text{(F.2)}$$

where the sum is over all N solvent molecules, being at a specific configuration $\mathbf{X}^N = \mathbf{X}_1, \ldots, \mathbf{X}_N$. Note that in (F.2), we do not include the solute–solute interaction. The average in (F.1) is over all the configurations of the solvent molecules, with the probability distribution of the "pure" solvent,[285] i.e.

$$P_0(\mathbf{X}^N) = \frac{\exp[-\beta U(\mathbf{X}^N)]}{\int \cdots \int d\mathbf{X}^N \exp[-\beta U(\mathbf{X}^N)]}. \qquad \text{(F.3)}$$

---

[284]As usual, we assume that the solutes are simple spherical particles and that the internal partition function of each solute is not affected by the presence of the solvent.
[285]"Pure" in the sense that the solvent does not include the two solute particles at $\mathbf{R}_1$ and $\mathbf{R}_2$. The solvent can be a mixture of any number of components.

In (F.1), we took the average in the T, V, N ensemble. If we take the average in the T, P, N ensemble, then the quantity on the left-hand side of (F.1) would be the Gibbs energy of solvation of the pair of solutes.

The average on the right-hand side of (F.1) may be viewed as an average over a product of two functions. Using (F.2) in (F.1) we write

$$\langle \exp[-\beta B_{SS}(\mathbf{R}_1, \mathbf{R}_2)]\rangle_0 = \langle \exp[-\beta B_1(\mathbf{R}_1)]\exp[-\beta B_2(\mathbf{R}_2)]\rangle_0.$$

$$(F.4)$$

We shall say that the two solutes are independently solvated if the average of the product can be factored into a product of two averages, in which case we can rewrite (F.4) as

$$\langle \exp[-\beta B_{SS}(\mathbf{R}_1, \mathbf{R}_2)]\rangle_0$$
$$= \langle \exp[-\beta B_1(\mathbf{R}_1)]\rangle_0 \langle \exp[-\beta B_2(\mathbf{R}_2)]\rangle_0$$
$$= \langle \exp[-\beta B(\mathbf{R}_0)]\rangle_0^2 = \exp[-\beta 2\Delta A_S^*]. \qquad (F.5)$$

Note carefully that the solvation quantity on the left-hand side of (F.1) depends only on the distance between the two solutes $R = |\mathbf{R}_2 - \mathbf{R}_1|$. However, if the factorization on the right-hand side of (F.5) is valid, then each of the average quantities is formally dependent on the location of one solute. Since we can choose the location of a single solute at $\mathbf{R}_1$ or $\mathbf{R}_2$ at any location in the liquid, the two averages on the right-hand side of (F.5) have equal magnitude (presuming that the two solute molecules are identical). Therefore, we have chosen $\mathbf{R}_0$ for the location of a single solute in the second equality on the right-hand side of (F.5). Furthermore, the value of the average of the quantity $\exp[-\beta B(\mathbf{R}_0)]$ does not depend on the specific location chosen for $\mathbf{R}_0$. The last equality follows from the definition of the solvation Helmholtz energy of a single solute S.

Thus, what we have, in effect, gotten is the equality

$$\Delta A_{SS}^{*}(\mathbf{R}_1, \mathbf{R}_2) = 2\Delta A_{S}^{*}, \qquad (F.6)$$

which means that the solvation Helmholtz (or Gibbs) energy of the pair of solutes at a distance R is the same as the solvation Helmholtz energy of two separate molecules.

The condition of independence of the solvation of the two solutes as given above is the same as the condition for uncorrelated random variables in probability theory.[286] We can reformulate the condition for independence of the solvation in terms of the pair correlation function. In general, for two solute particles at $\mathbf{R}_1$ and $\mathbf{R}_2$ we have the equality[287]

$$\begin{aligned}
\langle \exp[&-\beta \Delta A_{SS}^{*}(\mathbf{R}_1, \mathbf{R}_2)]\rangle_0 \\
&= \langle \exp[-\beta \Delta A_{S}^{*}]\rangle_0^2 \exp[-\beta W_{SS}(R) + \beta U_{SS}(R)] \\
&= \langle \exp[-\beta \Delta A_{S}^{*}]\rangle_0^2 \, g_{SS}(R) \exp[+\beta U(R)]. \qquad (F.7)
\end{aligned}$$

In (F.7), W_{SS} is the PMF for the two solutes, $U_{SS}(R)$ is the direct interaction energy between the two solutes, and $g_{SS}(R)$ is the pair correlation function between the two solutes. The product of the two factors on the right-hand side of (F.7) is sometimes denoted by

$$y_{SS}(R) = g_{SS}(R) \exp[\beta U_{SS}(R)], \qquad (F.8)$$

and is referred to as the indirect (or the solvent-induced) part of the pair correlation function. Thus, independence of the solvation of the two solutes at the distance R is equivalent to the condition that the distance R is large enough so that the indirect part of the pair correlation function is unity.

A more qualitative way of understanding the concept of independence of the solvation is to assume that each solute has

---

[286]See, for example, Ben-Naim (2008).
[287]For details, see Appendix C and Ben-Naim (1992, 2006).

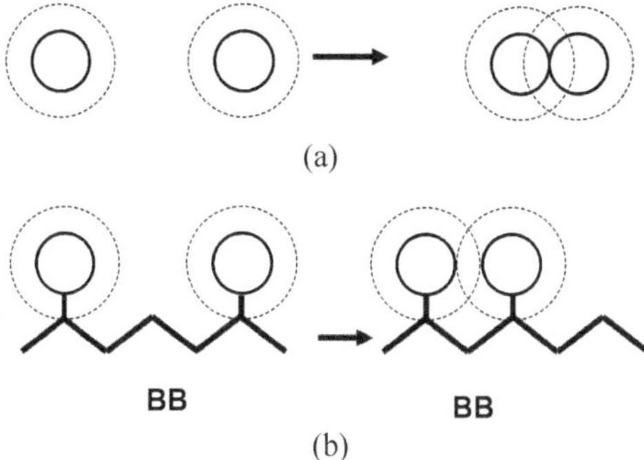

(a)

(b)

Fig. F.1 (a) Two solutes that are independently solvated on the left-hand side become dependently solvated; (b) two groups attached to a backbone become dependently solvated.

a "sphere of influence," or a "coordination sphere." This is a sphere around the solute in which the distribution of the solvent molecules is affected by the solute. When the two solute particles are at very large distances, their two spheres of influence are non-overlapping. In this case, the solvation of one solute does not affect the solvation of the second molecule, hence they are considered to be independently solvated. As they come to a close distance, their spheres of influence overlap, and therefore the solvation of one solute is affected by the presence of the second solute (Fig. F.1a).

The concept of independence of solvation of the two solutes can be extended to the case of two groups attached to a backbone (BB). In this case, if the two groups are far apart, their *conditional* solvations are independent. When they approach each other to a close distance, they become dependent. All the equations written above apply to this case, with interpretation of the averages as *conditional* averages. Instead of the probability distribution (F.3), we should use the conditional probability $P_0(X^N/BB)$, i.e. the probability distribution of the solvent configurations given the BB at some fixed location in the solvent. The qualitative argument provided above in terms of the overlapping of "spheres of influence" can also be carried out in the case of conditional solvation. Here, again, each group has its own "sphere of influence." Both of these are influenced by the backbone, but whenever they do not overlap, we can consider the conditional solvation of the two groups as being independent. Compare Figs. F.1a and F.1b.

Approximate Estimates of the *HϕI* Interaction between Two, Three and Four *HϕI* Groups at a Distance of 4.5 Å

In the first part of this appendix, we will present two approximate estimates of the strength of the *HϕI* interaction between two "arms" of the *HϕI* groups or molecules. The first is quick, based on incorrect argument, and provides an exaggerated value for the *HϕI* interaction. The second is lengthier, still very approximate, but provides a more realistic value of the *HϕI* interaction at some specific configuration of the two *HϕI* groups. In the second part of this appendix, we shall quote some values of the *HϕI* interaction between two, three and four *arms* of *HϕI* groups at the right configuration so that they can be bridged by a single water molecule.

Consider first the reaction shown in Fig. G.1, which can represent either an intramolecular or an intermolecular HB. We bring two arms of, say, two water molecules from infinite separation to the final distance of about 2.8 Å so that they form a direct HB. In this process, we assume that the HB energy is about $-27.2\,\mathrm{kJ\,mol^{-1}}$. The loss of the solvation Gibbs energy[288] of the two arms is about $2 \times 9.4 = 18.8\,\mathrm{kJ\,mol^{-1}}$. The net change in

---

[288]Ben-Naim (2009), Chapter 3.

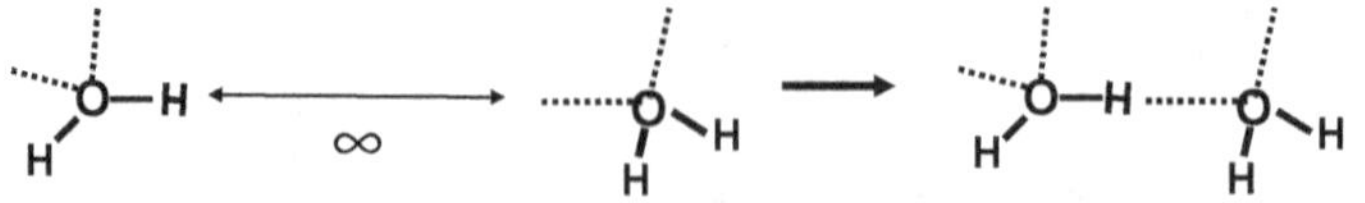

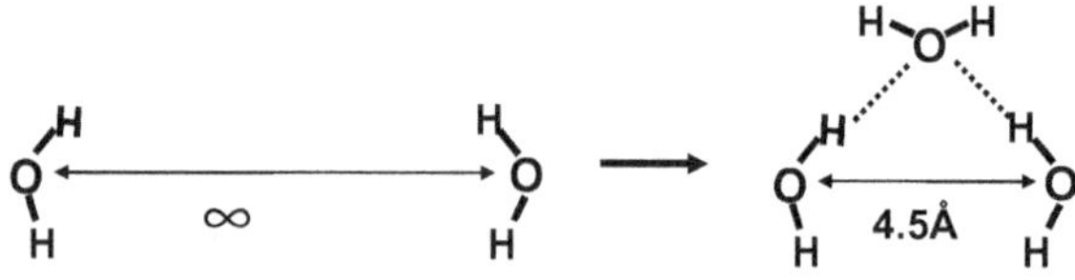

Fig. G.1 The process of bringing two water molecules (or two *HφI* groups) to form a direct HB.

Gibbs energy is therefore

$$\Delta G \text{ (Process in Fig. G.1)} \approx -27.2 + 18.8 = -8.4 \,\text{kJ mol}^{-1}.$$
$$(G.1)$$

The above calculation is based on a correct procedure except for the numerical values, which might not be correct. Nevertheless, I believe that this result is of the correct order of magnitude.

Inspired by the calculation of ΔG in (G.1), we now attempt to estimate the Gibbs energy change for the process of bringing two "arms" of either *HφI* groups or *HφI* molecules from infinite separation to the distance of about 4.5 Å, and properly oriented in such a way that the two arms can form a hydrogen-bonded bridge by a water molecule (Fig. G.2).

Assuming again that the two HBs are formed, contributing about $-2 \times 27.2 \,\text{kJ mol}^{-1}$, and that *four* solvation Gibbs energies of four arms are lost ($4 \times 9.4 \,\text{kJ mol}^{-1}$), we estimate

$$\delta G \text{ (Process in Fig. G.2)}$$
$$\approx -2 \times 27.2 + 4 \times 9.4 = -16.8 \,\text{kJ mol}^{-1}. \qquad (G.2)$$

Fig. G.2 The process of bringing two water molecules (or two *HφI* groups) to a distance of about 4.5 Å so that they can be bridged by a water molecule.

Note that here we estimate only the solvent-induced parts of the Gibbs energy change. We can assume that the direct interaction energy between the two *HφI* groups at this distance is much smaller than the indirect part of the Gibbs energy change.

The estimate in (G.2) is probably too high. The reason is that we have assumed that the water molecule *actually* forms two HBs with the two arms of the *HφI* group. However, the process that actually occurs when we bring the two *HφI* groups from infinity to the distance 4.5 Å is not the process depicted in Fig. G.2, i.e. we do not *form* a two-hydrogen-bond bridge, but instead, the two arms are *solvated* by water molecules in both the initial and in the final configuration. In other words, we need to take into account the statistical character of the solvation of the two arms in the final configuration. To do this, we need a more elaborate argument (see below).

We present here another heuristic argument which is based on the radial distribution function (RDF) of water (Fig. G.3).

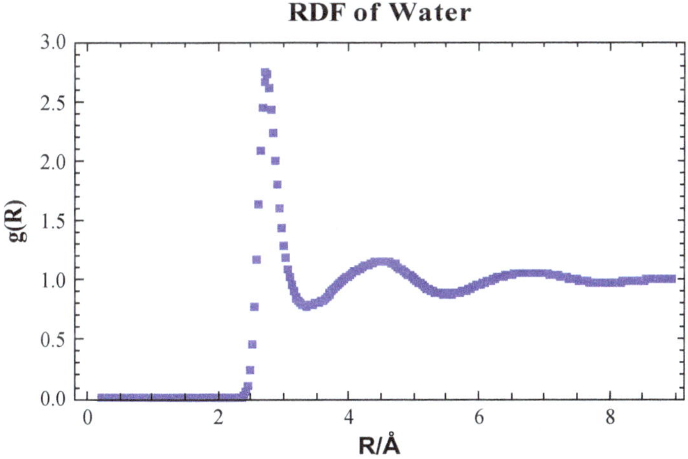

Fig. G.3   The radial distribution function of water.

First, note that the RDF is actually an average over all possible orientations of the two water molecules[289]

$$\overline{g}(R) = \frac{1}{(8\pi^2)^2} \int g(R, \mathbf{\Omega}_1, \mathbf{\Omega}_2) d\mathbf{\Omega}_1, d\mathbf{\Omega}_2. \qquad (G.3)$$

At the distance of $R_1 \approx 2.8\,\text{Å}$, we find a first peak in the RDF of height about 2.

We now wish to *de-average* the quantity $\overline{g}$, i.e. we want to find a value of $g(R_1, \mathbf{\Omega}_1, \mathbf{\Omega}_2)$ given $\overline{g}(R_1)$. Unfortunately, such a procedure is not possible (the word "de-average" does not even exist). Nevertheless, we shall create the word and proceed to execute the "de-averaging" process in the following sense:

We assume that the integrand $g(R_1, \mathbf{\Omega}_1, \mathbf{\Omega}_2)$ has only two values: A very high value when the orientations of the two molecules are such that a direct HB is formed between the two molecules, and a small value for all other orientations. Figure G.1 shows one example of an orientation for which the two water molecules form a HB.

Let x_H be the fraction of all orientations which are favorable for HBing. Hence, we can rewrite the average in (G.3) as

$$\overline{g}(R_1) = g_H x_H + g_L(1 - x_H). \qquad (G.4)$$

We can estimate the highest value of g_H as follows:

The work of bringing the two water molecules from infinite separation to the final distance R_1 with such orientations that a direct HB is formed is about

$$W(R_1) \approx -27.2 + 18.83 - 2.51 \approx -10.9\,\text{kJ mol}^{-1}. \qquad (G.5)$$

The three terms in (G.5) correspond to the energy of the HB formed by the two molecules, the loss of the solvation of two

---

[289]For more details see Ben-Naim (2009), Sec. 4.8.

arms, and some weak van der Waals interaction between the two water molecules at R_1. The value of $W(R_1)$ corresponds to the value of the pair correlation function of water at room temperature (25 °C):

$$g_H(R_1) \approx \exp[-\beta W(R_1)] \approx 77. \qquad (G.6)$$

The experimental value of the (angle average) pair correlation function of water at $R_1 \approx 2.8\,\text{Å}$ is about 2.2. This means that *averaging* over all orientations has reduced the value of the pair correlation function from the highest value of about 77 to the average value of about 2.2, i.e. reduction by a factor of about 35. We may infer that the "de-averaging" of the experimental value of $\overline{g}(R_1)$ should cause an increase in the value of the pair correlation function at the most favorable configuration for forming a HB by a factor of about 35.

Now, the heuristic — and very weakly justified — inference regarding the second peak of the pair correlation function of water at $R_2 = 4.5\,\text{Å}$, which is about

$$\overline{g}(R_2) \approx 1.2. \qquad (G.7)$$

"De-averaging" this value by multiplying it by 35, we may estimate the highest value of the pair correlation function at a specific configuration, such as that shown in Fig. G.2. Thus, we conclude that

$$g_H(R_2) \approx 1.2 \times 35 = 42, \qquad (G.8)$$

which corresponds at room temperature to

$$W_H(R_2) \approx -k_B T \ln g_H(R_2) \approx -9.4\,\text{kJ}\,\text{mol}^{-1}. \qquad (G.9)$$

This is, of course, a very crude estimate. It is much smaller than the value we calculated in (G.2). We believe that the real value is somewhat between these two values.

A more elaborate calculation of the $H\phi I$ interaction at the distance $R_2 = 4.5$ Å is given in Appendix I of Part I. The value obtained there is about $W(R_2 \approx 4.5) \approx -11.7\,\mathrm{kJ\,mol^{-1}}$, which is in good agreement with some experimental data[290] and simulated calculation.[291]

Thus, assuming that the direct interaction between the two arms at the configuration (G.2) is much smaller than the indirect, solvent-induced effect, we may conclude that

$$\delta G(1,2/LJ) \approx -11.7\,\mathrm{kJ\,mol^{-1}}. \qquad (G.10)$$

We quote here more estimates of the $H\phi I$ interactions from Ben-Naim (1992).

The $H\phi I$ interaction between three $H\phi I$ arms at the correct distances and orientations (Fig. G.4b) such that a water molecule can form three HBs with three $H\phi I$ groups is estimated to be[292]

$$\delta G(1,2,3/LJ) \approx -26.4\,\mathrm{kJ\,mol^{-1}}. \qquad (G.11)$$

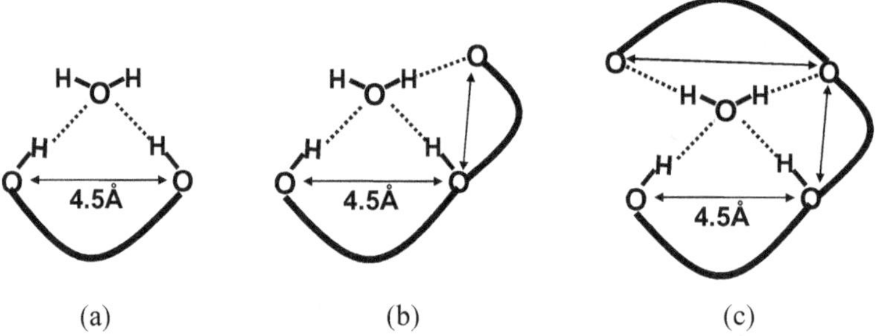

Fig. G.4 The formation of a one-water bridge connecting (a) two $H\phi I$ groups, (b) three $H\phi I$ groups and (c) four $H\phi I$ groups.

---

[290] See Part I, Sec. 4.8.
[291] Mezei and Ben-Naim (1990).
[292] Ben-Naim (1992) Appendix C.

The *HϕI* interaction between four *HϕI* arms at the correct distances and orientations (Fig. G.4c) such that a water molecule can form four HBs with these groups is estimated to be[293]

$$\delta G(1, 2, 3, 4/LJ) \approx -47.3 \, \text{kJ} \, \text{mol}^{-1}. \qquad \text{(G.12)}$$

---

[293] Ben-Naim (1992) Appendix C.

Evaluating The Inadequacy of Kauzmann's Model for the Role of the $H\phi O$ Effect in Protein Folding

We present here an extremely simple model of the "protein folding" process for which we can evaluate the adequacy of Kauzmann's model for the role of the $H\phi O$ effect in the driving force of protein folding.

Consider a linear polymer where hard spheres (HS) side chains are attached to the backbone (BB). We also assume that these side chains are independently solvated, i.e. they are far apart from each other so that their solvation spheres do not overlap (Fig. H.1). The BB is assumed to be a hard filament. In the folded form, we assume that the polymer forms a compact sphere. Therefore, the solvent-induced contribution to this process of protein folding is

$$\delta G(U \to F) = \Delta G_F^* - \Delta G_U^*$$

$$= \Delta G_F^{*HS} - \left(\Delta G_U^{*BB} + \sum_{i=1}^{M} \Delta G^{*i/BB} \right). \qquad (H.1)$$

Note that in this process, δG is the same as ΔG for the folding process. Thus, there are three contributions to δG. First, there is a loss of the solvation Gibbs energy of the BB of the unfolded form. Second, there is the loss of the conditional solvation Gibbs energies of the M side-chains. This is simply M

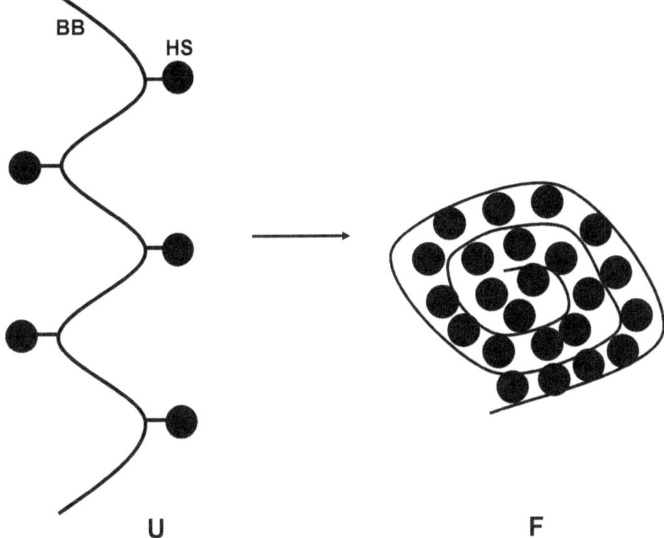

Fig. H.1 A hypothetical "protein" made up of a thin filament on which hard sphere (HS) side chains are attached. The solvation Gibbs energy of the F form is simply the work required to create a cavity to accommodate **F**.

times the conditional solvation Gibbs energy of one side chain. Finally, there is a gain of Gibbs energy of solvation of the folded form, which in our model is simply the work of formation of a cavity suitable to accommodate the **F** form.

Kauzmann's model would have used, for this particular process, the Gibbs energy change for transferring of one HS from water into a HS solvent (representing the organic liquid in the Kauzmann's model). We can write Kauzmann's estimate for the contribution of each side chain for the driving force as

$$\Delta G \text{ (per side chain)}$$
$$\approx \Delta G_{HS}^* \text{ (in HS solvent)} - \Delta G_{HS}^* \text{ (in water)}. \quad (H.2)$$

This is essentially the same as the Gibbs energy change for transferring of one HS from water into a HS solvent.

Clearly, the two components in (H.2) are not the same as the contributions in (H.1). First, instead of the solvation Gibbs energy of a HS in water, we need only the *conditional* solvation Gibbs energy of a HS attached to a BB. Second, there is no need to take the solvation Gibbs energy of a HS in a HS solvent. Although the environment of each side chain in the interior of the folded form is similar to the environment of a HS solvent, the solvation of a HS in a HS solvent does not feature in (H.1). Instead, we have the work of cavity formation of the entire F form. Once this is formed and taken into account in H.1, there is no need to "dig" a new cavity within an already existing cavity.

Here, we have discussed a caricature of a protein to highlight the inadequacy of Kauzmann's model, both with respect to the solvation of a $H\phi O$ molecule in water, and its solvation in an organic liquid.

APPENDIX I

The Cracks in the Hydrogen Bond Inventory Argument

The HB inventory argument is based on the following stoichiometric "reaction"

$$E\text{--}OH\cdots w \;+\; w\cdots O{=}C\text{--}S \;\rightarrow\; E\text{--}OH\cdots O{=}C\text{--}S \;+\; w\cdots w, \tag{I.1}$$

where E and S may be an enzyme (E) and a substrate (S), or any two FGs — one donor and one acceptor for hydrogen bonding.

Perhaps the first "reaction" of the form (I.1) was written by Schellman[294] in his study of the association between urea molecules in water. In that case, the urea molecule serves both as a donor and an acceptor for hydrogen bonding.

The term "HB inventory" was coined by Fersht.[295] Basically, it states that on the left-hand side of (I.1), the donor [E in (I.1)] and the acceptor [S in (I.1)] each make a HB with a water molecule. On the right-hand side of (I.1), the acceptor and the donor make a HB, and "the released water molecules each make a hydrogen bond with water, which is formally equivalent to binding to each other."[296]

---

[294]Schellman (1955).
[295]Fersht *et al.* (1985); Fersht (1987, 1999).
[296]Fersht (1999), page 337.

The HB inventory argument is basically the following[297]:

A crude, but very effective way of understanding the energetics of hydrogen bonding is to perform a hydrogen bond inventory, i.e. count the number and the nature of hydrogen bonds on each side of the chemical equation.

Fersht continued his argument, focusing on the enthalpy change in this reaction, and concludes: "Thus, hydrogen bonding should be relatively isoenthalpic."[298] However, "although the hydrogen bond inventory is zero, hydrogen bonding is energetically favorable in the formation of enzyme–substrate complexes because of the increase in entropy on the release of bound water molecules."[299]

In my opinion, neither of the conclusions about the enthalpy, nor about the entropy change in reaction (I.1) is correct. First, because the "counting" of HBs on the two sides of the Eq. (I.1) is not valid (see below). Second, the entropy change in this process cannot be predicted from just *looking* at reaction (I.1) and observing the release of water molecules. The entropy change associated with the process (I.1) could be positive or negative depending on the "structural changes" in the water induced by the formation of a HB between E and S.[300]

In this appendix, we shall first discuss the validity of the very writing of a stoichiometric reaction of the form (I.1). Then, we shall comment on the entropy and the enthalpy changes in this process.

Without specifying the thermodynamic quantity associated with this reaction, one can *see* from (I.1) that in the process of formation of a HB between E and S, two hydrogen bonds are

---

[297]Fersht (1999), page 338.

[298]This conclusion is similar to the one reached by Kauzmann (1959) based on Schellman's work. See Part I, Sec. 4.12 for more details.

[299]Fersht (1999), page 338.

[300]For more details, see Ben-Naim (2009) Chapter 3.

broken and two hydrogen bonds are *formed*. Assuming that the strength of the HB energy is approximately the same between a hydroxyl group and water, between a carbonyl group and water, and between two water molecules, one can reach the conclusion that the formation of a HB (inter or intramolecular) will have a minor effect on the energetics of such a reaction.

Indeed, this conclusion is correct for the specific reaction as written in (I.1). However, this is not the reaction that takes place when a HB is formed between E and S. The correct reaction is not (I.1) but (I.2).

$$(E-OH)_{\text{solvated}}+(O=C-S)_{\text{solvated}}\rightarrow(E-OH\cdots O=C-S)_{\text{solvated}}.$$
$$(I.2)$$

Thus, an enzyme E and a substrate S form a complex ES having a direct HB. All these three components are *solvated* by water.

The replacement of the (correct) reaction (I.2) with the (incorrect) reaction (I.1) reveals the two flaws that invalidate the HB inventory argument. First, the $H\phi I$ groups on the left hand side of Eq. (I.2) do not *form* a HB with a water molecule but are *solvated* by water molecules. The HB formed by E and S on the right-hand side is a *genuine* HB. Second, the water molecules "released" in the reaction (I.1) are at the same chemical potential before and after the association of E and S. Therefore, the effect of the released water molecules is taken into account in the *difference* in the solvation Gibbs energies of the species E, S and ES. Both the enthalpy and the entropy changes associated with this reaction cannot be predicted by mere inspection of the reaction (I.1). These quantities depend on the "structural changes" in the water induced by the reaction. These structural changes can contribute positively or negatively to the entropy and the enthalpy of the reaction, but because of the exact compensation (see Appendix O and below), these structural changes will have no effect on the Gibbs energy of the reaction.

If we are interested in the change in the Gibbs energy for a unit reaction (I.2), we should write

$$\Delta G \text{ (Reaction (I.2))} = \mu_{ES}^* - \mu_S^* - \mu_E^*, \qquad \text{(I.3)}$$

where the asterisk stands for any chosen standard state, or for the pseudo-chemical potential. The solvation of the species *ES*, *S* and *E* are taken into account through the coupling works that are part of the chemical potential.

Thus, the reaction that actually occurs is (I.2), and all the thermodynamic quantities associated with this reaction should be derived from the standard Gibbs energy change (I.3).

However, if one insists on examining stoichiometric reactions of the form (I.1), one must realize that the ellipsis, denoting a HB, does not have the same meaning on the right-hand side of this reaction as it does on the left-hand side.

Whereas the hydroxyl and the carbonyl groups are *solvated* by water molecules on the left-hand side of (I.1), there is a *genuine* HB between the hydroxyl and the carbonyl groups on the right-hand side of (I.1). Therefore, if one insists on examining stoichiometric reactions of the type (I.1), one must consider many possible "reactions" of this type. A few examples of these are shown in Fig. I.1 (here, the ellipses are genuine HBs).

Clearly, there are many stoichiometric reactions of the form (I.1). In some, the hydroxyl group may be engaged in zero, one, two or three HBs with water molecules. Similarly, the carbonyl group may be engaged in zero, one or two HBs. On the right-hand side of the reaction, a HB is formed by *E* and *S*. The water molecules released may or may not form a HB between themselves. Each of these reactions has a well-defined standard Gibbs energy.

Thus, if one prefers a specific stoichiometric reaction of the type (I.1), one should realize that this reaction does not represent what actually happens when a HB is formed between *E* and *S*. This is only one out of the many possible reactions of the type

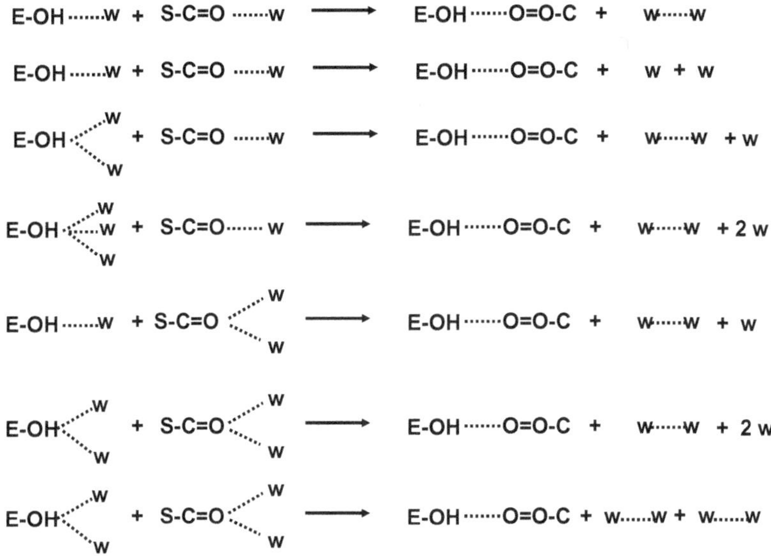

Fig. I.1   A few possible stoichiometric reactions involved in the process (I.2).

shown in Fig. I.1. The simpler approach is, of course to study the reaction (I.2), which takes into account all the possibilities of the reactions in Fig. I.1 through the *solvation* Gibbs energies of the species E, S and ES.[301]

To summarize, the stoichiometric reaction as written in (I.1) might lead to erroneous conclusions: First, because the HB formed on the right-hand side is a genuine HB and contributes an HB energy. On the other hand, the ellipsis on the left-hand side does not represent a genuine HB, but the solvation of a hydroxyl and a carbonyl group by water. Second, the water molecules released into the solvent are the same molecules that solvate the E and S molecule. They have the same chemical potential and therefore do not contribute a genuine HB energy as might be implied by the writing a reaction of the form (I.1).

---

[301] Note that in Fig. I.1, we listed only a very few of the possible reactions involving *real* HBs. In fact, one must also take reaction between E and S, which interact with water molecules through weak interactions. These would make the list in Fig. I.1 infinitely long.

Some authors discuss the effect of the release of the water molecules on the entropy and the enthalpy change in the process. This is also a potentially interesting problem. However, it is very difficult, if not impossible, to reach a quantitative estimate the contribution of the released water molecules on the entropy and the enthalpy change in a reaction such as (I.1). Furthermore, these changes are irrelevant to the Gibbs energy change of the process. We present a simple argument below based on the mixture model approach.

The exact entropy–enthalpy compensation was first shown in a two-structure model for water and later proved in the general form.[302] Here, we use a simple mixture model (MM) for water to show that the water release in the process (I.2) may affect the entropy and the enthalpy of the process, but cannot affect the Gibbs (or the Helmholtz) energy of the process.

Consider the following MM view of liquid water.[303] At each snapshot of the system, we can classify all water molecules into five groups and count the number of molecules in each group. The vector $(N_0, N_1, N_2, N_3, N_4)$ describes the composition of the mixture of species at a specific configuration of the system, where N_i is the number of water molecules engaged in i HBs. We can also define the *average* composition of the system by the vector $(\overline{N}_0, \overline{N}_1, \overline{N}_2, \overline{N}_3, \overline{N}_4)$. With this vector, one can define the concept of the structure of water,[304] but we shall not need this concept here.

Whenever a process occurs in aqueous solutions, there might be changes in the composition of the system (as well as in the *structure* of water).

---

[302] More details of the definition of the HB and the method of classification into five species may be found in Ben-Naim (2009).
[303] See Ben-Naim (2009).
[304] See Ben-Naim (2009).

Within the MM approach to water, we can write the chemical potential, the partial molar entropy and the partial molar enthalpy of any solute A as

$$\mu_A = \left(\frac{\partial G}{\partial N_A}\right)_{N_w}$$

$$= \left(\frac{\partial G}{\partial N_A}\right)_{\overline{N}_0,\dots,\overline{N}_4} + \sum_{i=0}^{4}\left(\frac{\partial G}{\partial \overline{N}_i}\right)\left(\frac{\partial \overline{N}_i}{\partial N_A}\right)_{N_w}, \qquad (\text{I.4})$$

$$\overline{S}_A = \left(\frac{\partial S}{\partial N_A}\right)_{N_w}$$

$$= \left(\frac{\partial S}{\partial N_A}\right)_{\overline{N}_0,\dots,\overline{N}_4} + \sum_{i=0}^{4}\left(\frac{\partial S}{\partial \overline{N}_i}\right)\left(\frac{\partial \overline{N}_i}{\partial N_A}\right)_{N_w}, \qquad (\text{I.5})$$

$$\overline{H}_A = \left(\frac{\partial H}{\partial N_A}\right)_{N_w}$$

$$= \left(\frac{\partial H}{\partial N_A}\right)_{\overline{N}_0,\dots,\overline{N}_4} + \sum_{i=0}^{4}\left(\frac{\partial H}{\partial \overline{N}_i}\right)\left(\frac{\partial \overline{N}_i}{\partial N_A}\right)_{N_w}. \qquad (\text{I.6})$$

All the derivatives are at constant P, T and number of any other solute molecules. The derivative at constant N_w means that the *total* number of water molecules $N_w = \overline{N}_0 + \overline{N}_1 + \overline{N}_2 + \overline{N}_3 + \overline{N}_4$ is kept constant, but each of the $\overline{N}_i$ may change. On the other hand, a derivative at constant $\overline{N}_0,\dots,\overline{N}_4$ means that we add $\overline{N}_A$ molecules of A in a hypothetical liquid where the conversion between all the species of water molecules is "frozen in."

Each partial molar quantity is written in two terms. First, we add N_A in a "frozen-in" (fr) system, then we release the constraint on fixed values of $\overline{N}_0,\dots,\overline{N}_4$ and let the system reach a new equilibrium composition. We may refer to the first term as the "frozen-in" (fr) partial molar quantity, and the second term as the relaxation term, or as the *structural* changes in the solvent induced by the process of adding $\overline{N}_A$. We rewrite (I.4) to (I.6) as[305]

$$\mu_A = \mu_A^{fr} + \sum_{i=0}^{4} \mu_i \left(\frac{\partial \overline{N}_i}{\partial N_A}\right)_{N_w}, \tag{I.7}$$

$$\overline{S}_A = S_A^{fr} + \sum_{i=0}^{4} \overline{S}_i \left(\frac{\partial \overline{N}_i}{\partial N_A}\right)_{N_w}, \tag{I.8}$$

$$\overline{H}_A = H_A^{fr} + \sum_{i=0}^{4} \overline{H}_i \left(\frac{\partial \overline{N}_i}{\partial N_A}\right)_{N_w}. \tag{I.9}$$

At equilibrium, we must have the equality of the chemical potentials

$$\mu_w = \mu_0 = \mu_1 = \mu_2 = \mu_3 = \mu_4, \tag{I.10}$$

where μ_w is the chemical potential of the water, and μ_i is the chemical potential of the species i of water molecules.

Using the equilibrium condition (I.10), we can rewrite the second term on the right-hand side of (I.7) as

$$\sum_{i=0}^{4} \mu_i \left(\frac{\partial \overline{N}_i}{\partial N_A}\right)_{N_w} = \mu_w \sum_{i=0}^{4} \left(\frac{\partial \overline{N}_i}{\partial N_A}\right)_{N_w}$$

$$= \mu_w \frac{\partial}{\partial N_A} \left(\sum \overline{N}_i\right) = 0. \tag{I.11}$$

Thus, we can rewrite (I.7) to (I.9) as

$$\mu_A = \mu_A^{fr},$$

$$\overline{S}_A = S_A^{fr} + \Delta S_A^{st},$$

$$\overline{H}_A = H_A^{fr} + \Delta H_A^{st}. \tag{I.12}$$

---

[305]Note that the bar on $\overline{S}_A$ and $\overline{H}_A$ is for partial molar quantity, whereas the bar on $\overline{N}_i$ is for average number of the species i.

This result is important. Structural changes in the solvent can affect the partial molar entropy, enthalpy, volume, etc. of a solute A, but not the chemical potential of water. The reason for this result is the chemical equilibrium condition (I.10).

The last result can be applied to any process occurring in water. For instance, for the isomerization reaction $A \to B$, we have

$$\Delta G(A \to B) = \mu_B - \mu_A = \mu_B^{fr} - \mu_A^{fr} = \Delta\mu^{fr}, \tag{I.13}$$

$$\Delta S(A \to B) = \bar{S}_B - \bar{S}_A = \Delta S^{fr} + \Delta S_B^{st} - \Delta S_A^{st}, \tag{I.14}$$

$$\Delta H(A \to B) = \overline{H}_B - \overline{H}_A = \Delta H^{fr} + \Delta H_B^{st} - \Delta H_A^{st}. \tag{I.15}$$

Thus, in any process occurring in water,[306] there can be a redistribution in the composition of the water, which we refer to as "structural changes" in water. These changes may affect the entropy, enthalpy, volume, etc. of that reaction. However, the change in Gibbs energy (or the "driving force") for the process is not affected.

Specifically, for the dimerization of urea in water, or the intramolecular hydrogen bonding in protein, some water molecules are released from the solvation sphere of the two $H\phi I$ groups that form a HB. However, this release of the water molecules occurs at a constant chemical potential of the water. Therefore, the water molecules that are released in the process cannot affect the "driving force" for the process.

Finally, it should be noted that neither the sign nor the magnitude of the quantities ΔS_α^{st} may be easily estimated. A process can cause a redistribution of the "composition" of the solvent (any solvent) when viewed as a mixture of species. This change of composition will cause changes in the entropy, enthalpy, volume, etc. of the system. To predict the sign of the

---

[306]Actually, this result applies for any liquid viewed as a mixture of species. See Ben-Naim (2006, 2009).

changes in these thermodynamic quantities, one must know the partial molar quantities of all the species involved, as well as the changes in the average number of the molecules of each species. An exact argument based on the Kirkwood–Buff theory of solution for the case of solvation of inert gas molecules in water is provided in Ben-Naim (2009).

Can "Statistical Potential," Derived from Protein Structures, be Interpreted as a Potential of Mean Force?

In Chapter 3, we dissected the Gibbs energy change for the process of protein folding into "smaller," more manageable components. We have also seen that one of these "components" is the conditional pair correlation function between two $H\phi O$ or $H\phi I$ groups, or more generally between any two side chains. These are essentially the Gibbs energy changes for the process of bringing two side chains from infinite separation (i.e. independently solvated) to some close separation, at which their "solvation spheres" are correlated. We have already discussed the difference between pair correlations and the conditional pair correlations in Appendix F. Also, we noted the relation between the total Gibbs energy of the process and the solvent-induced part of the Gibbs energy change, i.e.

$$W(\mathbf{X}_1, \mathbf{X}_2) = U(\mathbf{X}_1, \mathbf{X}_2) + \delta G(\mathbf{X}_1, \mathbf{X}_2). \qquad (\text{J.1})$$

Each of the quantities in (J.1) refers to the process of bringing two molecules or two groups from infinite separation to the final configuration $\mathbf{X}_1, \mathbf{X}_2$.

At present, there exists no reliable information on either the PMF or the conditional PMF between any two side chains of a real protein. However, there are some claims in the literature that one can extract such a PMF from the crystallographic data available in the protein data bank (PDB). These are sometimes referred to as "standard free energies of contact," "contact energies" or simply the "pair potential of mean force."[307]

There have been several attempts to extract information on the PMF from the PDB. Basically, one uses the statistical data on the distance distribution between two side chains, say α and β, to construct the pair correlation function $g_{\alpha\beta}(R)$, which is related to the probability of finding two side chains, of types α and β, at some distance between R and $R + dR$. These data and the corresponding pair correlation function are valuable statistics based on the available crystallographic structures of the protein. The question I have raised in my criticism is: To what extent can one use this data to obtain information on the PMF between the two species α and β? It is tempting to use the exact relation between the PMF and the pair correlation function

$$W_{\alpha\beta}(R) = -k_B T \ln g_{\alpha\beta}(R). \qquad (J.2)$$

Actually, one defines the PMF by this equation, then shows that the gradient of $W_{\alpha\beta}(R)$ is indeed the average force.[308]

Relation (J.2) is sometimes referred to as the Boltzmann law, probably because it resembles the relationship between probability and energy of the form (see also Appendix P).

$$\Pr(R) = C \exp[-\beta U(R)]. \qquad (J.3)$$

In (J.2), $g_{\alpha\beta}(R)$ represents the probability of finding the two side chains at a distance R, and the PMF in (J.2), which is a *free energy*, replaces the potential energy in (J.3).

---

[307] For more details on the literature, the method of calculating these quantities, and some critical comments, the reader is referred to Ben-Naim (1997b).
[308] For more details, see Ben-Naim (2006).

However, the Boltzmann law, either in the form (J.3) or in the form (J.2), is valid only when the probability is *determined* by the corresponding energy. The well-known example is the barometric distribution of the density (or of the pressure) in a vertical column of air which has the form[309]

$$\rho(h) = \exp[-mgh/k_B T] = \exp[-U(R)/k_B T]. \qquad (J.4)$$

Thus, if one measures the density of the air at some point in the column of air, one can convert from density into height (provided one also assumes that the gas is ideal and the temperature is uniform in the column of air, etc.). This conversion from density (or probability) into potential energy (or height) is valid only when the density is *determined* by the Boltzmann law, i.e. by Eq. (J.4). One can easily build up a column of air or a column of liquid or solid in which the density distribution is determined by some natural or artificial method.

Figure J.1a shows the mass density distribution as a function of the height on a mountain. Figure J.1b shows the density of all the mass as a function of height in a five-storey building. These are certainly meaningful density distributions. It is also clear that one cannot convert these density distributions into

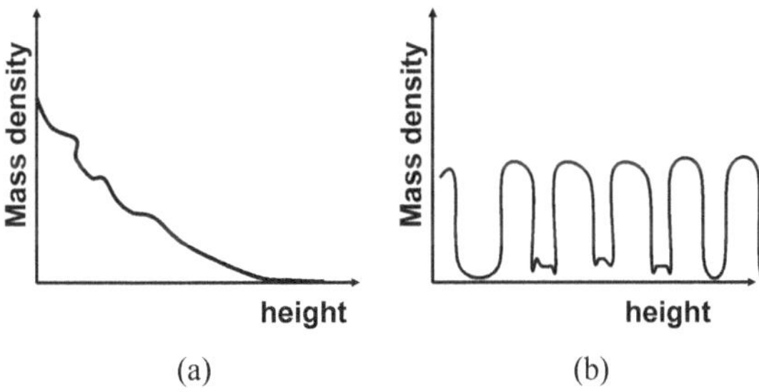

Fig. J.1 Hypothetical density distribution of mass on (**a**) a mountain, (**b**) in a five-storey building.

---

[309]For details on the derivation of this law, see Ben-Naim (1992, 2008).

heights using Boltzmann's law (J.4). The density of mass in the mountain or in the building was *determined* by some unknown force of Nature, or on the drawing board of an architect — but not by Boltzmann!

The situation is quite the same when we determine the distance distribution of side chains using all the available data on the structure of proteins.[310] The statistical distribution is valid data, perhaps also potentially of interest. However, one cannot convert this data into PMF using Eq. (J.2). The reason is as in the previous examples: These pair distributions were determined by evolution, not by Boltzmann! Therefore, one cannot interpret the quantity $W_{\alpha\beta}(R)$ *defined* in (J.2), as the potential of mean force.

---

[310]See, for example, Bahar and Jernigan (1997).

APPENDIX K

Work of Creating a Cavity and the Probability of Finding a Cavity in a Solvent

---

An important part of any solvation process is the work of cavity formation. For globular proteins in particular, there exists a reasonably good approximation for estimating the work of cavity formation. It is also important to be familiar with the relationship between this *work* on one hand, and the probability of finding a cavity at some *fixed* position on the other. This relationship is a particular example of a more general law discussed in Appendix P.[311]

One caveat before we start: "Formation of a cavity" always means a cavity at some *fixed position* in the liquid. We shall discuss this point later in this appendix.

The work associated with creating a cavity of diameter σ at some fixed position is of interest both in the study of the properties of pure liquids, and as a first step in calculating the solvation Gibbs or Helmholtz energy of a solute in a liquid.

We start with a spherical cavity.

A cavity of diameter σ at a point $\mathbf{R}_0$ in the fluid is defined as the spherical region of radius $\sigma/2$ centered at $\mathbf{R}_0$ from which

---

[311] In most textbooks on thermodynamics, the relationship is derived between the probability and the entropy of the system. A similar relationship exists between probability and the Gibbs or Helmholtz energy. See also Ben-Naim (2008).

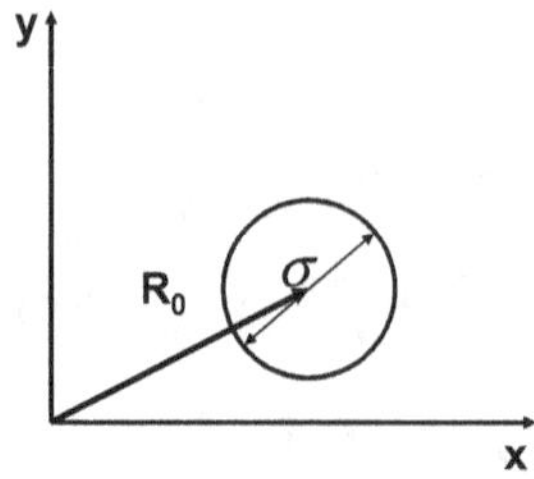

Fig. K.1 A cavity of diameter σ at some fixed position $\mathbf{R}_0$.

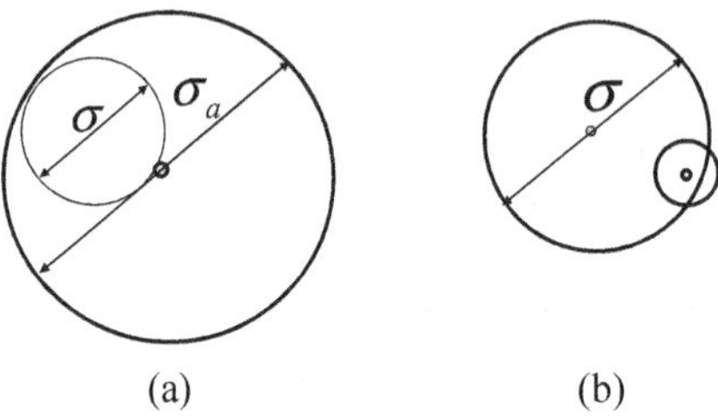

(a)   (b)

Fig. K.2 (a) A cavity of diameter σ is empty according to the definition although it is filled by a particle of diameter σ_α; (b) a "cavity" of diameter σ looks empty, but because the center of another particle is in it, it is not a cavity.

the *centers* of all other particles are excluded. Figure K.1 depicts such a cavity.[312]

It is important to realize that a cavity may seem to be filled in the conventional sense, yet be empty according to the *definition* given above. Two extreme cases are illustrated in Fig. K.2. In Fig. K.2a, we have a cavity of diameter σ which is completely *filled* by a molecule of diameter σ_a. Yet, since no center of a molecule falls in this cavity, it is empty, according our definition. In Fig. K.2b, we have a sphere of diameter σ at $\mathbf{R}_0$ which looks empty, but since the center of a molecule is in this "cavity," it is not a cavity according to our definition.

---

[312]Note that σ is used for the diameter, not the radius of the cavity.

The work required to create a cavity of diameter σ (keeping T, V, and N constant) is defined by

$$\Delta A_{cav}(\mathbf{R}, \sigma) = A(T, V, N; \mathbf{R}_0, \sigma) - A(T, V, N). \qquad \text{(K.1)}$$

Here, $A(T, V, N; \mathbf{R}_0, \sigma)$ stands for the Helmholtz energy of a system in the T, V, N ensemble having a cavity of diameter σ centered at $\mathbf{R}_0$. Clearly, in a homogenous fluid, all points in the system are considered to be equivalent (except for a negligible region near the boundaries of the system), and hence $\Delta A_{cav}(\mathbf{R}_0, \sigma)$ does not depend on the particular choice of $\mathbf{R}_0$. Nevertheless, we shall keep $\mathbf{R}_0$ in our notation to stress the fact that we are concerned with a cavity at a *fixed* position in the system.

From the definition of the concept of a cavity, it follows that all centers of the N particles are excluded from the spherical region of radius $\sigma/2$ centered at $\mathbf{R}_0$. Denoting this region by $V(\mathbf{R}_0, \sigma)$, we can write ΔA_{cav} in terms of the ratio of two partition functions.[313]

$$\exp[-\beta \Delta A_{cav}(\mathbf{R}, \sigma)]$$
$$= \frac{(q^N/\Lambda^{3N}N!) \int \cdots_{V-V(\mathbf{R},\sigma)} \int d\mathbf{R}^N \exp[-\beta U_N(\mathbf{R}^N)]}{(q^N/\Lambda^{3N}N!) \int \cdots_V \int d\mathbf{R}^N \exp[-\beta U_N(\mathbf{R}^N)]}.$$

$$\text{(K.2)}$$

Note that the only difference in the two integrals in (K.2) is in the limits of the integration.

The right-hand side of (K.2) can now be interpreted as the probability of observing a cavity of radius $\sigma/2$ at $\mathbf{R}_0$. We recall that the probability density of observing any specific

---

[313] For details, see Ben-Naim (1992, 2006).

configuration $\mathbf{R}^N$ is given by[314]

$$\Pr(\mathbf{R}^N) = \frac{\exp[-\beta U_N(\mathbf{R}^N)]}{\int \cdots \int d\mathbf{R}^N \exp[-\beta U_N(\mathbf{R}^N)]}. \qquad (K.3)$$

Suppose we choose a region S in our system. The probability of finding the centers of all the particles in S is obtained by summing over all the events (configurations) conforming to the requirement that all $\mathbf{R}_i(i = 1, 2, \ldots, N)$ are within S. Hence,

$$\Pr(S) = \int_S \cdots \int d\mathbf{R}^N \Pr(\mathbf{R}^N). \qquad (K.4)$$

As a particular case, if S coincides with the entire volume of the system, then $\Pr(V) = 1$. Similarly, if all the centers are to be excluded from $V(\mathbf{R}_0, \sigma)$, we have the case in which $S = V - V(\mathbf{R}_0, \sigma)$, and therefore

$$P_{cav}(\mathbf{R}_0, \sigma) = \int_{V - V(\mathbf{R}_0, \sigma)} \cdots \int d\mathbf{R}^N \Pr(\mathbf{R}^N). \qquad (K.5)$$

From (K.2) and (K.5), we arrive at the important relationship

$$P_{cav}(\mathbf{R}_0, \sigma) = \exp[-\beta \Delta A_{cav}(\mathbf{R}_0, \sigma)]. \qquad (K.6)$$

On the right-hand side of (K.6), we have the work required to form a cavity of radius $\sigma/2$ at $\mathbf{R}_0$, keeping T, V and N constant. On the left-hand side, we have the probability of observing such a cavity in a system in the T, V, N ensemble. It is quite straightforward to obtain similar relations in other ensembles. For instance, the probability of finding a cavity of radius $\sigma/2$ in a T, P, N ensemble is related to the Gibbs energy for creating such a cavity.

---

[314]See Appendix P and Ben-Naim (2006).

Cavity formation and the pseudo-chemical potential of a hard sphere

We derive here a useful relation between the work required to form a cavity and the pseudo-chemical potential of a hard-sphere solute. We shall still discuss only a system of spherical particles in the T, V, N ensemble. The results are also valid for more general systems and in other ensembles.

Consider a system of N particles with an effective hard-core diameter σ_α. By "effective," we mean that the two molecules at a distance σ_α from each other "feel" almost infinite repulsive forces. There is no unique method of determining the value of σ_α for a real molecule. Nevertheless, for, say, the Lennard-Jones particles, the parameter σ is a reasonable choice for an effective hard-core diameter.

In Appendix B, we obtained the pseudo-chemical potential for a one-component system. We repeat the same process here, but instead of adding the $(N + 1)$th particle, we add a hard-sphere particle of diameter σ_{HS} to a fixed position $\mathbf{R}_0$ in a system of particles having an effective hard-core diameter σ_α. The work associated with this process, keeping T, V, N constant, is given by

$$
\begin{aligned}
\exp\left(-\beta\mu^*_{HS}\right) \\
= \exp\{-\beta[A(T, V, N; \mathbf{R}_0, \sigma_{HS}) - A(T, V, N)]\} \\
= \frac{\int \cdots \int_V d\mathbf{R}^N \exp[-\beta U_N(\mathbf{R}^N) - \beta B_{HS}(\mathbf{R}_0)]}{\int \cdots \int_V d\mathbf{R}^N \exp[-\beta U_N(\mathbf{R}^N)]}.
\end{aligned}
\tag{K.7}
$$

Note that since the hard-sphere solute is presumed to have no internal structure, its addition to the system does not introduce an internal partition function. The quantity $B_{HS}(\mathbf{R}_0)$ is the total interaction energy between the hard sphere at $\mathbf{R}_0$ and N

molecules in the configuration $\mathbf{R}^N$.

$$B_{HS}(\mathbf{R}_0) = \sum_{i=1}^{N} U(\mathbf{R}_i, \mathbf{R}_0). \tag{K.8}$$

According to our assumption, σ_α is the hard-core diameter of the solvent particles. This statement means that

$$U(\mathbf{R}_i, \mathbf{R}_j) = \infty, \quad \text{for } \mathbf{R}_{ij} = |\mathbf{R}_j - \mathbf{R}_i| < \sigma_\alpha,$$

$$i, j = 1, 2, \ldots, N, \quad i \neq j. \tag{K.9}$$

For $R_{ij} \geq \sigma_\alpha$, the potential function can have any form. However, the interaction between a particle in the system and the hard-sphere solute is presumed to have the form

$$U(\mathbf{R}_i, \mathbf{R}_0) = \begin{cases} \infty & \text{for } R_{i0} < \dfrac{1}{2}(\sigma_{HS} + \sigma_\alpha) \\[2mm] 0 & R_{i0} \geq \dfrac{1}{2}(\sigma_{HS} + \sigma_\alpha) \end{cases} \quad i = 1, 2, \ldots, N. \tag{K.10}$$

The condition (K.10) can be rewritten in an equivalent form as

$$\exp[-\beta U(\mathbf{R}_i, \mathbf{R}_0)] = \begin{cases} 0 & \text{for } R_{i0} < \dfrac{1}{2}(\sigma_{HS} + \sigma_\alpha) \\[2mm] 1 & \text{for } R_{i0} \geq \dfrac{1}{2}(\sigma_{HS} + \sigma_\alpha). \end{cases} \tag{K.11}$$

Using (K.8) and (K.11) in (K.7), we obtain

$$\exp(-\beta\mu_{HS}^*) = \int_V \cdots \int d\mathbf{R}^N \Pr(\mathbf{R}^N) \exp[-\beta B_{HS}(\mathbf{R}_0)]$$

$$= \int_V \cdots \int d\mathbf{R}^N \Pr(\mathbf{R}^N) \prod_{i=1}^{N} \exp[-\beta U(\mathbf{R}_i, \mathbf{R}_0)]$$

$$= \int_{V-V(\mathbf{R}_0, \sigma)} \cdots \int d\mathbf{R}^N \Pr(\mathbf{R}^N). \tag{K.12}$$

The last equality in (K.12) follows from the property of the step function (K.11). Since we have a product of N such factors (each of which operates on one vector $\mathbf{R}_i$) in the integrand of (K.12), their effect is to reduce the region of integration for each $\mathbf{R}_i$ from V to $V - V(\mathbf{R}_0, \sigma)$. Comparing with (K.5) and (K.12), we arrive at

$$\exp\left(-\beta\mu^*_{HS}\right) = P_{cav}(\mathbf{R}_0, \sigma) = \exp[-\beta\Delta A_{cav}(\mathbf{R}_0, \sigma)], \quad \text{(K.13)}$$

with the condition

$$\sigma = \sigma_{HS} + \sigma_\alpha. \quad \text{(K.14)}$$

Thus, the work required to produce a cavity of diameter σ at $\mathbf{R}_0$ (which may be chosen at any point in the fluid) in a system of particles with an effective hard-core diameter σ_α is equal to the work required to introduce a hard-sphere particle of diameter σ_{HS} (given by $\sigma_{HS} = \sigma - \sigma_\alpha$) into a fixed position $\mathbf{R}_0$.

It is important to realize that the last statement is valid only when we refer to a *fixed* position for both processes. The equivalence of the two processes is clear on intuitive grounds. Creation of a cavity means imposing a restriction on the centers of all particles, keeping them out of a certain region. This constraint is achieved simply by putting a hard-sphere solute at the position $\mathbf{R}_0$, provided that we have properly chosen its diameter by relation (K.14). In other words, as far as the solvent (i.e. the particles of the system) is concerned, there is no difference if we create a cavity at $\mathbf{R}_0$ or put a hard-sphere solute there, provided that (K.14) is fulfilled.

The above reasoning breaks down if we remove the condition of *fixed* position for both of the processes. We recall that the work required to add a hard-sphere solute to the system (at T, V, N constant) is equal to the chemical potential of the solute, which can be written as

$$\mu_{HS} = \mu^*_{HS} + kT \ln\left(\rho_{HS}\Lambda^3_{HS}\right), \quad \text{(K.15)}$$

where $\rho_{HS} = 1/V$ is the (number) density of the solute and Λ_{HS}^3 its momentum partition function. We also recall that the second term on the right-hand side of (K.15) is the contribution to μ_{HS} associated with the release of the constraint on the *fixed* position for the solute. The analog of this contribution in the case of a cavity is not evident. First, a "free cavity" has no momentum partition function akin to the quantity Λ_{HS}^3 of a real solute. Second, there is no clear-cut definition of the (number) density of "free cavities of diameter σ." For these reasons, the equivalence of the process of introducing a hard-sphere solute and the creation of a cavity should be restricted to only a *fixed position* in the liquid.

Having the pseudo-chemical potential of a hard sphere in any liquid, we can always add the liberation Helmholtz energy to obtain the chemical potential of the hard sphere in the same system. The same cannot be done to the quantity ΔA_{cav}. Therefore, it is meaningful to speak about the chemical potential of hard spheres in a liquid, but there is no meaning to the analogous concept of the "chemical potential of cavities" in a liquid.

Generalization to non-spherical cavities

In any liquid or mixture of liquids, we can introduce a hard particle of any form at some fixed position and orientation. Associated with this process there is a well-defined pseudo-chemical potential. This hard particle by its definition excludes the centers of all other particles from some region which we denote by V^{EX}. Clearly, if the solvent particles are not spherical, or if they have different sizes, then the excluded volume of the hard particle will be different for different orientations of the solvent particles or for particles having different sizes. If we are interested in the work required to create a cavity which is equivalent to introducing a hard particle, we must take the intersection of all possible V^{EX} with respect to all orientations and all solvent species. The probability of finding such an excluded volume empty is simply obtained by generalization of (K.5), namely

$$\Pr(V_\alpha^{EX}) = \int \cdots \int_{V-V_\alpha^{EX}} \Pr(\mathbf{X}^N)d\mathbf{X}^N. \qquad (K.16)$$

Since the integrand is always positive, the integral will be larger when the region of integration increases or when the excluded region decreases. If the cavities are spherical, then a larger cavity implies a lower probability. In other words, Pr(cavity of radius R) is a decreasing function of R. This follows directly from the definition (K.16) and applies to any liquid.

In general, a larger *volume* of the cavity does not necessarily imply a lower probability. For two excluded volumes for which

$$V_{\alpha_1}^{EX} \supset V_{\alpha_2}^{EX}, \qquad (K.17)$$

we have

$$\Pr(V_{\alpha_1}^{EX}) < \Pr(V_{\alpha_2}^{EX}). \qquad (K.18)$$

(K.17) means that the region $V_{\alpha_2}^{EX}$ is *included* in the region $V_{\alpha_1}^{EX}$. Thus, for the two spheres of radii $\mathbf{R}_1 < \mathbf{R}_2$,

$$V_{\mathbf{R}_1}^{EX} < V_{\mathbf{R}_2}^{EX} \qquad (K.19)$$

and

$$V_{\mathbf{R}_1}^{EX} \subset V_{\mathbf{R}_2}^{EX}. \qquad (K.20)$$

Hence,

$$\Pr(V_{\mathbf{R}_1}^{EX}) > \Pr(V_{\mathbf{R}_2}^{EX}). \qquad (K.21)$$

If the two excluded volumes have the same shape, then the larger one also includes the smaller region, and hence (K.18) holds. If the two excluded volumes have different shapes, say a sphere and a rod, then in general, we cannot say anything on the relative probabilities even when we know their relative sizes.

As in the case of spherical cavities, there is a relation between the probability of finding a region V_α^{EX} empty and the corresponding Gibbs or Helmholtz energy change for the creation of such a cavity. In all cases, the relevant cavity must be at some *fixed* position in the system. It does not matter where we choose that fixed position (if the system is homogenous), and therefore we may omit the explicit notation of $\mathbf{R}_0$.

To see how important it is to fix the position of the cavity, here is a simple example. We consider a system of N hard spheres of diameter σ in a volume V at temperature T. We choose a density small enough so that the system's partition function is

$$Q(T, V, N) = \frac{V^N}{N!\Lambda^{3N}}. \tag{K.22}$$

The partition function of the same system with a cavity V^{EX} at some fixed position $\mathbf{R}_0$ is

$$\begin{aligned}
Q(T, V, N; \mathbf{R}_0) &= \frac{1}{N!\Lambda^{3N}} \int \cdots \int_{V-V^{EX}} d\mathbf{R}^N \\
&= \frac{(V - V^{EX})^N}{N!\Lambda^{3N}}.
\end{aligned} \tag{K.23}$$

The corresponding Helmholtz energy change is

$$\Delta A = A(T, V, N; \mathbf{R}_0) - A(T, V, N) = -k_B T \ln \frac{Q(T, V, N; \mathbf{R}_0)}{Q(T, V, N)}$$

$$= -k_B T \ln \left(1 - \frac{V^{EX}}{V}\right)^N = -k_B T \ln P_{cav}(\mathbf{R}_0). \tag{K.24}$$

In this particular case, ΔA is simply the work required to compress the system from volume V to $V - V^{EX}$. The probability of finding a cavity at $\mathbf{R}_0$ is simply

$$P_{cav}(\mathbf{R}_0) = \left(1 - \frac{V^{EX}}{V}\right)^N$$

$$\approx 1 - \frac{NV^{EX}}{V} = 1 - \rho V^{EX}. \qquad \text{(K.25)}$$

The last approximation is valid for any volume V^{EX} provided that it is small compared to the macroscopic volume V, i.e. $V^{EX}/V \ll 1$.

The above interpretation of P_{cav} becomes invalid if we talk about a cavity (of any form) at *any* point. Clearly, the probability of finding a cavity at *any* point in the system is unity (again we talk about a cavity of microscopic dimensions) — there is always some place within V where a small region of volume V^{EX} will be found empty of particles.

As with spherical particles, we can replace the cavity at $\mathbf{R}_0$ with a hard particle of a suitable form. Such a particle, when released, has a chemical potential of the form

$$\mu = \mu^* + k_B T \ln \rho \Lambda^3, \qquad \text{(K.26)}$$

where μ^* is the work required to place the hard particle at some fixed position and orientation. This work will be the same as the work required to form the corresponding cavity.

APPENDIX L

$H\phi I$ Interactions and Solubility of Isomeric Compounds

Perhaps the most dramatic difference in the solubility of two isomers is the case of cisbutenedioic acid (maleic acid) and trans-butenedioic acid (fumaric acid) (Fig. L.1).

Since the molecular weights of these two isomers are the same, we can compare the solubilities in either molar or weight concentrations. The solubilities of the two isomers are

$$\text{maleic acid} \quad 78\,\text{g}/100\,\text{ml} \quad \text{at } 25°\text{C}, \tag{L.1}$$

$$\text{fumaric acid} \quad 0.6\,\text{g}/100\,\text{ml} \quad \text{at } 25°\text{C}. \tag{L.2}$$

Note that the difference in the solubilities of the two isomers is more than two orders of magnitude. This is quite puzzling. Some have speculated than an intramolecular HB in maleic acid is responsible for the relatively high solubility.[315,316]

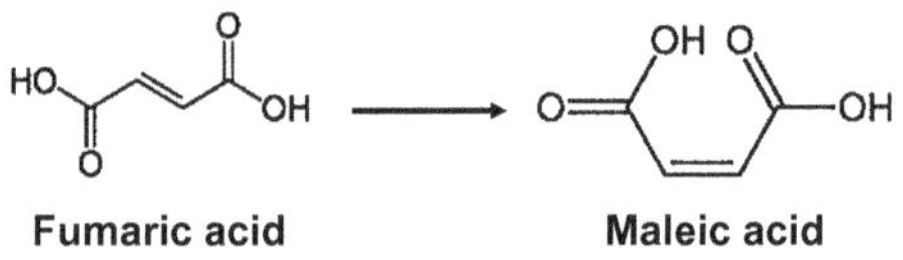

Fig. L.1 Fumaric and maleic acids. The difference in the solubilities of the two isomers may be interpreted as the indirect part of the Gibbs energy change for the process of transferring the carboxylic group from the trans to the cis position, Eq. (L.6).

---

[315]Frank *et al.* (1963).
[316]Apelblat *et al.* (2006).

However, there exists no convincing argument that explains why an intramolecular HB should cause such an increase in solubility.

We offer here an interpretation based on the solvent-induced correlation between the two carboxylic groups.

We assume that in fumaric acid (F), the two carboxylic groups are far apart, hence independently solvated. On the other hand, in maleic acid (M), the two carboxylic groups could be bridged by at least one water molecule.

The chemical potentials of the two isomers M and F at equilibrium are

$$\mu_M^P = \mu_M^* + k_B T \ln (\rho_M)_{\text{eq}} \Lambda^3, \tag{L.3}$$

$$\mu_F^P = \mu_F^* + k_B T \ln (\rho_F)_{\text{eq}} \Lambda^3. \tag{L.4}$$

Assuming that the chemical potentials of the two isomers are approximately equal in the pure state, we can write

$$-k_B T \ln \left(\frac{\rho_M}{\rho_F} \right)_{eq} \approx \mu_M^* - \mu_F^* \approx (\mu_M^* - \mu_M^{*ig}) - (\mu_F^* - \mu_F^{*ig})$$

$$= \Delta G_M^* - \Delta G_F^* = \delta G(F \to M). \tag{L.5}$$

Based on these assumptions we can interpret the ratio of the solubilities of the two isomers in terms of the solvent-induced driving force for the process of isomerization $F \to M$. Furthermore, assuming that in the fumaric isomer (F), the two carboxylic groups are far apart, hence, independently solvated (see Appendix F), we can interpret $\delta G(F \to M)$ in terms of $HϕI$ interaction between the two $HϕI$ groups at the configuration of maleic acid, i.e.

$$\delta G \ (HϕI \text{ in maleic acid}) \approx -k_B T \ln \frac{78}{0.6} \approx -12.2 \, \text{kJ} \, \text{mol}^{-1}$$

$$\tag{L.6}$$

This is close to the value of the $H\phi I$ interaction between two arms of two water molecules (see Sec. 1.7 and Appendix G). Note that here the two $H\phi I$ groups are attached to the backbone of the butene molecule. It seems that the effect of the BB on the $H\phi I$ interaction is relatively small compared to the effect on $H\phi O$ interaction (see Sec. 1.7).

This result supports the estimate of the $H\phi I$ interaction due to a one-water bridge between two $H\phi I$ groups, and at the same time provides a molecular interpretation of the relatively high solubility of the maleic acid.

Figure L.2 shows similar data for the transfer of a carboxylic group from position 4 on a benzene ring to position 2, given a carboxylic group at position 1 (Fig. L.3a). The indirect part of

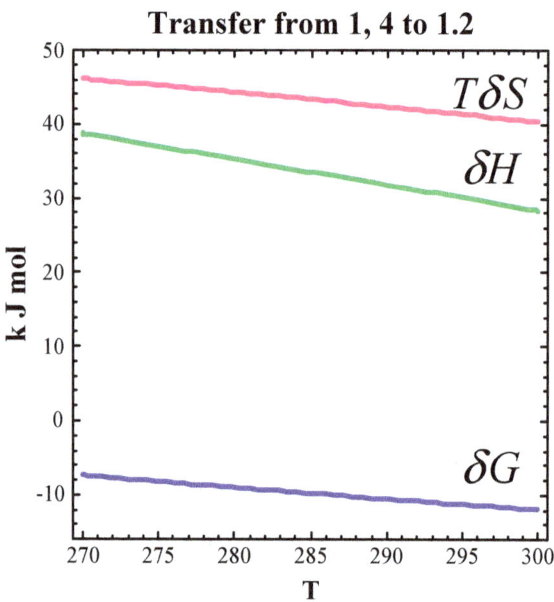

Fig. L.2   Values of δG, δH and $T\delta S$ for the process of transferring a carboxylic group from position 4 to position 2 relative to a group at position 1 (see Fig. L.3a).

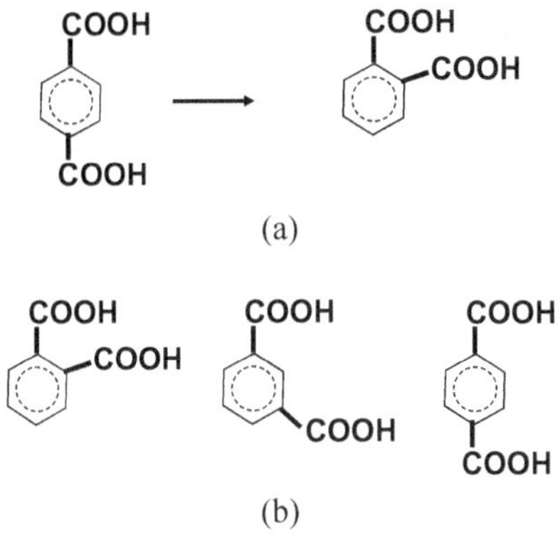

Fig. L.3 (a) The process of transferring a carboxylic from position 4 to position 2; (b) the three isomers of dicarboxylic benzene.

the Gibbs energy change for this process is

$$\delta G(1,4 \to 1,2) = \Delta G^*_{1,2} - G^*_{1,4} \approx \mu^*_{1,2} - \mu^*_{1,4}$$
$$= -k_B T \ln \left(\rho_{1,2}/\rho_{1,4} \right)_{eq}. \tag{L.7}$$

The assumption made here is that the solubility ratio is determined by the difference in the solvation Gibbs energies of the two isomers. We also neglect the difference in the rotational partition functions of the two isomers.

It is interesting to note that the value of δG for this "reaction" at 298 K is almost the same as the value for the reaction of transferring a carboxylic group from the trans to the cis isomers in Fig. L.1 provided in Eq. L.6.

Figure L.2 shows the values of δH and $T\delta S$ along with the values of δG for the same process as in Eq. L.6. Note that both δH and $T\delta S$ are positive and much larger in absolute magnitude than the values of δG.

Having examined the reaction shown in Fig. L.2, a few words of caution are in order. All the above conclusions were based on the assumption that the chemical potential in the pure state is insensitive to variation in the molecular structure of the solute. This is a gross simplification. It is also known that some of these compounds are hydrated in the solid state, a fact which might affect the entire argument given above.

Moreover, our association of δG with the ratio of the solubilities is valid provided that the solubilities are low enough so that Henry's law is valid. There is evidence that association by intermolecular HBs occurs in aqueous solutions. This again could affect our arguments and might invalidate our conclusion.

Figures L.4 and L.5 show the solubility data of the various benzene-polycarboxylic acids in water. As can be expected, the hexacarboxylic acid has the highest solubility. The 1,3- and 1,4-dicarboxylic acids have the lowest solubilities. Unexpectedly, the solubility of benzoic acid is much higher than the 1,3- and 1,4-dicarboxylic isomers, but slightly lower than the 1,2-isomer. This is surprising, since the dicarboxylic acids are certainly more hydrophilic than benzoic acid.

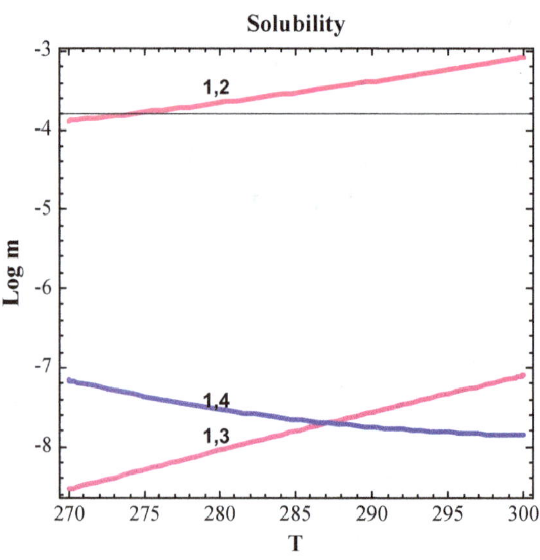

Fig. L.4   Solubility data for the three isomers in Fig. L.3b.

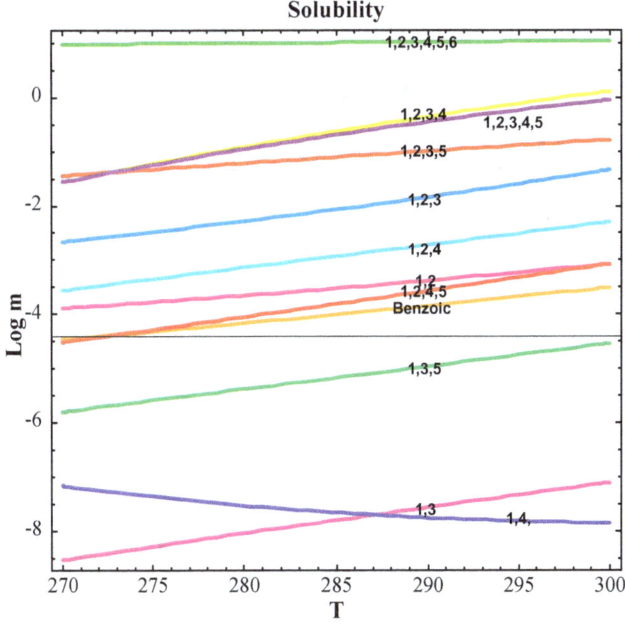

Fig. L.5 Solubility data for benzene polycarboxylic acids.[317]

It is not clear to what extent the trends shown in the solubility are mainly a result of the *solvation* of these acids. We believe, however, that the large difference between the 1,2-dicarboxylic and the 1,3- and 1,4-dicarboxylic is due to the *HφI* pairwise interaction at the 1,2 configuration. Perhaps, similar conclusions may be drawn from the solubilities of the 1,2,4- and the 1,3,5-tricarboxylic acid. Such a conclusion should await more detailed data on the nature of the solid states and the extent of association between these compounds in water.

---

[317]Data from Apelblat *et al.* (2006).

APPENDIX M

Further Inflating the Already Inflated Value of the $H\phi O$ Effect

---

In Chapters 1, 3 and Appendix H, we have discussed the inadequacy of the Kauzmann model of using the Gibbs energy of transferring a *$H\phi O$ molecule* from water into an organic liquid to estimate the contribution of a *$H\phi O$ group* transferred to the interior of a protein to the "driving-force" of protein folding. We have shown that the solvation Gibbs energy of a $H\phi O$ molecule such as methane or ethane is *much larger* than the conditional solvation Gibbs energy of the corresponding methyl or ethyl group attached to the backbone of the protein. We also have seen that the solvation of the molecule in the organic liquid representing the interior of a protein is not necessary (Appendix H).

Thus, the very use of the solvation Gibbs energy of a $H\phi O$ molecule to estimate the contribution to the driving force for the process of protein folding already gives an inflated value for the $H\phi O$ effect.

This appendix discusses some suggestions made in the literature that led to the further inflation of the value of the $H\phi O$ effect. There is also an important pedagogical moral spawned from the content of this appendix: Even a seemingly *exact derivation* could conceal a misleading result.

The solvation process and the corresponding solvation Gibbs energy were defined in Chapter 1. In earlier publications, various ways of measuring the solvation Gibbs energy were discussed.[318] The simplest and most straightforward method is to measure the densities of the molecule A in the liquid, and in a gaseous phase at equilibrium. If the gaseous phase may be considered to be an ideal gas, then we have

$$\mu_A^l = \mu_A^{*l} + k_B T \ln \rho_A^l \Lambda_A^3, \tag{M.1}$$

$$\mu_A^g = \mu_A^{*g} + k_B T \ln \rho_A^g \Lambda_A^3, \tag{M.2}$$

from which one obtains the exact result[319]

$$\Delta G_A^* = \mu_A^{*l} - \mu_A^{*g} = k_B T \ln (\rho_A^g / \rho_A^l)_{eq}. \tag{M.3}$$

As we have pointed out in several places throughout the book, the quantity that is needed in the driving force for protein folding is not ΔG_A^*, but the corresponding *conditional* Gibbs energy of solvation. From the available data, it seems that the conditional solvation energy could be one or two orders of magnitude *smaller* than the solvation Gibbs energy of the same group or molecule. Therefore, by using ΔG_A^* instead of the conditional quantity $\Delta G_{A/BB}^*$, one already commits an exaggeration of the $H\phi O$ effects in protein folding.

In 1991, Sharp *et al.* proposed a "corrected" method of calculating the solvation Gibbs energy of a molecule.[320] Instead of the measure of ΔG_A^* in (M.3), they proposed a corrected measure of the solvation Gibbs energy that reads

$$\Delta G_A^\circ = k_B T \ln (\rho_A^g / \rho_A^l)_{eq} + k_B T \ln \left(\frac{\overline{V}_A}{\overline{V}_B} - 1 \right), \tag{M.4}$$

---

[318]See, for example, Ben-Naim (1987, 2006, 2009).

[319]This is an exact result for a classical system, i.e. a system for which classical statistical mechanics applies.

[320]Sharp *et al.* (1991). It should be noted that the motivation for suggesting a corrected solvation Gibbs energy was not to obtain a new $H\phi O$ effect, but to reconcile the discrepancy between the macroscopic and the microscopic surface tension. The unwarranted reconciliation between the two "surface tensions" was criticized by Ben-Naim and Mazo (1993, 1997).

where $\overline{V}_A$ and $\overline{V}_B$ are the partial molar volumes of the solute A and the solvent B in an infinite dilution of A in B. Thus, the relation between the solvation Gibbs energy ΔG_A^* as calculated from (M.3) and the "corrected" solvation Gibbs energy calculated from (M.4) is

$$\Delta G_A^\circ = \Delta G_A^* + k_B T \left(\frac{\overline{V}_A}{\overline{V}_B} - 1 \right). \tag{M.5}$$

Clearly, whenever the partial molar volume of the solute is *larger* than the partial molar volume of the solvent, the modified quantity ΔG_A° will be larger than the solvation Gibbs energy. This has led Sharp *et al.* to conclude that the $H\phi O$ effect as measured by ΔG_A^* has been *underestimated*, and that the "real" $H\phi O$ effect as measured by the quantity ΔG_A° is much larger.

Recall that in Chapter 1, we concluded that ΔG_A^* is already an *overestimation* of the $H\phi O$ effect in protein folding. Therefore, the newly suggested quantity ΔG_A° is, in our opinion, an *over overestimation* of the contribution of the $H\phi O$ effect to protein folding.

The derivation by Sharp *et al.* was criticized by Ben-Naim and Mazo,[321] and by Holtzer.[322] We shall not discuss the details here. The main point of contention in our criticism was that relationship (M.3) is exact (within classical statistical mechanics), and the quantity ΔG_A^* is a well-defined quantity pertaining to the solvation process. With this criticism we felt that the issue of the correct measure of the *solvation* Gibbs energy was settled. Unfortunately, the controversy still lingered for some time. In 1994, Tunon *et al.*[323] published a paper claiming to have derived the Sharp *et al.* measure (Eq. M.4) *exactly*. This development

---

[321]Ben-Naim and Mazo (1993). See also Sharp *et al.* (1996) and Ben-Naim and Mazo (1997).

[322]Holtzer (1992, 1994).

[323]Tunon *et al.* (1994).

seems to support the suggested modified quantity ΔG_A°. The pendulum swung sharply toward favoring the suggestion from Sharp *et al.* Sharply indeed, but only apparently, and not for long. To see the pitfall, we repeat very briefly the derivation by Tunon *et al.*[324]

Tunon *et al.* started with the quantity G^* defined by

$$G^* = N_A \mu_A^* + N_B \mu_B^*, \tag{M.6}$$

where μ_A^* and μ_B^* are the pseudo-chemical potentials of the solute A and the solvent B, respectively. Differentiating (M.6) with respect to N_A leads to

$$\mu_A^r = \left(\frac{\partial G^*}{\partial N_A}\right)_{T,P,N_B} = \mu_A^* + N_A \frac{\partial \mu_A^*}{\partial N_A} + N_B \frac{\partial \mu_B^*}{\partial N_A}. \tag{M.7}$$

Here, all the pseudo-chemical potentials are for the liquid mixture of A and B. Using the general expression for the chemical potential as in (M.1) and (M.2) for both A and B, we get[325]

$$\frac{\partial \mu_A^*}{\partial N_A} = \frac{\partial \mu_A}{\partial N_A} - k_B T \frac{\partial \ln \rho_A^l}{\partial N_A} = \frac{\partial \mu_A}{\partial N_A} - \frac{k_B T}{N_A} + \frac{k_B T \overline{V}_A}{V}, \tag{M.8}$$

$$\frac{\partial \mu_B^*}{\partial N_A} = \frac{\partial \mu_B}{\partial N_A} - k_B T \frac{\partial \ln \rho_B^l}{\partial N_A} = \frac{\partial \mu_B}{\partial N_A} + \frac{k_B T \overline{V}_B}{V}. \tag{M.9}$$

Multiplying (M.8) by N_A and (M.9) by N_B and summing up the two, we obtain

$$\mu_A^r = \mu_A^* + \left[N_A \frac{\partial \mu_A}{\partial N_A} + N_B \frac{\partial \mu_B}{\partial N_A} \right]$$

$$+ \frac{k_B T}{V}(N_A \overline{V}_A + N_B \overline{V}_A) - k_B T. \tag{M.10}$$

---

[324]Tunon *et al.* (1994).
[325]All the derivatives are P, T constants and the number of molecules N_B.

The expression in the squared brackets is zero due to the Gibbs–Duhem equation. In addition, when A (the solute) is very diluted in B (the solvent), i.e. $N_A \ll N_B$, then

$$V = N_A \overline{V}_A + N_B \overline{V}_B \approx N_B \overline{V}_B. \qquad (M.11)$$

Hence, we obtain from (M.10) and (M.11)

$$\mu_A^r = \mu_A^* + k_B T \left(\frac{\overline{V}_A}{\overline{V}_B} - 1 \right), \qquad (M.12)$$

where $\overline{V}_A$ and $\overline{V}_B$ are the partial molar volumes of A and B, respectively, in the limit of very dilute solutions.

We now construct a new "solvation" Gibbs energy change by defining

$$\Delta G_A^r = \mu_A^{rl} - \mu_A^{rg} = \mu_A^{*l} - \mu_A^{*g} + k_B T \left(\frac{\overline{V}_A}{\overline{V}_B} - 1 \right). \qquad (M.13)$$

Note that in the ideal gas phase, the partial molar volumes of A and B are the same: $\overline{V}_A = \overline{V}_B = \frac{k_B T}{V}$, and $\mu_A^{rg} = 0$.

The expression (M.13) is the same as the expression (M.5) suggested by Sharp *et al.* (1991). Since relation (M.13) is indeed an exact result, it implies that our criticism of Sharp *et al.* quantity is invalid.

At this point it seems that the Sharp *et al.* "corrected" quantity is not only correct, but an exact result. But wait a minute — an exact result indeed, but of what? To answer this question, one must not only validate the derivation of Eq. (M.3), but also examine the *meaning* of the quantity denoted by ΔG_A^r. Tunon *et al.* do not discuss the *meaning* of this quantity, but only state that ΔG_A^r is more general than ΔG^*, in the sense that it reduces to ΔG_A^* whenever $\overline{V}_A = \overline{V}_B$, which is correct as can be seen from (M.13).

To examine the meaning of ΔG_A^r, or of μ_A^{rl}, we go back to the quantity G^* defined in (M.6), which we can rewrite as

$$G^* = N_A[\mu_A - k_B T \ln \rho_A \Lambda_A^3] + N_B[\mu_A - k_B T \ln \rho_B \Lambda_B^3]$$

$$= [N_A \mu_A + N_B \mu_B] - [N_A k_B T \ln \rho_A \Lambda_A^3$$

$$+ N_B k_B T \ln \rho_B \Lambda_B^3]. \tag{M.14}$$

The second term on the right-hand side of (M.14) is the Gibbs energy of an ideal gas mixture of composition N_A and N_B at a given temperature T and volume V. Therefore, we may interpret G^* as the change in the Gibbs energy for the process of "turning on" *all* the interactions among all the molecules in the system keeping T, V, N_A, N_B constants (Fig. M.1). Let us call G^* the total coupling work at constant T, V, N_A, N_B. When we take the derivative of G^* with respect to N_A at constant T, P, N_B, we get a quantity that measures the change in the *total coupling work* associated with the process of adding one particle of A at constant T, P, N_B. Therefore, the derived quantity reflects the changes in the solute–solute, the solute–solvent and the solvent–solvent interactions.

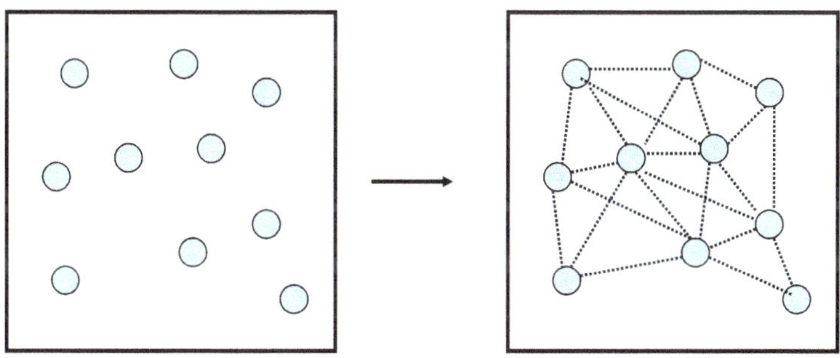

Fig. M.1   Schematic description of the process for which G^* in Eq. (M.14) is its Gibbs energy change. On the left-hand side, we have a system of N non-interacting particles in volume V. On the right-hand side, we have "turned on" the interactions among all the particles, keeping T and V constant. The interactions are depicted as dotted lines (not all interactions are shown).

In the study of the solvation process, we are interested in the coupling work of *one solute A* to the solvent when the solvent–solvent interactions are fully coupled. Clearly, this is very different from the quantity μ_A^r defined in (M.7). Thus, although the derivation of Eq. (M.13) is exact, it is a derivation of quantity that is totally irrelevant to the *solvation process*. The fact that ΔG_A^r is identical to the quantity ΔG_A° that was suggested by Sharp *et al.* means that ΔG_A°, as is ΔG_A^r, is also irrelevant to the solvation process, and therefore irrelevant to the $H\phi O$ effect.

The Anfinsen Dogma and the "Thermodynamic Hypothesis" Applied to the Process of Protein Folding

Ever since Anfinsen showed that the protein ribonuclease can be reversibly denatured/renatured in a test tube, the "thermodynamic hypothesis" has been invoked in the explanation of the protein folding process. In the words of Anfinsen:[326]

> "This hypothesis states that the three-dimensional structure of a native protein in its normal physiological milieu (solvent, pH, ionic strength, presence of other components such as metal ions or prosthetic groups, temperature, and other) is the one in which the Gibbs free energy of the whole system is lowest..."

In a recent review article, Dill and Chan wrote:[327]

> "The famous experiments of Christian B. Anfinsen and co-workers beginning in the early 1960s[328,329] showed that the protein can fold reversibly, implying that the native structures of some globular proteins

---

[326]Anfinsen (1973). See also Rose *et al.* (2006).
[327]Dill and Chan (1997).
[328]Anfinsen, Haber, Sela and White (1961).
[329]Anfinsen (1973).

are thermodynamically stable states, and therefore are conformations at the global minima of their accessible free energies."

It is true that the second law of thermodynamics states that the Gibbs energy of a system characterized by the variables T, P, N is at a minimum. The question is: Minimum with respect to what? Does the second law require the protein to attain a conformation at the global minimum?

In this appendix, we examine the meaning and the validity of the "thermodynamic hypothesis" when applied to the process of protein folding. In Appendix Q, we also comment on the validity of the usage of the concept of reversibility.

The second law of thermodynamics states that in any spontaneous process in an isolated system, the entropy always increases. Specifically, suppose we start with N particles in one compartment and remove a partition between the two compartments as in Fig. N.1. The gas initially confined to a volume V will spontaneously expand and will fill the entire volume $2V$. The change in entropy in this process is $k_B N \ln 2$, where k_B is the Boltzmann constant, and N is the number of particles. If we do the same process quasi-statistically (see Appendix Q), the change in entropy will still be the same, though we can also follow the change in entropy along the process. We can safely say that *at any* stage in the process, the entropy increases monotonically, and it will reach a *single* maximum at the equilibrium state. Similarly, if we start with N particles in the second compartment (Fig. N.1b) and do the same experiment, the entropy change will be the same, and the final equilibrium state will be the same. Clearly, if the entropy climbs monotonically to the maximum at the equilibrium value, there can be only *one* such a maximum. The existence of more than a single maximum would imply that the entropy could go both upwards as

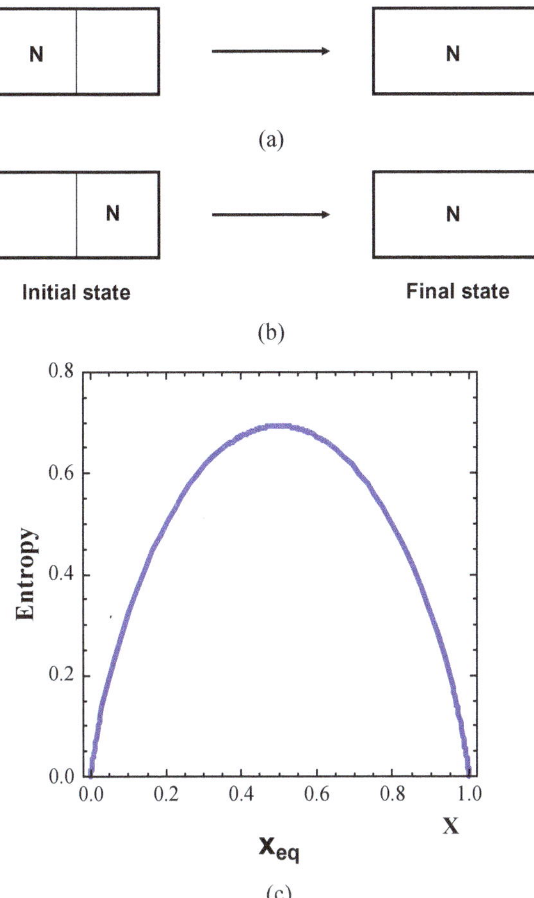

Fig. N.1   Schematic process of expansion. (a) Initially, all particles are in the left compartment; (b) initially, all particles are in the right compartment; (c) the entropy curve. From any initial condition, the mole fractions will change in such a way that it will reach an equilibrium value x_{eq} at which the entropy is maximum.

well as downwards. This is of course not allowed by the second law when applied to a spontaneous process in a *macroscopic* system.

The reason that we have a monotonic increase of the entropy, and that once it reaches a maximum value it stays there *forever*, is that the probability of attaining the equilibrium state is nearly one.[330] There exists a possibility of occurrence of small fluctuations away from the equilibrium state, but these are so small that we cannot observe them experimentally. On the other hand, larger fluctuations which could be observed have a negligible probability of occurrence.

Before moving on to the process of protein folding, let us reformulate the second law in terms of the Gibbs energy rather than in terms of the entropy. In a system characterized by the variables P, T, N, which are the most common variables that we control in the laboratory, any spontaneous process that occurs as a result of the removal of a constraint (e.g. removing a partition in the expansion or the mixing process) would result in a decrease in the Gibbs energy.

At the final equilibrium state, the Gibbs energy reaches its minimum value. Again, in a macroscopic system there can be no multiple minima. The reason is the same as provided above for the entropy. Also, the probability of finding measurable deviations from the equilibrium state is negligibly small.

All the statements that we made above apply to *macroscopic* systems. The main question is whether or not the so-called "thermodynamic hypothesis" applies to the case of a small system such as a protein (Chapter 3).

As we have seen in Sec. 1.2, when we release the two particles from an initial distance R, the system does not necessarily proceed monotonically towards the minimum. The same holds true for the protein. I believe that in applying the "thermodynamic hypothesis" to protein folding, there exists some confusion between two kinds of probabilities. One is the probability of finding the small system in some specific state. The other is the probability of finding the equilibrium *probability distribution* of

---

[330]See Ben-Naim (2008), Chapter 6.

the *states* of the protein. In the first, we deal with the probability (or the Gibbs energy) as a *function* of the state. In the second, we deal with the probability *functional*, or the Gibbs energy *functional* of the probability distribution of the states. To clarify this point, we consider the following examples:

(i) *A solute at a fixed position* $\mathbf{R}_0$

Consider an isolated system having a fixed energy U, volume V and number of particles N. Suppose we fix the location of one specific particle at $\mathbf{R}_0$ — we shall refer to this particle as the *solute*. It can be a molecule which differs from the other molecules in the system, or identical to all other $N-1$ molecules. However, by fixing the location of one particular molecule, we make it distinguishable from all other molecules in the system.[331]

For such a system, one can define the entropy $S(U, V, N; \mathbf{R}_0)$. Clearly, by fixing the location of one particle in the system, we have a constrained equilibrium state. Here, the parameter $\mathbf{R}_0$ is the microscopic parameter characterizing the constrained equilibrium state.

What happens if we remove the constraint and let the particle wander freely in the system? It is easy to calculate the change in entropy of the system upon releasing the constraint on a fixed location $\mathbf{R}_0$. One can also show that releasing the constraint always involves a positive change in entropy.[332] The entropy attains a new maximum value. But maximum with respect to what?

Here is a potential pitfall. We write the entropy *function* as $S(U, V, N; \mathbf{R})$, keeping U, V, N constant. We then release the constraint on $\mathbf{R}_0$, and we are inclined to conclude that the entropy, as a function of $\mathbf{R}$, increases towards a maximum.

---

[331]Ben-Naim (2008).
[332]Ben-Naim (2006).

Of course, no one has reached that conclusion for this particular example. But many have reached essentially the same conclusion for a more complicated case, which is discussed in the third example below.

The reason we naturally avoid falling into this trap is that even without any calculations, we feel that when the location of the particle is not constrained at $\mathbf{R}_0$, the entropy is not a monotonically increasing function of $\mathbf{R}$, and there exists no specific location, say, $R_{\max}$, for which the entropy is maximum. In fact, it is easy to prove using Shannon's theorem[333] that at equilibrium (after lifting the constraint on $\mathbf{R}$), the probability of finding the particle at any point $\mathbf{R}$ within the system is constant, independently of $\mathbf{R}$. Specifically, the probability density is

$$P_{eq} \to P_{eq}(\mathbf{R}) = \frac{1}{V}, \quad \text{for any } \mathbf{R}. \tag{N.1}$$

Thus, we are still left with the question posed earlier: Maximum with respect to what?

The answer is again given by Shannon's theorem. The entropy of the system (having fixed U, N, V) is maximum with respect to all possible *distributions* of *locations*.

Let $P(\mathbf{R})$ be any distribution, such that

$$\int_V P(\mathbf{R})d\mathbf{R} = 1, \tag{N.2}$$

i.e. the probability of finding the particle at *any* location within the volume V is one. Instead of the function $S(U, V, N; \mathbf{R})$, we look at the *functional* $S(U, V, N; P(\mathbf{R}))$. Shannon's theorem states that the entropy functional will be maximum with respect to the *function* $P(\mathbf{R})$. Thus, in the initial state, the entropy functional is written as $S(U, V, N; \delta(\mathbf{R} - \mathbf{R}_0))$, where $\delta(\mathbf{R} - \mathbf{R}_0)$ is the Dirac delta function. At equilibrium, there exists a single

---

[333]Ben-Naim (2008).

function, which we denoted by $P_{eq}(\mathbf{R})$, which maximizes the entropy, i.e. the condition for the maximum is that the functional derivative is zero

$$\frac{\delta S(P(\mathbf{R}))}{\delta P(\mathbf{R})} = 0. \tag{N.3}$$

The distribution function that maximizes the entropy functional is given in Eq. (N.1).

Before we turn to the next two examples, we reformulate the same principle in terms of probability and in terms of Gibbs energy.

In the probability formulation, there exists another pitfall that arises from mal-notation. The principle of maximum entropy may be stated as a principle of maximum probability. But care must be exercised in distinguishing between the probability of finding the particle at an element of volume $d\mathbf{R}$ around $\mathbf{R}$ [or the probability density which we denoted by $P(\mathbf{R})$] on the one hand, and the probability of finding a *specific probability distribution* $P(\mathbf{R})$ which we denote by $\Pr[P(\mathbf{R})]$ on the other. To avoid confusion, we refer to $P(\mathbf{R})$ as the probability distribution of locations, and to $\Pr$ as the super probability.[334] The super probability is a functional of the probability distribution, while $P(\mathbf{R})$ at equilibrium is constant independent of $\mathbf{R}$. The super probability has a single maximum at the same distribution that maximizes the entropy functional, i.e. at $P_{eq}(\mathbf{R})$.

The second equivalent formulation which we will discuss in more detail in the third example is in terms of Gibbs energy. For a system characterized by the variables P, T, N, we can write the Gibbs energy function of the system as $G(P, T, N; \mathbf{R})$.

If we start with a system having one particle at a fixed position, say, $\mathbf{R} = \mathbf{R}_0$, then release the constraint on $\mathbf{R}$, but keep T, P, N fixed, the system's Gibbs energy will always decrease by

---

[334]Ben-Naim (2008).

the amount

$$\Delta G = k_B T \ln \rho \Lambda^3 < 0. \qquad (N.4)$$

Note again that $G(P, T, N; \mathbf{R})$ is not a monotonic function of $\mathbf{R}$, and that there exists no value of $\mathbf{R}$ for which G is minimum. Instead, the *functional* $G(T, P, N; P(\mathbf{R}))$ has a minimum with respect to all possible distributions $P(\mathbf{R})$. There exists a distribution $P_{eq}(\mathbf{R})$, which is the same as (N.1), for which G is minimal.

In this example, we found that G as a function of $\mathbf{R}$ is constant. There exists no single value $\mathbf{R}_0$ at which G is minimal. In the next two examples, we shall encounter a Gibbs energy function that has one or more minima. The point to stress here is that these minima, when they occur, are not required by the second law. The second law requires that the entropy *functional* $S(U, V, N; P(\mathbf{R}))$ or the Gibbs energy functional $G(P, T, N; P(\mathbf{R}))$ will have a *single* extremum with respect to the probability distribution $P(\mathbf{R})$. The equilibrium distribution $P(\mathbf{R})$ may have any form, with or without minima or maxima.

(ii) *Ethane at a fixed conformation*

The second example brings us closer to the case of a protein where the thermodynamic principle of maximum entropy or minimum Gibbs energy is misapplied.

Consider a system characterized by (U, V, N_W, N_a), where N_w is the number of solvent molecules, say, water, and N_a is the number of ethane molecules. We also assume that $N_a \ll N_w$, so that the solute molecules do not interact with each other.

Suppose that we start with all the ethane molecules at a fixed dihedral angle ϕ, say, at $\phi_0 = 0$. Since ϕ is the only parameter that describes the configuration of the ethane molecule, we can write the entropy *function* for this system as $S(U, V, N_w, N_a; \phi)$.

If we now release the constraint on a fixed dihedral angle, the entropy of the system must increase, reaching a new maximum entropy. Again, the question we pose is: Maximum with respect to what?

In this particular example, we know that the probability distribution with respect to the parameter ϕ has three maxima and three minima. The same is true for the entropy function $S(U, V, N_w, N_a; \phi)$, or the Gibbs energy function $G(P, T, N_w, N_a; \phi)$. Thus, when we release the constraint on fixed ϕ, the system *does not* proceed to one of the maxima of the entropy. This statement sounds like a contradiction of the second law. However, this is not so. The second law does not require that the *function $S(U, V, N_w, N_a; \phi)$* have a single maximum at any particular value of ϕ. As we have seen in the previous example, $S(U, V, N; \mathbf{R})$ did not have any maximum with respect to $\mathbf{R}$. In the present example, we do have several minima and maxima of the function $S(U, V, N_w, N_a; \phi)$, but these are not the maxima that are dictated by the second law of thermodynamics. What the second law does say is that when we release the constraint on the fixed angle ϕ, the entropy *functional* will reach a maximum with respect to all possible *distributions $P(\phi)$*.

In other words, the initial distribution of angle was

$$P_{in}(\phi) = \delta(\phi - \phi_0). \tag{N.5}$$

The corresponding value of the entropy functional is $S(U, V, N_w, N_a; \delta(\phi - \phi_0))$. Upon releasing the constraint on the angular distribution, the system will attain an equilibrium distribution $P_{eq}(\phi)$ which has the form as in Fig. N.2, and for which the entropy functional is maximal, i.e. $S(U, V, N_w, N_a; P_{eq}(\phi))$ is a maximum with respect to all possible distribution functions $P(\phi)$.

Here is another potential pitfall. $P_{eq}(\phi)$ may or may not have maxima or minima. In the previous example, $P_{eq}(R)$ did

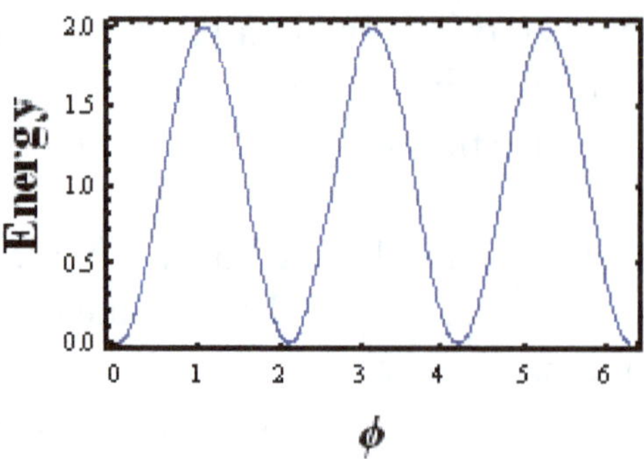

Fig. N.2   Schematic distribution angles ϕ in ethane.

not have any minimum or maximum. In the present example, $P_{eq}(\phi)$ has three minima and three maxima. In the next example we shall encounter many more minima and maxima in the probability distribution. What the second law states is that there exists a distribution function $P_{eq}(\phi)$ for which the super probability functional $\Pr[P(\phi)]$ has a *single* maximum with respect to all possible distributions $P(\phi)$. The equilibrium distribution $P_{eq}(\phi)$ that maximizes the entropy functional is also the distribution that maximizes the super probability functional.

We shall reformulate the same statement, but in terms of the Gibbs energy. The Gibbs energy functional $G(P, T, N_w, N_a; P(\phi))$ has a *single minimum* at a specific probability distribution $P_{eq}(\phi)$ over all possible distributions $P(\phi)$. The probability distribution $P(\phi)$ or the Gibbs energy function $G(P, T, N_w, N_a; \phi)$ may or may not have minima or maxima. The second law of thermodynamics applies to either the functional $G(P, T, N_w, N_a; P(\phi))$, or to the super probability $\Pr[P(\phi)]$, but not to the function $G(P, T, N_w, N_a; \phi)$ or to $P(\phi)$.

(iii) *Proteins and the Anfinsen Dogma*

The third example is essentially an extension of the previous example. Instead of one dihedral angle ϕ, proteins are described by the configuration of at least $2M$ angles

$\phi_1, \psi_1, \phi_2, \psi_2, \ldots \phi_M, \psi_M$, where M is the number of amino acids in the protein. The thermodynamic hypothesis applied to protein folding is essentially a generalization of the case discussed above. We denote by ϕ the conformational vector consisting of the $2M$ components. Let us denote by ϕ_U the conformation which we recognize as the unfolded or denatured form of the protein. This conformation is not a single configuration but a whole distribution of conformations $P_U(\phi)$. It is well known that the function $P_U(\phi)$ has many minima and maxima. The corresponding Gibbs energy *function* is $G(T, P, N; \phi)$, which is referred to as the Gibbs energy *landscape*. Next, we change the macroscopic parameters, say, reducing the temperature or removing a denaturing agent. The system's Gibbs energy will change to a new minimum. However, the minimum is not with respect to the conformation ϕ — it is the minimum of the Gibbs energy *functional* $G(T, P, N; P(\phi))$ with respect to all possible distributions $P(\phi)$. This is a single minimum as required by the second law, whereas the function $P(\phi)$ or $G(T, P, N; \phi)$ has multiple minima and maxima. An equivalent statement is in terms of the super probability functional denoted $\Pr[P(\phi)]$.

From the above examples, we can conclude that although the system's Gibbs energy is always at its minimum at equilibrium, the minimum is with respect to the *probability distributions of the states*. Thus, when we start from any initial conformation of the protein and change the environment (temperature, pressure of solvent composition), the protein might or might not proceed to some minimum in the energy landscape. The second law does not apply to the Gibbs energy *function* $G(T, P, N_w, N_a; \phi)$ or to the Gibbs energy landscape. The second law applies to the Gibbs energy *functional* $G(T, P, N_w, N_a; P(\phi))$.

The misinterpretation of Anfinsen's thermodynamic hypothesis has led to the conclusion that the Gibbs energy landscape should have a shape of a funnel. This is an unwarranted conclusion.

In the article by Dill and Chan (1997), mentioned in the beginning of this Appendix, six different shapes of funnels are described. The reader should be warned that these are purely hypothetical shapes. One can imagine hundreds or thousands of speculative shapes, none of which can be supported theoretically. The unwary reader might be misled into thinking that the funnel-like landscape is a fact, and that this fact follows from Anfinsen's thermodynamic hypothesis. As we have stressed in this Appendix, the Second Law states that there exists a single landscape which minimizes the Gibbs energy functional. The Second Law does not state anything regarding the shape of this landscape, whether it should or should not have one or more minima.

Entropy-Enthalpy Compensation; from an Exact Theorem to an Approximate Manifestation

The exact entropy–enthalpy compensation (EEC) theorem was first formulated in 1965 in terms of a mixture model for water.[335] It was proven later in more general terms, independently of the solvent or of the model used for that solvent.[336] An approximate version of the EEC principle was published by Lumry and Rajender.[337] Some aspects of both the exact and the approximate versions of the EEC were discussed in Part I. Here, we shall approach the EEC theorem in three steps: First, we shall consider an isomerization reaction in an ideal gas phase where the entropy and the energy of the reactions are, in general, uncorrelated. The next step is to carry out the same reaction in a solvent, in which we can identify the source of the exact EEC. Finally, we shall examine the conditions under which the empirical approximate EEC might be manifested for the same process. In this appendix we shall discuss the entropy–energy compensation rather than the entropy–enthalpy compensation. This is done for convenience only.[338] Also, we discuss the

---

[335]Ben-Naim (1965).

[336]Ben-Naim (1974).

[337]Lumry and Rajender (1970).

[338]See also Ben-Naim (1992, 2006).

simplest isomerization reaction. The conclusions can be applied straightforwardly to any process in a solvent.

We consider the simplest chemical reaction, the equilibrium between two isomers A and B:

$$A \rightleftarrows B. \tag{O.1}$$

We shall derive the equilibrium constant and the corresponding thermodynamics of this reaction for three cases. First, when A and B are two energy levels ε_A and ε_B, with degeneracies ω_A and ω_B, respectively. Second, when A and B correspond to any arbitrary division of the energy levels into two groups. Finally, we add the solvent and examine how the thermodynamics of this reaction change upon solvating the two components A and B.

(a) *Reaction in an ideal gas phase, when A and B are two energy levels ε_A and ε_B, with degeneracies ω_A and ω_B*

In an ideal gas phase, the chemical potentials of A and B are

$$\mu_A = k_B T \ln \rho_A \Lambda_A^3 q_A^{-1} = \mu_A^* + k_B T \ln \rho_A \Lambda_A^3, \tag{O.2}$$

$$\mu_B = k_B T \ln \rho_B \Lambda_B^3 q_B^{-1} = \mu_B^* + k_B T \ln \rho_B \Lambda_B^3, \tag{O.3}$$

where ρ_α is the number density, Λ_α^3 is the momentum partition function, and q_α is the internal partition function of the species α. μ_α^* is the pseudo-chemical potential of the species α.

Since A and B have the same mass, their momentum partition functions are equal, i.e.

$$\Lambda_A^3 = \Lambda_B^3. \tag{O.4}$$

At equilibrium, we have the condition

$$\mu_A = \mu_B. \tag{O.5}$$

From (O.2), (O.3), (O.4) and (O.5) we get

$$K^{ig} = \left(\frac{\rho_B}{\rho_A}\right)_{eq} = \exp[-\beta(\mu_B^* - \mu_A^*)]. \tag{O.6}$$

In this particular case, the internal partition functions are

$$q_A = \omega_A \exp[-\beta\varepsilon_A], \quad q_B = \omega_B \exp[-\beta\varepsilon_B]. \tag{O.7}$$

Hence,

$$K^{ig} = \left(\frac{q_B}{q_A}\right) = \frac{\omega_B}{\omega_A} \exp[-\beta(\varepsilon_B - \varepsilon_A)]. \tag{O.8}$$

Define the standard Helmholtz energy of the reaction by

$$\Delta A^\circ(A \to B) = \mu_B^* - \mu_A^* = -k_B T \ln K^{ig}$$

$$= (\varepsilon_B - \varepsilon_A) + k_B T \ln \frac{\omega_B}{\omega_A}. \tag{O.9}$$

The corresponding entropy and energy (or enthalpy) changes are

$$\Delta A^\circ(A \to B) = -\frac{\partial \Delta A^\circ}{\partial T} = k_B \ln \frac{\omega_B}{\omega_A}, \tag{O.10}$$

$$\Delta E^\circ(A \to B) = \Delta H^\circ(A \to B) = \varepsilon_B - \varepsilon_A. \tag{O.11}$$

The most important aspect of this simple reaction is that the energy change and the entropy change are completely independent of each other. The energy change is simply the difference in the energy levels, and the entropy change is related to the ratio of the degeneracies of the two energy levels. As we shall see below, this neat separation between energy levels and degeneracies no longer holds when a solvent is present.

We can rewrite the equilibrium constant for this reaction

$$K^{ig} = \exp[-\beta(\Delta E^\circ - T\Delta S^\circ)]. \tag{O.12}$$

Assuming that $\varepsilon_B - \varepsilon_A > 0$ and $\omega_B/w_A > 1$, i.e. $\Delta E° > 0$ and $\Delta S° > 0$, we can predict the following behavior of the equilibrium constants as we change the temperature.

At very low temperatures $T \to 0$, the term $T\Delta S°$ will be negligible. The equilibrium constant will depend only on the difference in the energy levels

$$K^{ig} = \exp[-\beta\Delta E°] \to 0. \qquad (O.13)$$

Hence, the relative population of the B level will tend to zero. At very high temperatures $T \to \infty$, the term $T\Delta S°$ will become larger than $\Delta E°$. At this limit, the equilibrium constant will be determined by the ratio of the degeneracies

$$K^{ig} \approx \exp\left[\frac{\Delta S°}{k}\right] = \frac{\omega_B}{\omega_A}. \qquad (O.14)$$

(b) Generalization of the reaction (O.1) in an ideal gas phase, when A and B are defined in terms of two groups of molecular states

We start with a one-component system in an ideal gas phase. Each molecule has a series of states $i = 1, 2, 3, \ldots, n$. The internal partition function of each molecule is

$$q = \sum_i \exp[-\beta\varepsilon_i]. \qquad (O.15)$$

Note that the sum in (O.15) is over all the states of the molecule. We now divide the total states of the molecules into two groups, A and B (see Fig. 3.4). The internal partition function is divided into a sum of two terms

$$q = \sum_{all\ i} \exp[-\beta\varepsilon_i] = \sum_{i \in A} \exp[-\beta\varepsilon_i] + \sum_{i \in B} \exp[-\beta\varepsilon_i]$$

$$= q_A + q_B. \qquad (O.16)$$

Thus, a molecule in a state i belonging to a group A is referred to as an A molecule, and similarly, a molecule in a state $i \in B$ is referred to as a B molecule. The corresponding internal partition functions are q_A and q_B, respectively.

The relations (O.1) to (O.6) apply to this case. The equilibrium constant for this case is

$$K^{ig} = \left(\frac{\rho_B}{\rho_A} \right)_{eq} = \exp[-\beta(\mu_B^* - \mu_A^*)], \tag{O.17}$$

where now the pseudo-chemical potentials are

$$\mu_A^* = -k_B T \ln q_A = -k_B T \ln \sum_{i \in A} \exp[-\beta \varepsilon_i], \tag{O.18}$$

$$\mu_B^* = -k_B T \ln q_B = -k_B T \ln \sum_{i \in B} \exp[-\beta \varepsilon_i]. \tag{O.19}$$

The pseudo partial molar entropy and energy (or enthalpy) are, in this case,

$$-S_A^* = \frac{\partial \mu_A^*}{\partial T} = -k_B \ln q_A - k_B T \frac{\partial \ln q_A}{\partial T}, \tag{O.20}$$

$$E_A^* = \mu_A^* + T S_A^*. \tag{O.21}$$

Denote by $p(i/A)$ the conditional probability of finding the state i given that the molecule is in one of the states in group A, i.e.

$$p(i/A) = \frac{\exp[-\beta \varepsilon_i]}{q_A}. \tag{O.22}$$

With this definition, we can rewrite the entropy S_A^* and the energy E_A^* in (O.20) and (O.21) as

$$S_A^* = k_B \ln q_A + \frac{\sum_{i \in A} \exp[-\beta \varepsilon_i]}{q_A T}$$

$$= -k_B \sum_{i \in A} p(i/A) \ln p(i/A) \tag{O.23}$$

$$E_A^* = \sum_{i \in A} \varepsilon_i p(i/A). \qquad (O.24)$$

Hence, the standard thermodynamic quantities for this case are

$$\Delta A^\circ(A \to B) = \mu_B^* - \mu_A^*$$
$$= \Delta E^\circ(A \to B) - T\Delta S^\circ(A \to B), \qquad (O.25)$$
$$\Delta S^\circ(A \to B) = S_B^* - S_A^*$$
$$= -k_B \sum_{i \in B} p(i/B) \ln p(i/B)$$
$$+ k_B \sum_{i \in A} p(i/A) \ln p(i/A), \qquad (O.26)$$
$$\Delta E^\circ(A \to B) = E_B^* - E_A^* = \sum_{i \in B} \varepsilon_i p(i/B) - \sum_{i \in A} \varepsilon_i p(i/A)$$
$$= \bar{\varepsilon}_B - \bar{\varepsilon}_A. \qquad (O.27)$$

Compare these relationships with (O.9), (O.10) and (O.11). Clearly, when each group consists of only one energy level, the last three relationships reduce to the corresponding relationships (O.9)–(O.11). Note also that the entropy difference is simply the difference in the Shannon's measure of information (SMI) defined on the distributions $\{p(i/A)\}$ and $\{p(i/B)\}$, respectively.[339] One can also generalize the last relationships for the case of continuous distribution of states. The interpretation of the various terms on the right-hand side of (O.26) and (O.27) does not change.[340]

In applying these results to the conversion of the unfolded to the folded form of a protein, the energy difference in (O.27)

---

[339]Shannon (1948), Ben-Naim (2008).

[340]Note that in the continuous case, each of the quantities S_A^* and S_B^* diverges. However, the difference $S_B^* - S_A^*$ is not divergent. For more details, see Ben-Naim (2008).

is positive (i.e. the average energy level in the folded form is lower than the average of the unfolded form). Also, the entropy difference is positive (i.e. the Shannon's measure of information of the unfolded form is much larger than that of the folded form). It is often said that the number of states in the unfolded form is much larger than the number of states in the folded form. This is true, of course, however, the entropy difference is equal to the difference in the SMI and not to the ratio of the number of states. It is only in the simple case when all the states of the unfolded form have the same energy, and all the states of the folded form have the same energy, that the difference in (O.26) reduces to expression (O.10), which depends only on the ratio of the degeneracies.

(c) *The same reaction in a solvent*

We use either case (a) or (b) discussed above and we add a solvent denoted w.[341] We also assume that the solvent does not affect the energy levels or the degeneracies, i.e. the internal partition function is the same as in either case (a) or (b). We are interested in the effect of the solvent on the chemical equilibrium. To do that, we write the chemical potentials of A and B as[342]

$$\mu_A = W(A|w) + k_B T \ln \rho_A \Lambda_A^3 q_A^{-1} = \mu_A + k_B T \ln \rho_A \Lambda_A^3,$$
$$\tag{O.28}$$

$$\mu_B = W(B|w) + k_B T \ln \rho_B \Lambda_B^3 q_B^{-1} = \mu_B + k_B T \ln \rho_B \Lambda_B^3.$$
$$\tag{O.29}$$

The quantity $W(\alpha|w)$ is the work associated with the insertion of an α molecule to a fixed position in the solvent. This

---

[341] Here, w is any solvent. It will stand for water in the next subsection (d).
[342] For details, see Ben-Naim (2006).

work is sometimes referred to as the coupling work of α to the solvent w.

At equilibrium we have the condition

$$\mu_A = \mu_B. \qquad (O.30)$$

Hence, the chemical equilibrium in the solvent is

$$K = \left(\frac{\rho_B}{\rho_A}\right)_{eq} = \frac{q_B}{q_A} \exp[-\beta\{W(B|w) - W(A|w)\}]$$

$$= K^{ig} \exp[-\beta\{W(B|w) - W(A|w)\}]. \qquad (O.31)$$

Thus, we see that the new equilibrium constant K is related to the equilibrium constant in an ideal gas and the difference in the coupling works of A and B to the solvent. The latter is also equal to the difference in the solvation Helmholtz (or Gibbs) energies of A and B, i.e.

$$K = K^{ig} \exp[-\beta(\Delta A_B^* - \Delta A_A^*)], \qquad (O.32)$$

where ΔA_α^* is the Helmholtz energy of solvation of an α molecule.

The thermodynamic quantities associated with the reaction are now

$$\Delta A^\circ(A \to B) = \mu_B^* - \mu_A^* = \Delta A^{oig} + (\Delta A_B^* - \Delta A_A^*), \qquad (O.33)$$

$$\Delta S^\circ(A \to B) = S_B^* - S_A^* = \Delta S^{oig} + (\Delta S_B^* - \Delta S_A^*), \qquad (O.34)$$

$$\Delta E^\circ(A \to B) = E_B^* - E_A^* = \Delta E^{oig} + (\Delta E_B^* - \Delta E_A^*). \qquad (O.35)$$

Note carefully that μ_α^* is the pseudo-chemical potential of a molecule α. S_α^* and E_α^* are the corresponding entropy and the energy associated with the pseudo-chemical potential. These quantities are related to the process of inserting α to a fixed position in the liquid. The quantities denoted ΔA^{oig}, ΔS^{oig} and ΔE^{oig} are the ideal-gas values of the Helmholtz energy, entropy and energy of the reaction. They are different for cases (a) and (b) discussed above.

The quantities $\Delta A_\alpha^*, \Delta S_\alpha^*$ and ΔE_α^* are the *solvation* Helmholtz energy, entropy and energy of the α molecule.

In case (a), we found that ΔS° and ΔE° were completely independent of each other — one depends on the ratio of the degeneracies and the other depends on the difference in the energy levels. In case (b), we saw that ΔS° and ΔE° become the difference in *average* quantities — the first is the difference in SMI defined on the two distributions $\{p(i/A)\}$ and $\{p(i/B)\}$, the second is the difference in the average energies of the two species. The two quantities are connected through the distribution functions with which the averages are taken, but they seem to have no common terms. The situation is quite different when we write down explicitly the quantities ΔS° and ΔE° in (O.34) and (O.35)[343]

$$\Delta A^\circ(A \to B) = \Delta A^{oig} + [W(B|w) - W(A|w)], \qquad (\text{O.36})$$

$$T\Delta S^\circ(A \to B) = T\Delta S^{oig} + [W(B|w) - W(A|w)]$$

$$+ [\langle B_B \rangle_B - \langle B_A \rangle_A] + [\langle U_N \rangle_B - \langle U_N \rangle_A], \qquad (\text{O.37})$$

$$\Delta E^\circ(A \to B) = \Delta E^{oig} + [\langle B_B \rangle_B - \langle B_A \rangle_A] + [\langle U_N \rangle_B - \langle U_N \rangle_A]. \qquad (\text{O.38})$$

As we can see, both $T\Delta S^\circ$ and ΔE° include the common terms. Hence, when we form the difference $\Delta E^\circ - T\Delta S^\circ$ these common terms cancel out, and we are left with an expression (O.36) that does not include these common terms. Thus, in the case (c) when a solvent is present we have an exact

---

[343]Here, we have written the entropy–energy compensation theorem. A similar entropy–enthalpy theorem exists and is discussed in Ben-Naim (1992, 2009).

entropy–energy compensation.[344] This theorem is important in the understanding the entropy and the energy (or enthalpy) change for any process occurring in aqueous solutions.[345]

The meaning of the various terms in (O.37) and (O.38) is as follows:

$\langle B_\alpha \rangle_\alpha$ means a conditional average of the binding energy of α to the solvent, averaged over all the configurations of the solvent molecules, given the *condition* that an α molecule is at a fixed position in the liquid. The quantity $\langle U_N \rangle_\alpha$ is the conditional average of the *total* interaction energy among the N solvent molecules, given an α molecule at some fixed position in the liquid. The difference $\langle U_N \rangle_B \langle -U_N \rangle_A$ may also be reinterpreted as the "structural changes" in the solvent induced by the process $A \to B$.[346] We shall not be interested in this particular interpretation in the present discussion. The important point to be emphasized here, is that whenever we form the combination $\Delta E^\circ - T\Delta S^\circ$, there are some terms which may be interpreted as "structural changes" in the solvent that cancel out.

Let us rewrite (O.37) and (O.38) as

$$T\Delta S^\circ(A \to B) = T\Delta S^{oig} - [W(B|w) - W(A|w)] + T\Delta S^{st},$$

$$(O.39)$$

$$\Delta E^\circ(A \to B) = \Delta E^{oig} + \Delta E^{st}, \qquad (O.40)$$

where the "structural changes" (st) parts are defined in (O.39) and (O.40). Thus, we have the exact entropy–enthalpy compensation

$$T\Delta S^{st} = \Delta E^{st}. \qquad (O.41)$$

This EEC is valid for any solvent w.

---

[344] Ben-Naim (2009).
[345] Ben-Naim (2009).
[346] Ben-Naim (2009).

(d) *The empirical EEC principle*

The EEC compensation in (O.41) is an exact theorem valid for any process in any solvent. There is another result which is also referred to as the EEC law, or the EEC principle, but is not an exact relationship. Lumry and Rajender[347] compiled an impressive body of experimental data on standard entropy and enthalpy of many processes in aqueous solutions. They found that a plot of $\Delta H°$ as a function of $\Delta S°$ of a series of processes is linear with a slope approximately equal to experimental temperature.

From these findings, Lumry and Rajender concluded:

"It is proposed that a linear enthalpy–entropy relationship be used as a diagnostic test for participation of water in protein process."

Although I agree with the general "spirit" of this conclusion, I disagree with the details. First, it is not clear that processes in other solvents, not necessarily water, will not show EEC. Second, there could be processes where relationship (O.41) is valid, but the EEC, in the sense defined by Lumry and Rajender, might not occur.

Therefore, although the EEC, in the Lumry–Rajender sense, is very likely to occur in aqueous solutions, it is not clear to what extent it might be used as a "diagnostic test for the participation of water."

We shall now discuss a possible condition under which the empirical Lumry–Rajender EEC might follow from the exact EEC (O.41).

Suppose one takes a series of reactions of the type $A \rightarrow B$ that differ by some parameter i, say, a series of similar proteins, or the same protein folding process but occurring in slightly different aqueous solutions (say, by addition of small amount of

---

[347]Lumry and Rajender (1970).

different solutes). Let ΔA_i° (or ΔG_i°) be the Helmholtz energy change for the process i. Then, we can write the exact relationship, for each i

$$\Delta E_i^\circ = \Delta A_i^\circ + T\Delta S_i^\circ. \tag{O.42}$$

Now, suppose that ΔS_i^{st} and ΔE_i^{st} in (O.39) and (O.40) dominate the entropy and the energy of the processes, and that ΔA_i° does not change much with i, in which case we can rewrite (O.42) in an approximate form

$$\Delta E_i^\circ \approx \Delta A^\circ + T\Delta S_i^\circ, \tag{O.43}$$

where ΔA° is approximately constant independent of i, and is small relative to ΔE_i° and $T\Delta S_i^\circ$. If ΔE_i^{st} and ΔS_i^{st} dominate the corresponding ΔE_i° and ΔS_i°, then we expect that a plot of ΔH_i° versus ΔS_i° for different i, will result in a linear curve with an approximate slope T.

Probability of Finding a Specific Configuration of a Protein and the Work Required to Obtain that Configuration

Consider a system of N solvent molecules and M solute molecules. To distinguish between a solute and a solvent molecule, we assume that the solute molecules are simple and spherical, hence the configuration of the M solutes is given by $\mathbf{R}^M = (\mathbf{R}_1, \ldots, \mathbf{R}_M)$. The configuration of the N solvent molecules is given by $\mathbf{X}^N = (\mathbf{X}_1, \ldots, \mathbf{X}_N)$, where $\mathbf{X}_i$ includes both locational and orientational components of the configuration of the ith solvent molecule.

A fundamental principle of statistical mechanics states that the probability density of finding any configuration of the entire system $(\mathbf{R}^M, \mathbf{X}^N)$ is given by

$$\Pr(\mathbf{R}^M, \mathbf{X}^N)$$
$$= \frac{\exp[-\beta U(\mathbf{R}^M) - \beta U(\mathbf{X}^M) - \beta U(\mathbf{X}^N|\mathbf{R}^M)]}{\int \cdots \int d\mathbf{R}^M d\mathbf{X}^N \exp[-\beta U(\mathbf{R}^M) - \beta U(\mathbf{X}^M) - \beta U(\mathbf{X}^N|\mathbf{R}^M)]}.$$
$$(\text{P.1})$$

This is simply the Boltzmann distribution law. The integrations in the denominator are over all possible configurations of all the molecules in the system. $U(\mathbf{R}^M)$ is the total interaction energy among the solute molecules, $U(\mathbf{X}^N)$ is the total interaction energy among the solvent molecules, and $U(\mathbf{X}^N|\mathbf{R}^M)$ is the total interaction energy between solute and solvent molecules. A schematic description of these three terms is shown in Fig. 3.14.

By integrating (P.1) over all the configurations of the solvent molecules, we get the probability density of finding the configuration $\mathbf{R}^M$ of the solute molecules, i.e.

$$\Pr(\mathbf{R}^M)$$

$$= \int \cdots \int \Pr(\mathbf{R}^M, \mathbf{X}^N) d\mathbf{X}^N$$

$$= \frac{\int \cdots \int d\mathbf{X}^N \exp[-\beta U(\mathbf{R}^M) - \beta U(\mathbf{X}^M) - \beta U(\mathbf{X}^N|\mathbf{R}^M)]}{\int \cdots \int d\mathbf{R}^M d\mathbf{X}^N \exp[-\beta U(\mathbf{R}^M) - \beta U(\mathbf{X}^M) - \beta U(\mathbf{X}^N|\mathbf{R}^M)]}.$$

$$(\text{P.2})$$

Next, we choose some reference configuration for the solvent molecules, say, all solute molecules being far apart from each other. We denote this configuration by $(\mathbf{R}^M = \infty)$. The corresponding probability density is

$$\Pr(\mathbf{R}^M = \infty)$$

$$= \frac{\int \cdots \int d\mathbf{X}^N \exp[-\beta U(\mathbf{X}^W) - \beta U(\mathbf{X}^N|\mathbf{R}^M = \infty)]}{\int \cdots \int d\mathbf{R}^M d\mathbf{X}^N \exp[-\beta U(\mathbf{X}^N) - \beta U(\mathbf{R}^M) - \beta U(\mathbf{X}^N|\mathbf{R}^M)]}.$$

$$(\text{P.3})$$

Taking the ratio of the probability densities in (P.2) and (P.3), we have

$$\frac{\Pr(\mathbf{R}^M)}{\Pr(\mathbf{R}^M = \infty)}$$

$$= \frac{\int \cdots \int d\mathbf{X}^N \exp[-\beta U(\mathbf{R}^M) - \beta U(\mathbf{X}^N) - \beta U(\mathbf{X}^N|\mathbf{R}^M)]}{\int \cdots \int d\mathbf{X}^N \exp[-\beta U(\mathbf{X}^N) - \beta U(\mathbf{X}^N|\mathbf{R}^M = \infty)]}$$

$$= \exp[-\beta(A(\mathbf{R}^M) - A(\mathbf{R}^M = \infty))]$$

$$= \exp[-\beta \Delta A(\mathbf{R}^M = \infty \to \mathbf{R}^M)]. \qquad (\text{P.4})$$

On the left-hand side of (P.4), we have essentially a correlation function. On the right-hand side of (P.4), we have the work (at T, V, N constants) for the process of bringing the M solutes from fixed positions but at infinite separation to the final configuration $\mathbf{R}^M$. A similar relationship holds for a system at constant T, P, N, i.e.

$$\frac{\Pr(\mathbf{R}^M)}{\Pr(\mathbf{R}^M = \infty)} = \exp[-\beta \Delta G(\mathbf{R}^M = \infty \to \mathbf{R}^M)]. \qquad (P.5)$$

The Many Faces of Reversibility and Irreversibility

The process of denaturation–renaturation of protein is said to be reversible. Such statements appear in almost any discussion of the problem of protein folding. Surprisingly, I could not find a single article in which the meaning of the reversibility of protein folding is discussed.

The term *reversible process* is used quite often in the textbooks on thermodynamics. Even in the context of thermodynamics, there are at least four senses assigned to the term "reversible process." Here are the four senses that I can identify as distinct:

R1: The mechanical sense.

R2: A thermodynamic process for which the entropy change is zero.

R3: A thermodynamic process $A \to B$ that proceeds along a dense sequence of equilibrium states.

R4: A thermodynamic process $A \to B$ that can be reversed $B \to A$.

Before we discuss the four definitions of the term "reversibility" in thermodynamics, it is instructive to see that this term could have several different meanings even when used colloquially.

Suppose a person walks downhill from point A to point B. After that, we are told that the same person went back to point A. One might be wondering how this process was *reversed*.

Did the person walk backwards, as would have been seen by rewinding the movie showing the person going forwards and backwards? This reversal of the motion would be fun to watch, but not a realistic process.

Another reversal process that one can imagine, but is still almost unrealistic, is that the person went back uphill in such a way that after the completion of the cycle A $\rightarrow$ B $\rightarrow$ A, everything in the universe had returned to the initial state. Clearly, this is not a realistic process, but still an imaginable one.

Another reversal, now more realistic, is that the person simply went back from A to B along the same path. Although the same person retraced the same path going downwards and upwards, some changes in both the person and in the hill had necessarily occurred.

The fourth possibility is the simplest: The person returned to A, not necessarily along the same path, and of course changes must have occurred in both the person and in the entire universe.

Bearing these examples in mind, let us move on to discuss the various meanings of "reversibility" as used in thermodynamics.

(i) The sense R1 is used in mechanics in connection with the reversibility of the equations of motion. A process is said to be reversible if by inversion of the velocities of all particles, the process is reversed. This sense of the term is not usually used in thermodynamics. It is sometimes used in statistical mechanics in connection with the apparent conflict between the reversibility of the equations of motion of the atoms and molecules, and the irreversibility [in the sense discussed in (iv) below] of the spontaneous thermodynamic process.

(ii) The sense R2 is sometimes used in connection with the second law of thermodynamics. It states that in any spontaneous

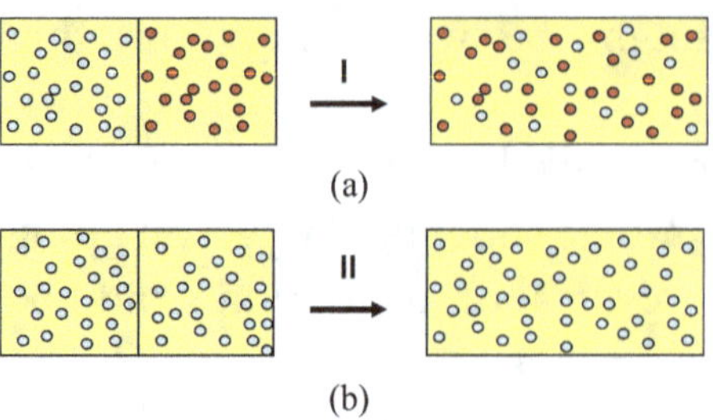

Fig. Q.1   (a) Mixing of two different gases; (b) "mixing" of the same gas.

process occurring in an isolated system, the entropy can never decrease.[348] A process for which the entropy increases is called an irreversible process. A process along a constant-entropy path is referred to as a reversible process. This nomenclature is used by Callen (1985) to distinguish between reversible and quasi-static processes. For instance, removing a partition separating two different gases will result in a spontaneous mixing, and the entropy increases (Fig. Q.1a). This specific process is deemed to be irreversible,[349] not because it cannot be *reversed*, but because it cannot be brought back to the initial state without any change in the entropy of the entire universe (the latter is vague, but it means that the entropy of the system and the surroundings must increase). Strictly, reversible processes in the sense R.2 rarely exist. Examples are reversible mixing of ideal gases,[350] or change of the shape of a *macroscopic* system, say from a sphere to a cube or from a cube to a sphere (Fig. Q.2). The latter is reversible in the R2 sense provided one can neglect surface effects.

(iii) The sense R3 is most commonly used in thermodynamics. It is sometimes confused with the sense R2. The difference

---

[348]For equivalent formulation of the second law, see Appendix N.

[349]Note that there exist mixing processes that do not involve positive change in entropy. See Ben-Naim (2008).

[350]See example in Ben-Naim (2008).

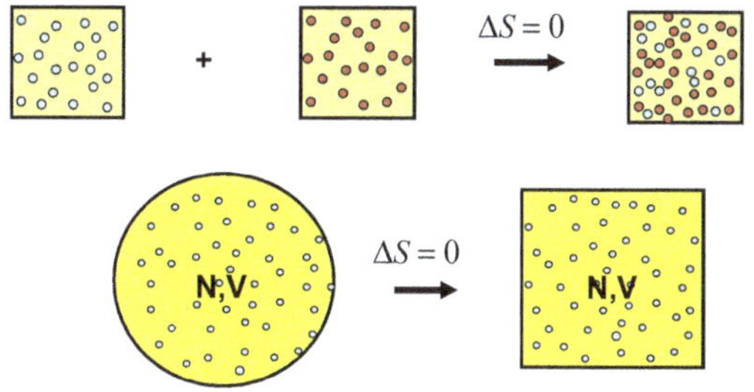

Fig. Q.2   Reversible processes in the sense R2.

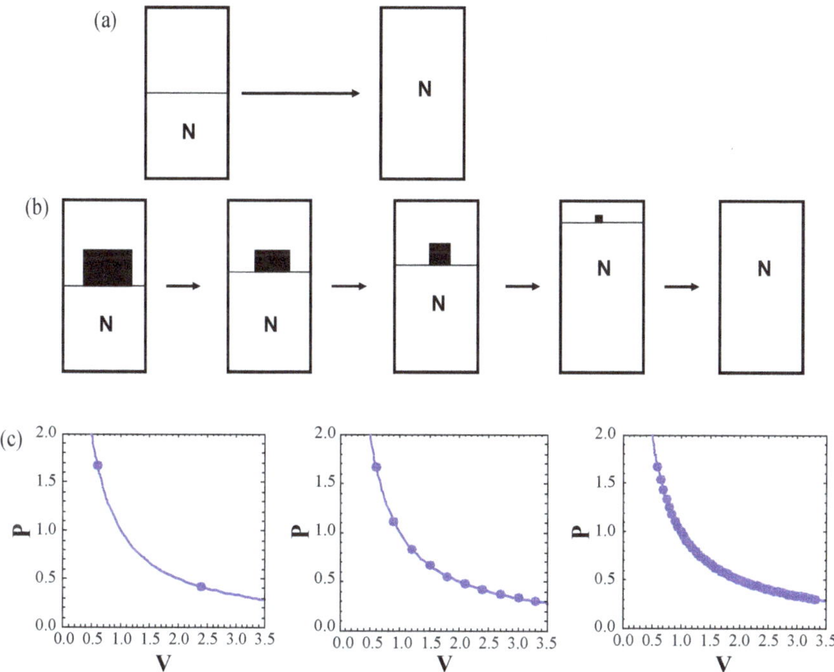

Fig. Q.3   (a) Spontaneous expansion; (b) expansion by small steps; (c) the PV diagram of the expansion process in different number of steps.

between the two can be clarified by the following simple example.

Consider a spontaneous process of expansion depicted in Fig. Q.3. Initially, all the N molecules are confined to one compartment. Removal of the partition causes a spontaneous process of expansion. This is a typical irreversible process in the sense discussed in (ii). The initial and the final states of the system may be described by two points in the PV diagram (Fig. Q.3c).

Clearly, the process is not reversible in the sense R2. But it is also not reversible in the sense R3, since we cannot trace back the path of the process in the PV diagram. This is so simply because the thermodynamic states of the system *during* the expansion process are not well defined, and it is therefore meaningless to talk about reversal along the thermodynamic path from the initial to the final state.

Suppose next that we change the weight on the piston gradually, each time reducing the weight by ΔM, waiting for the system to reach an equilibrium state, then reducing the weight again by ΔM, and so on. In this process, we can draw all the equilibrium points in the PV diagram (Fig. Q.3b). We can say that we know, say, ten points on the path from the initial to the final states, but we do not know the exact path between any two equilibrium points.

Next, we can imagine that at each step, we remove only an infinitesimal weight dM on the piston. In the limit, when we perform this process in infinitesimal steps, we can draw an almost continuous curve in the PV diagram (Fig. Q.3c). Clearly, in this limiting process, there is a *path* in the PV diagram leading from the initial to the final states. Therefore, it is meaningful to speak about the thermodynamic path from the initial to the final state, as well as about the reversed path from the final to the initial state.

To distinguish between this reversible process from the reversible process in the sense R2, the former is sometimes referred to as a quasi-static process.[351] A quasi-static process

---

[351]See Callen (1985).

is simply a process which is carried out in very small steps so that the system effectively goes through an almost continuous series of equilibrium states. Because each equilibrium state is well defined thermodynamically, it follows that the path of the process is well defined. It is therefore meaningful to talk about reversing the path, or reversing the process.

It should be stressed that a quasi-static process, i.e. a reversible process in the sense R3, does not imply that the process is reversible in the sense R2. It is therefore advisable to use two different terms to distinguish between the two.

Note also that reversible in the sense R3 is not reversible in the sense R1. The thermodynamic path is reversed, but the molecular trajectories are not.

It should be noted that in textbooks on thermodynamics, one writes

$$(dS)_{\text{rev}} = \frac{dQ}{T}, \qquad (Q.1)$$

where dQ is a small amount of heat flow and T is the absolute temperature. Clearly, the "rev" in the notation (Q.1) is in the sense R3, i.e. the equality is valid for a quasi-static flow of heat. It is not in the sense R2, since the process of flow of heat is spontaneous, and the change in entropy is positive even when the process is performed quasi-statistically.

(iv) The weakest form of reversibility is when a process $A \rightarrow B$ can be reversed $B \rightarrow A$. No requirement that the reversal will be along the same thermodynamic path (sense R3), or that the entropy will not change (sense R2). It is the weakest since it is difficult to find a thermodynamic process that cannot be reversed in this sense. We exclude from this discussion processes of life and death which, at least at the present level of our knowledge, seem to be completely irreversible in all the senses mentioned above. An example which is given quite often is boiling an egg. Such a process cannot be reversed (un-boil the egg?). However, in thermodynamics, we are discussing processes from one well-defined *equilibrium state* to another *equilibrium* state. It is far from clear whether an egg is in an equilibrium state either before or after the cooking. And besides, perhaps one day we could devise a process that could "un-boil" the egg.

Any chemical reaction that proceeds from one equilibrium state to another equilibrium state is, in principle, reversible in the R4 sense. Protein folding is one example of a chemical process for which reversibility in the R4 sense is applied. If that is the case, then perhaps it is superfluous to attach the adjective "reversible" to the protein folding process. Some proteins seem to denature "irreversibly." However, it is far from clear whether these proteins *cannot* be renatured; perhaps we have not been able to find the right conditions under which the process can be reversed.

Finally, it is instructive to mention a stronger "irreversible" process that is "more absolute" than the irreversibility of the mixing process in Fig. Q.1a. Gibbs, who pioneered the study of the process of mixing (Fig. Q.1a), compared that process with the mixing of "two gas masses of the same kind" (Fig. Q.1b). Gibbs concludes that process I is irreversible (in the sense R2, i.e. $\Delta S > 0$), but can be reversed (in the sense of R4). However, the mixing in process II cannot be reversed — that is "entirely impossible." This conclusion, and a similar one reached by Maxwell, are obviously correct. There is no way we can separate the two gases that are mixed in process Q.1b. Nevertheless, reversal of process Q.1b is not only *possible*, but it is also trivial, and can be done effortlessly. For further discussion of this topic see Ben-Naim (2008).

APPENDIX R

Cooperativity in Protein Folding?

The term cooperativity is used in many senses, both inside and outside the sciences. In most cases, when the term is utilized in the biochemical literature, one means one particle (or one system) *cooperates* with another to reach some result. I shall first present two examples of the usage of this term, and then examine its adequacy when applied to the process of protein folding.

Perhaps, the most common (and meaningful) usage of the term cooperativity is in the process of binding of ligands to sites on an adsorbent molecule. Suppose that each adsorbent molecule has two binding sites. If the binding of a ligand on one site is independent of the state of occupation of the second site, we get the typical Langmuir isotherm (Fig. R.1a). The analytical form of this binding curve is

$$\theta = \frac{N}{M} = \frac{K\rho}{1 + K\rho},\tag{R.1}$$

where ρ is the density of the ligand (in the gaseous or the liquid phase), θ is the fraction of occupied sites, and K is the binding constant.

If on the other hand, binding on one site affects the probability (as well as the Gibbs energy) of binding to a second site, we say that the process of binding is cooperative.[352] The simplest case of cooperative binding occurs when the two ligands occupying the two sites interact directly.

---

[352]Ben-Naim (2001).

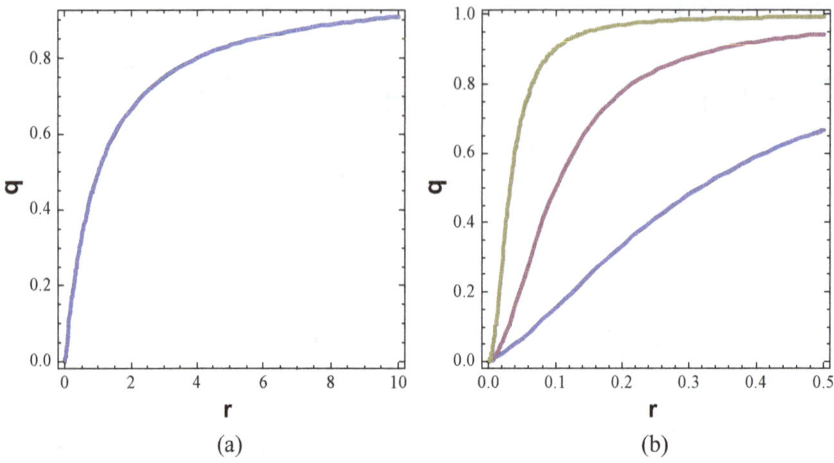

Fig. R.1   Lanmuir Isotherm.

In this case, the binding isotherm has the form

$$\theta = \frac{K_1\rho + K_1^2 g\rho^2}{1 + 2K_1\rho + K_1^2 g\rho^2},$$

(R.2)

where g is a correlation function.[353] For attractive interactions, $g > 1$, and for repulsive interactions, $g < 1$. We say that there is positive and negative cooperativity in these two cases, respectively. For the case of positive cooperativity (i.e. binding on one site enhances the probability of binding to a second site), the binding curve becomes an S shape. This change of the shape of the curve is more pronounced when there are more than two sites with positive cooperative binding, e.g. the binding of oxygen to hemoglobin. Thus, a change in the curve from Fig. R.1a to Fig. R.2b is a result of the *dependence*, or cooperation between ligands occupying two sites. In qualitative terms, one ligand absorbed on a site "helps" another ligand, enhancing its probability of binding to other sites.

---

[353] See Ben-Naim (2001).

Another rather different meaning of cooperativity is applied to hydrogen bonding. There is evidence that the binding energy of a third water molecule to a pair of HBed molecules is stronger than the HB energy between two water molecules.[354] Here, the term cooperativity has a different meaning: A water molecule approaching another water molecule can "sense" whether the second water molecule is already hydrogen-bonded to a third molecule. In a sense, each water molecule "cooperates" with the two other molecules to determine the strength of the newly formed HB.

The term cooperativity as used in connection with the protein folding process seems to have quite a different meaning. Whenever a sharp denaturation curve occurs, having a sigmoidal shape, the process is deemed to be highly cooperative. It seems to me that this term is used not because the mechanism of folding/unfolding is cooperative, but simply because of the shape of the denaturation curve, which is reminiscent of a cooperative phenomenon.

Of course, one can think of a folding process such that the various regions or side chains of the protein "cooperate" with each other leading to the folded form. However, this kind of cooperation does not necessarily follow from the shape of the denaturation curve. As we have seen in Chapter 3, a sharp transition between two isomers can occur without any type of cooperation. Therefore, the very fact that the transition is sharp (sometimes said to be an all-or-none transition) does not imply a cooperative process.

---

[354]See, for example, Ben-Naim (2009).

Local Densities of Water Molecules near *HϕI* Groups

In this appendix, we present some very rough estimates on the conditional densities of water molecules at some specific configuration $\mathbf{X}_w$ given one or more *HϕI* groups in the neighborhood of $\mathbf{X}_w$.

Let $g(\mathbf{R}_2, \mathbf{R}_1)$ be the (orientational average) pair correlation function of water. The local density of water molecules at $\mathbf{R}_2$ given a water molecule at $\mathbf{R}_1$ is given by

$$\rho(\mathbf{R}_2/\mathbf{R}_1) = \rho_w g(\mathbf{R}_1, \mathbf{R}_2) = \rho_w g(R), \qquad (S.1)$$

where $R = |\mathbf{R}_2, \mathbf{R}_1|$ is the distance between the two locations $\mathbf{R}_1$ and $\mathbf{R}_2$, and ρ_w is the number density of the bulk of water molecules at room temperature and one atmospheric pressure

$$\rho_w \approx 5.55 \times 10^{-2} \text{ mol/cm}^3 \qquad (S.2)$$

Experimentally, the value of $g(R)$ for water at $R \approx 2.8\,\text{Å}$ is about $g(R \approx 2.8\,\text{Å}) \approx 2$. This is also the value of the pair correlation function for water molecules near a *HϕO* molecule such as methane, at a distance of about $(\sigma_w + \sigma_M)/2$, where σ_w and σ_M are the diameters of water and methane, respectively.

The density ρ_w is the bulk density of water molecules at any point $\mathbf{R}$ in the liquid. Let $\rho^{(1)}(\mathbf{X})$ be the density of water molecules at a specific configuration $\mathbf{X}$, i.e. a specific location $\mathbf{R}$

and orientation $\boldsymbol{\Omega}$. The normalization for $\rho^{(1)}(\mathbf{X})$ is

$$\int \rho^{(1)}(\mathbf{X})d\mathbf{X} = N, \tag{S.3}$$

where the integration is carried out over all the configurations (i.e. locations and orientations) of the water molecule.

In a homogenous and isotropic fluid, we expect that

$$\rho^{(1)}(\mathbf{X}) = C, \tag{S.4}$$

i.e. the density at $\mathbf{X}$ is a constant C independent of the specific location $\mathbf{R}$ and the specific orientation $\boldsymbol{\Omega}$ (of course, the density can still be dependent on the temperature and the pressure). Therefore, integrating over all possible locations and orientations, we get

$$\int \rho^{(1)}(\mathbf{X})d\mathbf{X} = CV(8\pi^2) = N, \tag{S.5}$$

where V is the total volume of the system (obtained by integrating over all locations of the molecule), and $8\pi^2$ is obtained by integration over all orientations of a water molecule.[355]

$$\int d\mathbf{R} = V, \tag{S.6}$$

$$\int d\boldsymbol{\Omega} = \int_0^{2\pi} d\phi \int_0^{\pi} \sin\theta d\theta \int_0^{2\pi} d\psi = 8\pi^2. \tag{S.7}$$

Therefore, the relationship between the density $\rho^{(1)}(\mathbf{X})$ and the bulk density ρ_w is

$$\rho^{(1)}(\mathbf{X}) = C = \frac{N}{V(8\pi^2)} = \frac{\rho_w}{8\pi^2}. \tag{S.8}$$

Note that $\rho^{(1)}(\mathbf{X})$ is independent of $(\mathbf{X})$, but we retain this notation to distinguish this density from the bulk density.

---

[355]For more details, see Ben-Naim (1992), Chapter 1.

The probability of finding a water molecule in a small "volume" of configurations $d\mathbf{X}$ about $\mathbf{X}$ is

$$\Pr(\mathbf{X})d\mathbf{X} = \rho^{(1)}(\mathbf{X})d\mathbf{X} = \frac{\rho_w}{8\pi^2}d\mathbf{X}. \qquad (S.9)$$

We next turn to estimate the local density of water at $\mathbf{X}_w$ given one or more molecules or a group at some specific configurations.

(i) Local density at $\mathbf{X}_w$ given a $H\phi I$ group at $\mathbf{X}_1$.

The conditional local density of water molecules at $\mathbf{X}_w$ given a water molecule or a $H\phi I$ group at $\mathbf{X}_1$ is

$$\rho(\mathbf{X}_w/\mathbf{X}_1) = \frac{\rho_w}{8\pi^2}g(\mathbf{X}_w,\mathbf{X}_1)\exp[-\beta W(\mathbf{X}_w,\mathbf{X}_1)], \qquad (S.10)$$

where $W(\mathbf{X}_w,\mathbf{X}_1)$ is the PMF at $(\mathbf{X}_w,\mathbf{X}_1)$. A rough estimate of the PMF at a configuration $(\mathbf{X}_w,\mathbf{X}_1)$ such that one arm of the water can form a HB with the arm of the group at $\mathbf{X}_1$ (Fig. S.1a) is

$$W(\mathbf{X}_w,\mathbf{X}_1) = U(\mathbf{X}_w,\mathbf{X}_1) + \delta G(\mathbf{X}_w,\mathbf{X}_1)$$

$$= (\varepsilon_{HB} + \varepsilon_{LJ}) + (\delta G^{LJ} + \delta G^{HB/LJ}), \qquad (S.11)$$

where the direct interaction energy consists of two contributions, a Lennard-Jones (LJ) energy of interaction and a HB

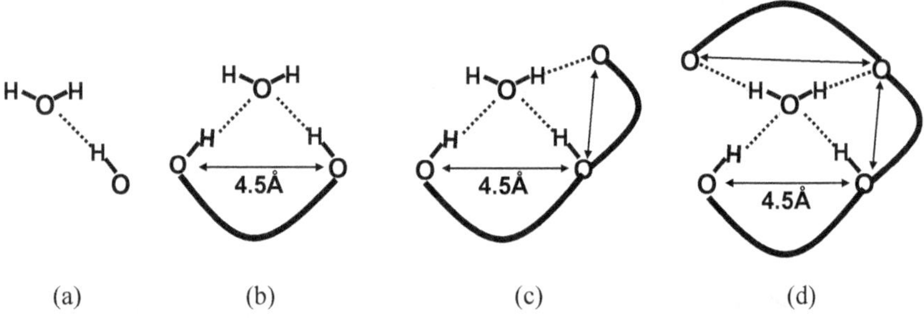

Fig. S.1 The configurations of the $H\phi I$ groups required to produce high local density of water molecules. The double arrow represents a distance of 4.5 Å.

energy. The quantity $\varepsilon_{LJ} + \delta G^{LJ}$ is the PMF for two LJ particles in water. This quantity was computed by Ravishanker *et al.*[356] based on simulation of aqueous solutions of LJ particles. They got a value of about $-3.39\,\text{kJ}\,\text{mol}^{-1}$. The quantity $\delta G^{HB/LJ}$ is the solvent-induced part of the work required to bring the two arms from infinite separation to the final configuration $\mathbf{X}_w, \mathbf{X}_1$. This quantity was taken as twice the loss of the conditional solvation Gibbs energy of one arm, i.e. about $9.4\,\text{kJ}\,\text{mol}^{-1}$.

Thus, the estimate of $W(\mathbf{X}_w, \mathbf{X}_1)$ is

$$W(\mathbf{X}_w, \mathbf{X}_1) \approx -27.2 - 3.35 + 2 \times 9.4$$

$$= -11.75\,\text{kJ}\,\text{mol}^{-1}. \qquad (\text{S}.12)$$

Using this value in (S.10), we get an estimate of the conditional density

$$\rho(\mathbf{X}_w/\mathbf{X}_1) \approx 114 \frac{\rho_w}{8\pi^2}. \qquad (\text{S}.13)$$

Thus, the local density at $\mathbf{X}_w$ given an arm at $\mathbf{X}_1$ (at the right configuration to form a HB) is about 110 times larger than the bulk density of water molecules at $\mathbf{X}_w$.

(ii) Local density at $\mathbf{X}_w$ given two *HφI* groups at $\mathbf{X}_1, \mathbf{X}_2$

The quantity required is the conditional density of water molecules at $\mathbf{X}_w$ given two *HφI* groups at such configuration $\mathbf{X}_1, \mathbf{X}_2$ that the water molecule can form two HBs with the two *HφI* groups (Fig. S.1b). We write this as

$$\rho(\mathbf{X}_w/\mathbf{X}_1, \mathbf{X}_2) = \frac{\rho_w}{8\pi^2} g(\mathbf{X}_w | \mathbf{X}_1, \mathbf{X}_2)$$

$$= \frac{\rho_w}{8\pi^2} \exp[-\beta W(\mathbf{X}_w) | \mathbf{X}_1, \mathbf{X}_2]. \qquad (\text{S}.14)$$

Here, $W(\mathbf{X}_w | \mathbf{X}_1, \mathbf{X}_2)$ is the work required to bring a water molecule from infinity to the final configuration $\mathbf{X}_w$ given the

---

[356]Ravishanker *et al.* (1982).

configuration of the two $H\phi I$ groups at $\mathbf{X}_1, \mathbf{X}_2$. We estimate this quantity as

$$W(\mathbf{X}_w|\mathbf{X}_1, \mathbf{X}_2) \approx 2\varepsilon_{HB} + 2\varepsilon_{LJ} + \delta G^{LJ} + \delta G^{HB/LJ}$$

$$\approx -2 \times 27.2 - 2 \times 3.35 + 2 \times 9.4 + 29.2$$

$$= -12.5 \, \text{kJ} \, \text{mol}^{-1}. \tag{S.15}$$

Thus, the first term is twice the HB energy, the second term is twice the estimate of $\varepsilon_{LJ} + \delta G^{LJ}$ as we have done in (S.12), the third term is due to the loss of the solvation Gibbs energy of the two arms of the water molecule, and the last term on the right-hand side of (S.15) is due to the loss of solvation of the two arms of the $H\phi I$ group which are correlated (this is the same as the loss of the solvation of the two arms 2×9.4, and the loss or the correlation between the two arms which we take as $10.5 \, \text{kJ} \, \text{mol}^{-1}$). Thus, the local density in (S.14) is estimated to be

$$\rho(\mathbf{X}_w/\mathbf{X}_1, \mathbf{X}_2) \approx 155 \frac{\rho_w}{8\pi^2}. \tag{S.16}$$

(iii) Local density at $\mathbf{X}_w$ given three $H\phi I$ groups at $\mathbf{X}_1, \mathbf{X}_2, \mathbf{X}_3$

Again, we assume that the configuration of the three $H\phi I$ groups is such that they form three HBs with one water molecule (Fig. S.1c). The local density is

$$\rho(\mathbf{X}_w/\mathbf{X}_1, \mathbf{X}_2, \mathbf{X}_3) = \frac{\rho_w}{8\pi^2} \exp[-\beta W(\mathbf{X}_w|\mathbf{X}_1, \mathbf{X}_2, \mathbf{X}_3)]. \tag{S.17}$$

In this case, we do not know the value of the $H\phi I$ interaction between three $H\phi I$ groups at the right configuration to form three HBs with a water molecule. We estimate that this is at least twice as large as in the case of two $H\phi I$ groups, i.e. $2 \times 10.5\,\text{kJ}\,\text{mol}^{-1}$.[357] Therefore, we estimate

$$W(\mathbf{X}_w|\mathbf{X}_1,\mathbf{X}_2,\mathbf{X}_3)$$
$$\approx -3 \times 27.2 - 3 \times 3.35 + 6 \times 9.4 + 2 \times 10.5$$
$$\approx -14.25\,\text{kJ}\,\text{mol}^{-1}, \qquad\qquad\qquad (\text{S}.18)$$

and the corresponding local density

$$\rho(\mathbf{X}_w/\mathbf{X}_1,\mathbf{X}_2,\mathbf{X}_3) = 314\frac{\rho_w}{8\pi^2}. \qquad\qquad (\text{S}.19)$$

(iv) Finally, we estimate the local density of $\mathbf{X}_w$ given four $H\phi I$ groups at the vertices of a regular tetrahedron of edge length 4.5 Å such that the water molecule at $\mathbf{X}_w$ can form four HBs with the $H\phi I$ groups (Fig. S.1d).

$$\rho(\mathbf{X}_w/\mathbf{X}_1,\mathbf{X}_2,\mathbf{X}_3,\mathbf{X}_4) = \frac{\rho_w}{8\pi^2}\exp[-\beta W(\mathbf{X}_w|\mathbf{X}_1,\mathbf{X}_2,\mathbf{X}_3,\mathbf{X}_4)],$$
$$(\text{S}.20)$$

and the corresponding work of bringing a water molecule to $\mathbf{X}_w$ given the configuration $\mathbf{X}_1,\mathbf{X}_2,\mathbf{X}_3,\mathbf{X}_4$ is estimated

$$W(\mathbf{X}_w|\mathbf{X}_1,\mathbf{X}_2,\mathbf{X}_3,\mathbf{X}_4)$$
$$\approx -4 \times 27.2 - 4 \times 3.35 + 8 \times 9.4 + 3 \times 10.5$$
$$\approx -15.5\,\text{kJ}\,\text{mol}^{-1}. \qquad\qquad\qquad (\text{S}.21)$$

Note again that we form four HBs, and we lose solvation of eight arms. The rationale for the choice of the last term on the right-hand side of (S.21) is that bringing two $H\phi I$ groups involves $-10.5\,\text{kJ}\,\text{mol}^{-1}$, and bringing a third and fourth $H\phi I$

---

[357] The qualitative rationale for this choice is that when we bring two $H\phi I$ groups together, the $H\phi I$ interaction is about $-10.5\,\text{kJ}\,\text{mol}^{-1}$. When a third $H\phi I$ group is brought to the position where it forms an equilateral triangle of edge length 4.5 Å, the additional $H\phi I$ interaction is of the same order of magnitude as $-10.5\,\text{kJ}\,\text{mol}^{-1}$.

group also involves $H\phi I$ interaction of similar order of magnitude. Therefore, the loss of $H\phi I$ interaction is estimated to be about $-3 \times 10.5\,\mathrm{kJ\,mol^{-1}}$.

The local density at $\mathbf{X}_w$ is now estimated to be

$$\rho(\mathbf{X}_w/\mathbf{X}_1, \mathbf{X}_2, \mathbf{X}_3, \mathbf{X}_4) \approx 520\frac{\rho_w}{8\pi^2}. \qquad (\text{S.22})$$

All the estimates carried out above are very qualitative. However, we see, as expected, that there is a dramatic increase in the local density near $H\phi I$ groups, and the larger the number of $H\phi I$ groups, the higher the local densities.

APPENDIX T

What Drives the "Driving Force?"

We have already discussed in Chapter 1 the distinction between the *force* and the *thermodynamic driving force*. We have seen that what drives the motion of the protein towards its equilibrium distribution is the *force* acting on each of the groups or atoms of the protein. We have also seen that the *driving force* defined either by $\Delta S > 0$ or $\Delta G < 0$ (depending on the thermodynamic variable used to characterize the system) certainly does not *drive* the molecular process such as protein folding.

Yet, the "driving force" holds a central place in the literature dealing with protein folding. Why? To clarify this problem, let us consider the following simple experiment.

An ideal gas is initially confined to a volume V. We remove a partition separating the two compartments as in Fig. T.1, and we observe a spontaneous expansion of the gas. At the new equilibrium state, the gas occupies the entire new volume $2V$. We can calculate the change in entropy associated with this process, either by thermodynamics or by the methods of statistical

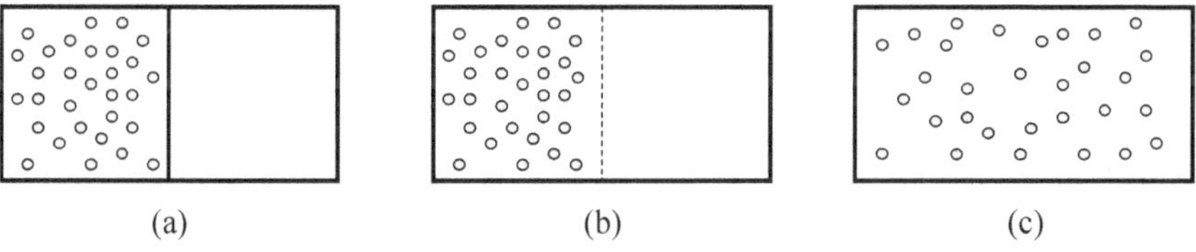

(a)   (b)   (c)

Fig. T.1 A spontaneous expansion of an ideal gas.

mechanics, and we find that

$$\Delta S = k_B N \ln \frac{2V}{V} = k_B N \ln 2 > 0. \qquad (T.1)$$

Here is a potential pitfall: We know that in any spontaneous process occurring in an isolated system, the change in entropy of the system is *positive*. The "reverse" of this statement is obviously not true. Knowledge of the positive change in entropy does not *drive* the process. In our example depicted in Fig. T.1, we can calculate the change in entropy between the initial and final equilibrium states and find the value of ΔS, as in Eq. (T.1). However, the process of expansion will not occur unless we remove the partition between the two compartments. Therefore, we cannot claim that $\Delta S > 0$ *drives* the spontaneous process of expansion. Similar experiments lead to the same conclusion. For instance, we can prepare a mixture of hydrogen and oxygen. If we add a catalyst, a spontaneous reaction will occur, and we can calculate the entropy change to be positive. However, as long as the catalyst is absent, we have an equilibrium mixture of two non-interacting gases. We can calculate the change of the entropy between the two equilibrium states before and after adding the catalyst and find $\Delta S > 0$. But this positive entropy change does not *drive* the process of chemical reaction. Notwithstanding the conclusion stated above, why do we still find statements associating the positive change in entropy with the driving force for that process? Where is the pitfall?

Let us examine again the simple process of expansion. The second law of thermodynamics states that in an isolated system (i.e. a system characterized by a fixed energy, volume and number of particles), when a constraint is removed, e.g. removing the partition or adding a catalyst, the system's entropy will always increase. The second law does not state that the entropy change drives the spontaneous process. The change in entropy is a *result* of the occurrence of the spontaneous process, not the *cause* of the process. If we accept the last statement, then what *is the cause* of the spontaneous process? The answer to this question is: Probability.[358]

Before we discuss the role of probability in the specific process of expansion, we need to remove another pitfall which is hidden in the concept of probability.

The probability of an event A is always a *conditional probability*.[359] If Ω is the sample space, i.e. the set of all possible outcomes of an experiment, the probability of Ω is one. This means that we perform an experiment, and one of the outcomes has occurred, hence $P(\Omega) = 1$. Therefore, when we write $P(A)$, we actually mean $P(A/\Omega)$, i.e. the conditional probability of an event A, *given the occurrence* of Ω. We usually suppress Ω in our notation and simply write $P(A)$.

Back to our experiment in Fig. T.1. In the initial equilibrium state (a), the probability of finding all the particles in V is *one*. Similarly, in the final state (c), the probability of finding all the particles in $2V$ is also *one*! Clearly, as long as the partition is in place, we cannot say that the system will evolve from a lower probability state to a higher probability state. This is so simply because we did not perform the experiment. Thus, before we remove the partition, ΔS is still positive, but the conditional probability to go from (a) to (c) is zero.

However, at the very moment we *remove* the partition between the two compartments (Fig. T.1b), we *can* talk of the probability of finding all the particles in any region of the system.

---

[358]Craig (2005) suggested that we *drive* the "driving force" out of our vocabulary in chemistry. His reasoning is that the term "driving force" suggests that Newtonian mechanics *drives* the process.

[359]Lindley (1965).

Assuming that the particles are independent in the sense that the probability of each particle being either in the right or in the left compartment does not depend on the state of all the other particles, then the probability of a single particle to be found in V is $1/2$, and the probability of finding *all* the particles in V is $(1/2)^N$. On the other hand, the probability of finding all the particles in the entire volume $2V$ is *one*. Thus, the ratio of the two probabilities is

$$\frac{\text{Pr (of all particles in } 2V)}{\text{Pr (of all particles in } V)} = \frac{1}{(1/2)^N} = 2^N. \qquad \text{(T.2)}$$

This is a very large number for N of the order of Avogadro's number.

If we now adopt the relative frequency interpretation of probability,[360] we can say that if we do the experiment, i.e. we remove the partition between the two compartments and follow the evolution of the states of the system (how many particles in each compartment), we shall find that the relative frequency of occurrence of the final state (each particle is either in the left or in the right compartment) is far larger than the relative frequency of finding all the particles in the initial state. Within this relative-frequency-of-occurrence meaning of probability, we can claim that the probability is the "driving force" for the expansion. This simply means that states that are expected to occur more frequently *will* occur more frequently. A more detailed account of the same expansion process carried out in quasi-static steps is described in Ben-Naim (2008).

Note carefully that after removing the partition, it is meaningful to ask about the *probability* of each particle being in one compartment, or in either one of the two compartments. On the other hand, the entropy of the system is defined only for

---

[360]See, for example, Papoulis and Pillai (2002).

an equilibrium state. Therefore, it is meaningless to talk about the entropy of the initial state right after removing the partition. Instead, one can do the expansion in very small steps such that at each step, the system reaches equilibrium, for which the entropy is definable. If we perform the expansion quasi-statically, say, by opening a very small window on the partition, then the entropy will go through a series of increasing values, and the total change in entropy will be $\Delta S = k_B N \ln 2$.

As we have seen, a positive change in entropy is not sufficient for a spontaneous process to occur. On the other hand, a higher probability for a final state means that the system will visit that state more frequently. If the probability of the final state is almost one, then the system will *always* be in that state.

The conclusion is that the thermodynamic "driving force" as measured by ΔS does not really *drive* the spontaneous process in an isolated system. A similar conclusion can be reached for the Gibbs energy: Negative change in Gibbs energy does not *drive* the process under T and P constants. What drive the process are the relative probabilities of the initial and the final state after a constraint has been removed (e.g. removing a partition or introducing a catalyst).

The aforementioned conclusion holds for any spontaneous process. However, there is a general relationship between the probability of the state of the system and Shannon's measure of information (SMI).[361] For the expansion process, at each stage of the expansion, let x be the mole fraction of the particle in one compartment, then Shannon's measure of information for this particular distribution $(x, 1 - x)$ is

$$H(x, 1 - x) = -N[x \log x + (1 - x) \log (1 - x)], \qquad \text{(T.3)}$$

---

[361] Ben-Naim (2008).

and the probability, denoted Pr, of finding this distribution is related to SMI for large N by[362]

$$\Pr(x, 1 - x) \approx \left(\frac{1}{2}\right)^N \frac{2^{NH(x,1-x)}}{\sqrt{2\pi Npq}}. \tag{T.4}$$

Since for any equilibrium state, the SMI becomes identical with the entropy of the system, Eq. (T.4) provides a relationship between the entropy of a state described by $(x, 1 - x)$, and the probability of occurrence of that state. Thus, though the two concepts are very different, they are connected by Eq. (T.4), valid in the limit of $N \to \infty$. Furthermore, one can show that the probability of finding the system at the equilibrium state is almost one.[363] In the particular case of expansion, the equilibrium state is characterized by $x_{eq} = 1/2$. This state maximized both the SMI and the probability Pr with respect to the variable x. At this state, the value of the SMI is the same as the value of the entropy, except for a multiplicative constant and a choice of the base of the logarithm.

---

[362] Ben-Naim (2008).
[363] Ben-Naim (2008).

References

Amzel LM. (1997) *Proteins* **28**: 144.

Amzel LM. (1997) Loss of translational entropy in binding, folding and catalysis. *Proteins* **29**: 1.

Anderson PW. (1983) *Proc Natl Acad Sci USA* **80**: 3386.

Anderson PW. (1990) *Phys Today* **43**: 9.

Anfinsen CB. (1973) Principles that govern the folding of protein chains. *Science* **181**: 223.

Anfinsen CB, Haber E, Sela M, White FH Jr. (1961) *Science* **181**: 223.

Anfinsen CB, Haber E, Sela M, White FH Jr. (1961) *Proc Natl Acad Sci* **47**: 1309.

Anfinsen CB, Scheraga HA. (1975) *Adv Protein Chem* **29**: 205.

Antonini E, Brunori M. (1971) *Hemoglobin and Myglobin in their Reactions with Ligands*, North-Holland, Amsterdam.

Apelblat A, Manzurola E, Balal NA. (2006) *J Chem Thermodyn* **38**: 565.

Arai S, Chatake T, Ohara T, *et al.* (2005) *Nucleic Acids Res* **33**: 3017.

Arakawa T, Timasheff SN. (1984) *Biochemistry* **23**: 5924.

Arthur JW, Haymet ADJ. (1999) *J Chem Phys* **110**: 5873.

Auton M, Bolen DW. (2004) *Biochemistry* **43**: 1329.

Auton M, Bolen DW. (2005) *Proc Natl Acad Sci USA* **102**: 15065.

Auton M, Bolen DW, Rosgen J. (2008) *Proteins* **73**: 802.

Auton M, Marcelo L, Holthauzen F, Bolen DW. (2007) *Proc Natl Acad Sci USA* **104**: 15317.

Avbelj F, Baldwin RL. (2006) *Proteins* **63**: 282.

Bahar I, Jernigan RL. (1994) *Biophys J* **66**: 454.

Bahar I, Jernigan RL. (1997) *J Mol Biol* **266**: 195.

Baker EN, Hubbard RE. (1984) *Prog Biophys Mol Biol* **44**: 97.

Baker BM, Murphy KP. (1997) *J Mol Biol* **268**: 557–569.

Baldwin RL, Eisenberg D. (1987) Protein stability. In: Oxender DL, Fox CF (eds), *Protein Engineering*. Alan R. Liss, New York, pp. 127.

Baldwin RL. (1994) Matching speed and stability. *Nature* **369**: 183.

Baldwin RL. (1995) The nature of protein folding pathways: The classical versus the new view. *J Biomol NMR* **5**: 103.

Baldwin RL. (2003) *J Biol Chem* **278**: 17581.

Baldwin RL. (2006) *Adv Protein Chem* **72**: ix–xi.

Baldwin RL, Rose GD. (1999) *Trends Biochem Sci* **24**: 77.

Ball P. (2008) *Chemphyschem* **9**: 2677.

Banavar JR, Hoang TX, Maritan A, *et al.* (2004) *Phys Rev E Stat Nonlin Soft Matter Phys* 70: 041905.

Behe MJ, Lattman EE, Rose GD. (1991) *Proc Natl Acad Sci USA* 88: 4195.

Ben-Naim A. (1965) *J Phys Chem* 69: 3245.

Ben-Naim A. (1974) *Water and Aqueous Solutions, Introduction to Molecular Theory.* Plenum Press, New York.

Ben-Naim A. (1978) *J Phys Chem* 82: 792.

Ben-Naim A. (1980) *Hydrophobic Interactions.* Plenum Press, New York.

Ben-Naim A. (1989) *J Chem Phys* 90: 7412.

Ben-Naim A. (1990a) *Biopolymers* 29: 567.

Ben-Naim A. (1990b) *J Phys Chem* 95: 1473.

Ben-Naim A. (1991) *J Chem Phys* 93: 8196.

Ben-Naim A. (1992) *Statistical Thermodynamics for Chemists and Biochemists.* Plenum Press, New York.

Ben-Naim A, (1994) *Biophys Chem* 51: 203.

Ben-Naim A. (1997a) *Pure App Chem* 69: 2239.

Ben-Naim A. (1997b) *J Chem Phys* 107: 3698.

Ben-Naim A. (2001) *Cooperativity and Regulation in Biochemical Processes.* Kluwer Academic/Plenum Publishers, New York.

Ben-Naim A. (2002) *Biophys Chem* 101: 309.

Ben-Naim A. (2006) *Molecular Theory of Solutions.* Oxford University Press, Oxford.

Ben-Naim A. (2008) *A Farewell to Entropy, Statistical Thermodynamics Based on Information.* World Scientific, Singapore.

Ben-Naim A. (2009) *Molecular Theory of Water and Aqueous Solutions, Part I: Understanding Water.* World Scientific, Singapore.

Ben-Naim A. (2011) *Molecular Theory of Water and Aqueous Solutions, Part II.* World Scientific, Singapore.

Ben-Naim A, Mazo R. (1993) *J Phys Chem* 97: 10829.

Ben-Naim A, Mazo R. (1997) *J Phys Chem B* 101: 11221.

Ben-Naim A, Ting KL, Jernigan RL. (1989a) *Biopolymers* 28: 1309.

Ben-Naim A, Ting KL, Jernigan RL. (1989b) *Biopolymers* 28: 1327.

Ben-Naim A, Ting KL, Jernigan RL. (1990) *Biopolymers* 29: 901.

Birshtein TM, Ptitsyn OB. (1966) *Conformations of Macromolecules.* Interscience Publishers, New York.

Bolen DW, Rose GD. (2008) Structure and energetics of the hydrogen-bonded backbone in protein folding. *Annu Rev Biochem* 77: 339.

Brooks CL, Karplus M, Pettit BM. (1988) *Proteins: A Theoretical Perspective of Dynamics, Structure and Thermodynamics.* John Wiley and Sons, New York.

Bruge F, Fomili SL, Palma-Vittorelli MB. (1994) *J Chem Phys* 101: 2407.

Bryngelson JD, Onuchic JN, Socci ND, Wolynes PG. (1995) Funnels, pathways, and the energy landscape of protein folding: A synthesis. *Proteins* 21: 167.

Bryngelson JD, Wolynes PG. (1987) *Proc Natl Acad Sci USA* 84: 7524.

Bryngelson JD, Wolynes PG. (1989) Intermediates and barrier crossing in a random energy model (with applications to protein folding). *J Phys Chem* 93: 6902.

Cabani S, Gianni P, Mollica V, Lepori L. (1981) *J Sol Chem* 10: 563.

Cantor CR, Schimmel PR. (1980) *Biophysical Chemistry*, WH Freeman and Co., San Francisco.

Chalikian TV, Breslauer KJ. (1998) *Biopolymers* **48**: 264.

Chalikian TV, Sarvazyan AP, Plum GE, Breslauer KJ. (1994) *Biochemistry* **33**: 2394.

Chalikian TV, Volker J, Srinivasan AR, Olson WK, Breslauer KJ. (1999) *Biopolymers* **50**: 459.

Chang R. (2005) *Physical Chemistry for the Biosciences*. University Science Books, USA.

Chaplin M. (2006) *Nat Rev Mol Cell Biol* **7**: 861.

Chotia C. (1980) In: Jaenicke R (ed), *Protein Folding*. Elsevier, Amsterdam, pp. 583.

Chotia C. (1992) *Nature* **357**: 543.

Chotia C, Finkelstein AV. (1990) *Annu Rev Biochem* **59**: 1007.

Cohen C, Parry DAD. (1990) *Proteins* **7**: 1.

Cohn EJ, Edsall JT. (1943) Chapter 9 In: Cohn EJ, Edsall JT (eds), *Proteins, Amino Acids and Peptides as Ions and Dipolar ions*. Reinhold, New York.

Colombo MF, Rau DC, Parsegian VA. (1992) **256**: 655.

Corey RB, Pauling L. (1953) *Proc R Soc Lond B Biol Sci* **141**: 10.

Cornett JL, Cease KB, Margalit H, *et al.* (1987) *J Mol Biol* **195**: 659.

Covell DG, Wallqvist A. (1997) *J Mol Biol* **269**: 281.

Craig NC. (2005) *J Chem Educ* **82**: 827.

Creamer TP. (1998) *Proteins* **33**: 218.

Creighton TE. (1978) *Prog Biophys Mol Biol* **33**: 231.

Creighton TE. (1980) *J Mol Biol* **144**: 521.

Creighton TE. (1984) *Proteins, Structure and Molecular Principles*. WH Freeman and Co., New York.

Creighton TE. (1985) *J Phys Chem* **89**: 2452.

Creighton TE. (1993) *Proteins, Structure and Molecular Principles*, 2nd ed. WH Freeman and Co., New York.

Crovetto R, Fernandez-Prini R. (1982) *J Chem Phys* **76**: 1077.

Dawkins R. (1987) *The Blind Watchmaker*. Norton, New York.

De Grado WF. (1988) *Adv Protein Chem* **39**: 51.

De Grado WF, Wasserman ZR, Lear JD. (1989) *Science* **243**: 622.

Delvin MT. (2006) *Textbook of Biochemistry with Clinical Correlations*, 6th ed. Wiley-Liss, Hoboken, New Jersey.

Di Cera E. (1996) Thermodynamic theory of site-specific binding processes. In: *Biological Macromolecules*. Cambridge University Press, Cambridge.

Dill KA. (1985) *Biochemistry* **24**: 1501.

Dill KA. (1987) The stabilities of globular proteins. In: Oxender DL, Fox CF (eds), *Protein Engineering*. Alan R. Liss, New York, pp. 187–192.

Dill KA. (1990) *Biochemistry* **29**: 7133.

Dill KA. (1999) *Protein Science* **8**: 1166.

Dill KA, Chan HS. (1997) *Nat Struct Biol* **4**: 10.

Dill KA, Ozcan SB, Shell MS, Weikl TR. (2008) *Annu Rev Biophys* **37**: 289.

Dill KA, Ozcan SB, Weikl TR, *et al.* (2007) *Curr Opin Struct Biol* **17**: 342.

Drew HR, Dickerson RE. (1981) *J Mol Biol* **151**: 535.

Durell SR, Brooks BR, Ben-Naim A. (1994) *J Phys Chem* **98:** 2198.

Eisenberg D, McLachlan AD. (1986) *Nature* **319:** 199.

Eisenberg D, Wilcox W, Eshita SM, Pryciak, *et al.* (1986) *Proteins* **1:** 16.

Fasman GD. ed. (1989) *Prediction of Protein Structure and The Principles of Protein Conformation.* Plenum Press, New York.

Fersht AR. (1984) *Trends Biochem Sci* **9:** 145.

Fersht AR. (1987) *Trends Biochem Sci* **12:** 301.

Fersht AR, Serrano L. (1993) *Curr Opin Struct Biol* **3:** 75.

Fersht AR, Shi JP, Knill-Jones J, *et al.* (1985) *Nature* **314:** 325.

Finney JL. (1979) The organization and function of water in protein crystals, Chapter 2 In: Franks F (ed), *Water a Comprehensive Treatise,* Volume 6. Plenum Press, New York.

Finney JL. (1984) *Journal de Physique* **45:** C7–197.

Finney JL, Gellatly BJ, Golton IC, Goodfellow J. (1980) *Biophys J* **32:** 17.

Forslind E, Jacobsson A. (1975) In: Franks F (ed), *Water, A Comprehensive Treatise.* Volume 5. Plenum Press, New York.

Gekko K, Timasheff SN. (1981) *Biochemistry* **20:** 4677.

Gertsman BS, Chapagain PP. (2008) *Prog Mol Biol Transl Sci* **84:** 1.

Ginsburg A, Carroll WR. (1965) *Biochemistry* **4:** 2159.

Goldberg ME. (1985) *Trends Biochem Sci* **10:** 388.

Greer J, Erickson JW, Baldwin JJ, Varney MD. (1994) Application of the three-dimensional structures of protein target molecules in structure-based drug design. *J Med Chem* **37:** 1035.

Gutte B, Dauminger M, Wittschieber E. (1979) *Nature (London)* **281:** 649.

Haber E, Anfinsen CB. (1961) *J Biol Chem* **236:** 422.

Haberfield P, Kivuls J, Haddad M, Rizzo T. (1984) *J Phys Chem* **88:** 1913.

Hahn KW, Klis WA, Stewart JM. (1990) *Science* **248:** 1544.

Hard T, Lundback T. (1996) *Biophysical Chemistry* **62:** 121.

Head-Gordon T, Stillinger FH, Wright MH, Gay DM. (1992) *Proc Natl Acad Sci USA* **89:** 11513.

Hecht MH, Richardson JS, Richardson DC, Ogden RC. (1990) *Science* **249:** 884.

Helene C, Lancelot G. (1982) *Prog Biophys Mol Biol* **39:** 1.

Henderson D, Lozada-Cassou M. (1986) *J Colloid Interface Sci* **114:** 180.

Hermans J, Wang L. (1997) *J Am Chem Soc* **119:** 2707.

Hill CP, Anderson DH, Wesson L, *et al.* (1990) *Science* **249:** 543.

Hill TL. (1956) *Statistical Mechanics, Principles and Selected Applications.* McGraw-Hill, New York.

Hill TL. (1960) *Introduction to Statistical Thermodynamics.* Addison Wesley, Reading, Massachusetts.

Hine J. (1972) *J Am Chem Soc* **94:** 5766.

Hirschfelder JO, Curtiss CF, Bird RB. (1954) *Molecular Theory of Gases and Liquids.* John Wiley, New York.

Hoang TX, Trovato A, Seno F, *et al.* (2004) *Proc Natl Acad Sci USA* **101:** 7960.

Hochachka PW, Somero GN. (2002) *Biochemical Adaptation.* Oxford University Press, Oxford.

Holtzer A. (1992) *Biopolymers* **32:** 711.

Holtzer A. (1994) *Biopolymers* **34**: 315.

Honig B, Yang AS. (1995) *Adv Protein Chem* **46**: 27.

Hu CY, Lynch GC, Kokubo H, Pettitt BM. (2010) *Proteins* **78**: 695.

Hughes WL, Saroff HA, Carney AL. (1949) *J Am Chem Soc* **71**: 2476.

Imai K. (1982) *Allosteric Effects in Haemoglobin.* Cambridge University Press, Cambridge.

Israelachvili JN. (1985) *Intermolecular and Surface Forces.* Academic, London.

Israelachvili JN. (1987) *Acc Chem Res* **20**: 415.

Israelachvili JN, McGuiggan PM. (1988) *Science* **241**: 795.

Jaenicke R. (1987) Folding and association of proteins. *Prog Biophys Mol Biol* **49**: 117.

Janin J. (1995) *Proteins* **21**: 30–39.

Janin J, Chotia C. (1990) The structure of protein–protein recognition sites. *J Biol Chem* **265**: 16027.

Jencks WP. (1969) *Catalysis in Chemistry and Enzymology.* McGraw-Hill, New York.

Jesior JC. (2000) *J Protein Chem* **19**: 93.

Karplus M, Janin J. (1999) *Protein Engineering* **12**: 185.

Karshikoff A. (2006) *Non-Covalent Interactions in Proteins.* Imperial College Press, London.

Kauzmann W. (1959) *Adv Protein Chem* **14**: 1.

Kemeny JG, Snell JL. (1960) *Finite Markov Chains.* Van Nostrand Comp. Int., New York.

Kirkwood JG. (1935) *J Chem Phys* **3**: 300.

Kirkwood JG. (1954) In: McElroy WD, Glass B (eds), *A Symposium on the Mechanism of Enzyme Action.* Johns Hopkins University Press, Baltimore, MD.

Klapper MH. (1973) *Adv Chem Phys* **23**: 55.

Klotz IM, Franzen JS. (1962) *J Am Chem Soc* **84**: 3461.

Kolata G. (1986) *Science* **233**: 1037.

Kopka ML, Fratini AV, Drew HR, Dickerson RE. (1983) *J Mol Biol* **163**: 129.

Koshland DE. (1958) *Proc Natl Acad Sci USA* **42**: 98.

Kumar MD, Bava KA, Gromiha MM, *et al.* (2006) *Nucleic Acids Res* **34**: D204.

Kuntz ID. (1992) Structure-based strategies for drug design and discovery. *Science* **257**: 1078.

Kuznetsova N, Rau DC, Parsegian VA, Leiken S. (1997) *Biophys J* **72**: 353.

Lazaridis T, Archontis G, Karplus M. (1995) *Adv Protein Chem* **47**: 231.

Lee JC, Timasheff SN. (1981) *J Biol Chem* **256**: 7193.

Leszcynski JF, Rose GD. (1986) *Science* **234**: 849.

Levinthal C. (1968) Are there pathways for protein folding? *J Chim Phys* **65**: 44.

Levinthal C. (1969) In: Debrunner T, Tsibris JCM, Munck E (eds), *Mossbauer Spectroscopy in Biological Systems.* University of Illinois Press, Urbana, pp. 22.

Levy Y, Onuchic JN. (2004) *Proc Natl Acad Sci USA* **101**: 3325.

Lewin S. (1967) *J Theor Biol* **17**: 181.

Lightner DA, Gawronski JK, Wijekoon WMD. (1987) *J Am Chem Soc* **109**: 923.

Liljas A, Liljas L, Piskur J, *et al.* (2009) *Textbook of Structural Biology.* World Scientific, Singapore.

Lin TY, Timashef SN. (1994) *Biochemistry* **33**: 12695.

Lindley DV. (1965) *Introduction to Probability and Statistics*. Cambridge University Press, Cambridge.

Luo H, Sharp K. (2002) *Proc Natl Acad Sci USA* **99**: 10399.

Makhatadze GI, Privalov PL. (1993a) *J Mol Biol* **232**: 639.

Makhatadze GI, Privalov PL. (1993b) *J Mol Biol* **232**: 660.

Makhatadze GI, Privalov PL. (1995) *Adv Protein Chem* **47**: 307.

Marcelo L, Holthauzen F, Rosgen J, Bolen DW. (2010) *Biochemistry* **49**: 1310.

Mazo RM. (2002) *Brownian Motion, Fluctuations, Dynamics and Applications*. Clarendon Press, Oxford.

Mello CC, Barrick D. (2004) *Proc Natl Acad Sci USA* **101**: 14102.

Mezei M, Ben-Naim A. (1990) *J Chem Phys* **92**: 1359.

Minasov G, Tereshko V, Egli M. (1999) *J Mol Biol* **291**: 83.

Mirsky AE, Pauling L. (1936) *Proc Natl Acad Sci USA* **22**: 439.

Moglich A, Joder K, Kiefhaber T. (2006) *Proc Natl Acad Sci USA* **103**: 12394.

Morozov AV, Kortemme T. (2005) *Adv Protein Chem* **72**: 1.

Mutter M, Hersperger R, Gubernator K, Muller, K. (1989) *Proteins* **5**: 13.

Mutter M, Vuilleumier S. (1989) *Angew Chem Int Ed* **28**: 535.

Myers JK, Oas TG. (2002) *Annu Rev Biochem* **71**: 783.

Myers JK, Pace CN. (1996) *Biophys J* **71**: 2033.

Na GC, Timasheff SN. (1981) *J Mol Biol* **151**: 165.

Nozaki Y, Tanford C. (1963) *J Biol Chem* **238**: 4074.

Nozaki Y, Tanford C. (1965) *J Biol Chem* **240**: 3568.

Nozaki Y, Tanford C. (1971) *J Biol Chem* **246**: 2211.

Onuchic JN, Wolynes PG, Luthey-Schulten Z, Socci ND. (1995) Towards an outline of the topography of a realistic protein folding funnel. *Proc Natl Acad Sci USA* **92**: 3626.

Pace CN. (2009) *Nat Struct Mol Biol* **16**: 681.

Page MI, Jencks WP. (1971) *Proc Natl Acad Sci USA* **68**: 1678.

Pangali C, Rao M, Berne BJ. (1979) *J Chem Phys* **71**: 2975.

Papoulis A, Pillai SU. (2002) *Probability Random Variables and Stochastic Processes*, 4th ed. McGraw-Hill, New York.

Parsegian VA, Gingelli D. (1973) In: Lee LH (ed), *Recent Advances in Adhesions*. Gordon and Breach, London.

Pauling L. (1960) *The Nature of Chemical Bond*, 3rd ed. Cornell University Press, Ithaca, New York.

Perun TJ, Propst CL. (1989) *Computer-Aided Drug Design*. Marcel Dekker, New York.

Perutz MF. (1990) *Mechanisms of Cooperativity and Allosteric Regulation in Proteins*. Cambridge University Press, Cambridge.

Plotkin SS, Onuchic JN. (2002) *Q Rev Biophys* **35**: 111.

Plotkin TD, Onuchic JN. (2002) Dill and Chan (1997).

Pollard TD, Earnshaw WC. (2002) *Cell Biology*. Saunders, Philadelphia.

Predota M, Ben-Naim A, Nezbeda I. (2003) *J Chem Phys* **118**: 6446.

Privalov PL, Dragan AI, Crane-Robinson C, *et al.* (2007) *J Mol Biol* **365**: 1.

Privalov PL, Dragan AI, Crane-Robinson C. *Trends Biochem Sci* **690**: 1.

Privalov PL, Makhatadze GI. (1992) *J Mol Biol* **224**: 715.

Privalov PL, Makhatadze GI. (1993) *J Mol Biol* **232**: 660.

Privalov PL, Makhatadze GI. (1995) *Adv Protein Chem* **47**: 307.

Propst CL, Perun TJ. (1992) *Nucleic Acid Targeted Drug Design*. Marcel Dekker, New York.

Ptashne M. (1967) *Nature* **214**: 232.

Rebek J. (1994) *Scientific American* **271**: 34.

Rebek J, Askew B, Ballester P, Costero A. (1988) *J Am Chem Soc* **110**: 923.

Reichmann D, Phillip Y, Carmi A, Schreiber G. (2008) *Biochemistry* **47**: 1051.

Reiss H. (1966) *Adv Chem Phys* **9**: 1.

Richardson JM, Lopez MM, Makhatadze GI. (2005) *Proc Natl Acad Sci USA* **102**: 1413.

Roerdink FH, Kroon AM. (1989) *Drug Carrier Systems*. John Wiley and Sons, New York.

Saenger W, Hunter WN, Kennard O. (1986) *Nature* **324**: 385.

Sainsbury RM. (2009) *Paradoxes*, 3rd ed. Cambridge University Press, Cambridge.

Sanchez-Ruiz JM. (2006) Private communication.

Santos A. Private communication.

Schellman JA. (1955) *CR Trav Lab Carlsberg Ser Chim* **29**: 223.

Schellman JA. (1978) *Biopolymers* **17**: 1305.

Scholtz JM, Marqusee S, Baldwin RL, *et al.* (1991) *Proc Natl Acad Sci USA* **88**: 2854.

Schwabe JW. (1997) *Curr Opin Struct Biol* **7**: 126.

Schneider HJ. (2009) *Angew Chem Int Ed* **48**: 3924.

Schulz GE, Schirmer RH. (1979) *Principle of Protein Structure*. Springer Verlag, Berlin.

Sear RP. (2006) *Curr Opin Colloid Interface Sci* **11**: 35.

Searle MS, Williams DH, Gerhard U. (1992) Partitioning of free energy contributions in the estimation of binding constants; Residual motions and consequences for amide–amide hydrogen bond strengths. *J Am Chem Soc* **114**: 10697.

Sharp KA, Nicholls A, Fine RF, Honig B. (1991) *Science* **252**: 106.

Sharp KA, Nicholls A, Friedman R, Honig B. (1991) *Biochemistry* **30**: 9686.

Shaw GS, Hodges RS, Sykes BD. (1990) *Science* **249**: 280.

Shirley BA, Stanssens P, Hahn U, Pace CN. (1992) *Biochemistry* **31**: 725.

Shui X, Fail-Isom L, Hu GG, Williams LD. (1998) *Biochemistry* **37**: 8341.

Siebert X, Amzel LM. (2004) *Proteins* **54**: 104.

Sitkoff D, Sharp KA, Honig B (1994) *Biophys Chem* **51**: 397.

Spolar RS, Record MT. (1994) Coupling of local folding to site-specific binding of proteins to DNA. *Science* **263**: 777.

Stapley BJ, Creamer TP. (1999) *Protein Sci* **8**: 587.

Steinberg IZ, Scheraga H. (1963) *J Biol Chem* **238**: 172.

Stillinger FH, Wasserman Z. (1978) *J Phys Chem* **82**: 929.

Street TO, Bolen DW, Rose GD. (2006) *Proc Natl Acad Sci USA* **103**: 13997.

Stryer L. (1988) *Biochemistry*, 3rd ed. WH Freeman and Co., New York.

Sundaralingam M, Sekharudu YC. (1989) *Science* **244**: 1333.

Sundaralingam M, Sekharudu YC. (1989) *Biological Structure, Dynamics, Interactions and Expressions*. Volume 2. Sarma RS, Sarma MH (eds), Adenine Press.

Susi H, Ard JS. (1969) *J Phys Chem* **73**: 2440.

Tabushi I, Mizutani T. (1987) *Tetrahedron* **43**: 1439.

Tamura A, Privalov PL. (1997) *J Mol Biol* **273**: 1048.

Tanford C. (1962) *J Am Chem Soc* **84**: 4246.

Tanford C. (1964) *J Am Chem Soc* **86**: 2050.

Tanford C. (1968) *Adv Protein Chem* **23**: 121.

Tanford C. (1970) *Adv Protein Chem* **24**: 1.

Tanford C. (1973) *The Hydrophobic Effect: Formation of Micelles and Biological Membranes*. Wiley-Interscience, New York.

Tanford C. (1978) *Science* **200**: 1012.

Tanford C. (1980) *The Hydrophobic Effect*. Wiley, New York.

Tanford C, Reynolds J. (2001) *Nature's Robots, A History of Proteins*. Oxford University Press, Oxford.

Thanki N, Thornton JM, Goodfellow J. (1988) *J Mol Biol* **202**: 637.

Thomas DJ. (1992) *FEBS Lett* **307**: 10.

Tuñon I, Silla E, Pascual-Ahuir JL. (1994) *J Phys Chem* **98**: 377.

Uversky VN, Fink AL. (2002) *FEBS Lett* **515**: 79.

Villa J, Strajbl M, Glennon TM, *et al.* (2000) *Proc Natl Acad Sci USA* **97**: 11899.

Voet DJ, Voet JG, Pratt CW. (2008) *Principles of Biochemistry*, 3rd ed. John Wiley and Sons, Hoboken, New Jersey.

Volkenstein MV. (1977) *Molecular Biophysics*. Academic Press, New York.

von Hippel PH. (1979) In: Goldberger RF (ed), *Biological Regulation and Development*. Plenum, New York, pp. 279.

von Hippel PH. (1994) *Science* **263**: 769.

von Hippel PH, Berg OG. (1986) *Proc Natl Acad Sci USA* **83**: 1608.

Wang H, Ben-Naim A. (1996) *J Med Chem* **39**: 1531.

Wang H, Ben-Naim A. (1997) *J Phys Chem B* **101**: 1077, 1086.

Weinkam P, Zong C, Wolynes PG. (2005) *Proc Natl Acad Sci USA* **102**: 12401.

Whitford D. (2005) *Protein, Structure and Function*. John Wiley and Sons.

Wiggins PM. (1997) *Physica A* **238**: 113.

Wilf J, Ben-Naim A. (1994) *J Phys Chem* **98**: 8594.

Wolfenden R. (1983) *Science* **222**: 1087.

Wolfenden R, Anderson PM, Cullis C, Southgate CB. (1981) *Biochemistry* **20**: 849.

Wolynes PG. (2005) *Philos Trans R Soc Lond A* **363**: 453.

Wolynes PG, Onuchic JN, Thirumalai D. (1995) Navigating the folding routes. *Science* **267**: 1619.

Wu H. (1931) *Chin J Physiol* **5**: 321.

Wyman J, Gill SJ. (1990) *Binding Linkage: Functional Chemistry of Biological Macromolecules*. University Science Book, Mill Valley, CA.

Xu D, Lin SL, Nussinov R. (1997) *J Mol Biol* **265**: 68.

Yaacobi M, Ben-Naim A. (1974) *J Phys Chem* **78**: 175.

Yu YB, Privalov PL, Hodges RS. (2001) *Biophys J* **81**: 1632.

Yue K, Dill KA. (1992) *Proc Natl Acad Sci USA* **89**: 4163.

Zwanzig R, Szabo A, Bagchi B. (1992) Levinthal's paradox. *Proc Natl Acad Sci USA* **89**: 20.